D1670912

DEADLY LESSONS
UNDERSTANDING LETHAL SCHOOL VIOLENCE

Case Studies of School Violence Committee

Mark H. Moore, Carol V. Petrie, Anthony A. Braga, and
Brenda L. McLaughlin, editors

Committee on Law and Justice and Board on Children, Youth, and Families

Division of Behavioral and Social Sciences and Education

NATIONAL RESEARCH COUNCIL
INSTITUTE OF MEDICINE
OF THE NATIONAL ACADEMIES

THE NATIONAL ACADEMIES PRESS
Washington, D.C.
www.nap.edu

THE NATIONAL ACADEMIES PRESS • 500 Fifth Street, N.W. • Washington, D.C. 20001

NOTICE: The project that is the subject of this report was approved by the Governing Board of the National Research Council, whose members are drawn from the councils of the National Academy of Sciences, the National Academy of Engineering, and the Institute of Medicine. The members of the committee responsible for the report were chosen for their special competences and with regard for appropriate balance.

The study was supported by Grant No. S184U000010 between the National Academy of Sciences and the U.S. Department of Education. Any opinions, findings, conclusions, or recommendations expressed in this publication are those of the author(s) and do not necessarily reflect the view of the organizations or agencies that provided support for this project.

Library of Congress Cataloging-in-Publication Data

Deadly lessons : understanding lethal school violence : case studies of School Violence Committee / Mark H. Moore ... [et al.], editors ; Committee on Law and Justice and Board on Children, Youth, and Families, Division of Behavioral and Social Sciences and Education, National Research Council and Institute of Medicine.
p. cm.
Includes bibliographical references and index.
ISBN 0-309-08412-1 (hardcover)
1. School violence—United States—Case studies. I. Moore, Mark Harrison. II. National Research Council (U.S.). School Violence Committee. III. National Research Council (U.S.). Committee on Law and Justice. IV. National Research Council (U.S.). Board on Children, Youth, and Families.
LB3013.32 .D43 2002
371.7'82'0973—dc21

2002011776

Additional copies of this report are available from The National Academies Press, 500 Fifth Street, N.W., Box 285, Washington, DC 20055.

Call (800) 624-6242 or (202) 334-3313 (in the Washington metropolitan area)

This report is also available online at http://www.nap.edu

Printed in the United States of America

Suggested citation: National Research Council and Institute of Medicine. (2003) *Deadly Lessons: Understanding Lethal School Violence.* Case Studies of School Violence Committee. Mark H. Moore, Carol V. Petrie, Anthony A. Braga, and Brenda L. McLaughlin, editors. Division of Behavioral and Social Sciences and Education. Washington, DC: The National Academies Press.

THE NATIONAL ACADEMIES

Advisers to the Nation on Science, Engineering, and Medicine

The **National Academy of Sciences** is a private, nonprofit, self-perpetuating society of distinguished scholars engaged in scientific and engineering research, dedicated to the furtherance of science and technology and to their use for the general welfare. Upon the authority of the charter granted to it by the Congress in 1863, the Academy has a mandate that requires it to advise the federal government on scientific and technical matters. Dr. Bruce M. Alberts is president of the National Academy of Sciences.

The **National Academy of Engineering** was established in 1964, under the charter of the National Academy of Sciences, as a parallel organization of outstanding engineers. It is autonomous in its administration and in the selection of its members, sharing with the National Academy of Sciences the responsibility for advising the federal government. The National Academy of Engineering also sponsors engineering programs aimed at meeting national needs, encourages education and research, and recognizes the superior achievements of engineers. Dr. Wm. A. Wulf is president of the National Academy of Engineering.

The **Institute of Medicine** was established in 1970 by the National Academy of Sciences to secure the services of eminent members of appropriate professions in the examination of policy matters pertaining to the health of the public. The Institute acts under the responsibility given to the National Academy of Sciences by its congressional charter to be an adviser to the federal government and, upon its own initiative, to identify issues of medical care, research, and education. Dr. Harvey V. Fineberg is president of the Institute of Medicine.

The **National Research Council** was organized by the National Academy of Sciences in 1916 to associate the broad community of science and technology with the Academy's purposes of furthering knowledge and advising the federal government. Functioning in accordance with general policies determined by the Academy, the Council has become the principal operating agency of both the National Academy of Sciences and the National Academy of Engineering in providing services to the government, the public, and the scientific and engineering communities. The Council is administered jointly by both Academies and the Institute of Medicine. Dr. Bruce M. Alberts and Dr. Wm. A. Wulf are chair and vice chair, respectively, of the National Research Council.

www.national-academies.org

CASE STUDIES OF SCHOOL VIOLENCE COMMITTEE

MARK H. MOORE (*Chair*), John F. Kennedy School of Government, Harvard University
PHILIP J. COOK, Public Policy Studies, Duke University
THOMAS A. DISHION, Department of Psychology, University of Oregon
DENISE C. GOTTFREDSON, Department of Criminology and Criminal Justice, University of Maryland
PHILIP B. HEYMANN, Harvard Law School, Harvard University
JAMES F. SHORT, JR., Department of Sociology, Washington State University
STEPHEN A. SMALL, Department of Child and Family Studies, University of Wisconsin-Madison
LEWIS H. SPENCE, Department of Social Services, Boston, Massachusetts
LINDA A. TEPLIN, Department of Psychiatry and Behavioral Science, Northwestern University

CAROL PETRIE, *Study Director*
ANTHONY A. BRAGA (*Consultant*), John F. Kennedy School of Government, Harvard University
BRENDA MCLAUGHLIN, *Research Assistant*
LECIA QUARLES, *Project Assistant* (until March 2002)
MICHELLE AUCOIN MCGUIRE, *Project Assistant* (after March 2002)

COMMITTEE ON LAW AND JUSTICE

BOARD ON CHILDREN, YOUTH, AND FAMILIES

Contents

Preface ix

Executive Summary 1

1 Introduction 9

Part I: Case Studies of Lethal School Violence 17

2 The Copycat Factor: Mental Illness, Guns, and the Shooting Incident at Heritage High School, Rockdale County, Georgia 25
Mercer L. Sullivan and Rob T. Guerette

3 Bad Things Happen in Good Communities: The Rampage Shooting in Edinboro, Pennsylvania, and Its Aftermath 70
William DeJong, Joel C. Epstein, and Thomas E. Hart

4 A Deadly Partnership: Lethal Violence in an Arkansas Middle School 101
Cybelle Fox, Wendy D. Roth, and Katherine Newman

5 No Exit: Mental Illness, Marginality, and School Violence in West Paducah, Kentucky 132
David Harding, Jal Mehta, and Katherine Newman

6 Shooting at Tilden High: Causes and Consequences 163
John Hagan, Paul Hirschfield, and Carla Shedd

7 What Did Ian Tell God?: School Violence in East New York 198
Mindy Thompson Fullilove, Gina Arias, Moises Nunez, Ericka Phillips, Peter McFarlane, Rodrick Wallace, and Robert E. Fullilove III

8 A Cross-Case Analysis 247

Part II: Understanding and Preventing Lethal School Violence 285

9 Lethal School Violence in Statistical Context 287
10 Literature Review 302
11 Response Strategies: Observations on Causes, Interventions, and Research 330

References 343

Appendixes

A Case Study Methodology and the Study of Rare Events of Extreme Youth Violence: A Multilevel Framework for Discovery 351
Mercer L. Sullivan and Mindy Thompson Fullilove
B Biographical Sketches 364

Index 375

Preface

As the chair of the committee on studies of school violence, I would like to say on behalf of the committee that we felt privileged to have been given both the challenge and the opportunity to apply our collective knowledge to understanding as best we could the nature of the problem that the nation faced in dealing with the spate of school shootings that occurred in the late 1990s. Like all citizens, we were concerned about this apparently new and frightening phenomenon. As social scientists, we felt challenged by the task of making sense of something that seemed so new and so important. We hope we have used available knowledge of trends, literatures on violence, and the cases we commissioned to help the country as a whole understand the phenomenon we were asked to address. If we have done a good job, it is because we received some outstanding encouragement and help and we would like to acknowledge that fact here.

First, we would like to thank Representative James Greenwood for challenging the committee to address itself to this issue, even though it involved tasks that are unusual for a National Academies Committee to undertake. We are grateful for his leadership in insisting that this work be done, and that we experiment with the use of detailed case studies as a way of illuminating the problem he wanted us to address.

Second, we would like to thank the Committee on Law and Justice and the Board on Children, Youth, and Families for having the courage to approve the creation of a study committee with the unusual mandate that we had. We would also like to thank these committees as well as the

National Academies for their support of our adherence to rigorous scientific standards as we did our work in reviewing the trends, examining the relevant literatures, and commissioning six new cases, and drawing tentative inferences from a cross-case analysis.

Third, we would also like to thank the executive and professional staff of the Division of Behavioral and Social Sciences and Education (DBASSE), National Research Council (NRC), which acted on a day-to-day basis to ensure that we were both responsive to our mandate and acting consistently within the highest NRC standards. Barbara Torrey and Eugenia Grohman provided invaluable guidance as we went along with our work.

Those thanked above helped to create the opportunity for us to do the work and set the high standards that challenged us to do it well. We are also deeply grateful to those who helped us meet those standards.

We are particularly grateful to our colleagues who took the responsibility for developing the cases: William DeJong, Mindy Fullilove, John Hagan, Katherine Newman, and Mercer Sullivan. Mercer and Mindy were involved in this effort from the beginning, and it was partly their deep understanding of what the case study method brings to scientific inquiry that gave us all the courage to go ahead with this unusual project. They not only developed outstanding cases, but also kept us all focused on the important scientific methodological research involved in the use of cases for scientific inquiry. Mindy also served the committee exceptionally well by continuing to insist that we keep the spate of suburban school shootings in the context of the larger issues of youth and school violence in general.

Katherine Newman took the enormous responsibility of producing two cases rather than just one, and of organizing teams of students to accomplish this work. The quality of the cases her team produced showed that this strategy could be successful not only in developing first-rate cases, but also in developing the skills and interests of the students who worked on this project with her. It was also her extraordinary energy and the skills of her team that produced the earliest and most comprehensive "template" we used in deciding what facts had to be gathered across the sites, and how the cases would be analyzed as a combined "dataset." At a critical stage toward the end of our process, Bill DeJong stepped up to the plate to help us develop a simpler version of the "template" that all were comfortable in using for the cross-case analysis.

We think the cases developed for this project are the most important product of our effort, and we owe them entirely to the efforts of these colleagues. The case writing colleagues, although not members of the committee, were also extremely helpful in informing the committee deliberations: reminding us when facts they had gathered stood opposed to what we were thinking, suggesting new lines of thought about possible causes, and noting both significant variation and subtle consequences of certain kinds of interventions.

Finally, the committee depended crucially on the extraordinary effort and skills of the professional team that worked at the heart of the process that produced our report. Carol Petrie has for many years now been the backbone of the NRC committees in the domain of law and justice. Her tact, her persistence, her organizational skills, her good humor, her great substantive knowledge, and her writing and editorial skills make her invaluable to the production of first-rate reports. Anthony Braga has shown himself over and over again to be an outstanding researcher in the criminal justice field. He is meticulous with concepts and data, encyclopedic in his knowledge of the literature on crime and violence, and both subtle and creative as he considers what the data and the literature are trying to tell him about how the world actually works. He is also indefatigable and infinitely resourceful when there is work to be done. Brenda McLaughlin went far beyond the usual expectations for an NRC staff member working on a project like this. She engaged in very resourceful and creative original research as she traced down information about the trends that formed the basis of our analysis in Chapter 9. She also made it possible for us to look easily across different studies to determine which incidents were common to which studies and which were different. The field will owe her an enormous debt well into the future as we continue to work with the dataset she created. Lecia Quarles and Michelle AuCoin McGuire served as project assistants to the committee. Lecia organized the meetings both in Washington and Cambridge and took care of other logistics such as travel and producing and mailing out drafts of the report. Michelle took over the production of the final manuscript and was of invaluable assistance in developing the committee's response to reviewers. We are also grateful to Christine McShane, of the DBASSE editorial staff, who applied her superb editorial skills to the final draft.

While the committee as a whole owes much to these individuals, I have to say as the chair that I owe much to the committee. The committee was a diverse group, assembled quickly, assigned to a task that would have been tough for an individual researcher, let alone a collective group that had not previously worked together. They did a terrific job of addressing themselves seriously to the problem at hand, and deploying their great individual knowledge and analytic powers to develop insights, ideas, and hypotheses. They also did a terrific job of listening to one another and developing frameworks that all found useful. It was a pleasure to work with such an outstanding committee on such a challenging and novel task. We hope that we have made a useful contribution to society's future capacity to understand and respond to violence in general, but also that particular kind of violence that we came to call "school rampages," a very rare but very scary form of youth and school violence.

This report has been reviewed in draft form by individuals chosen for their diverse perspectives and technical expertise, in accordance with procedures approved by the Report Review Committee of the National Research Council (NRC). The purpose of this independent review is to provide candid and critical comments that will assist the institution in making the published report as sound as possible and to ensure that the report meets institutional standards for objectivity, evidence, and responsiveness to the study charge. The review comments and draft manuscript remain confidential to protect the integrity of the deliberative process.

We wish to thank the following individuals for their review of this report: Mark Anderson, National Center for Injury Prevention and Control, U.S. Centers for Disease Control and Prevention; Jane Costello, Developmental Epidemiology Program, Duke University School of Medicine; John Modell, Department of Sociology, Brown University; John Monahan, School of Law, University of Virginia; Vivian Reznik, Department of Pediatrics, University of California San Diego School of Medicine; Lawrence Steinberg, Department of Psychology, Temple University; Kate Stetzner, Butte Public Schools, Montana; and Marcelo Suarez-Orozco, Graduate School of Education, Harvard University.

Although the reviewers listed above have provided many constructive comments and suggestions, they were not asked to endorse the conclusions or recommendations nor did they see the final draft of the report before its release. The review of this report was overseen by Kenneth A. Shepsle, Department of Government, Harvard University, and Elaine L. Larson, School of Nursing, Columbia University. Appointed by the National Research Council, they were responsible for making certain that an independent examination of this report was carried out in accordance with institutional procedures and that all review comments were carefully considered. Responsibility for the final content of this report rests entirely with the authoring panel and the institution.

Mark H. Moore, Chair
Case Studies of School Violence Committee

DEADLY LESSONS

UNDERSTANDING LETHAL SCHOOL VIOLENCE

Executive Summary

In the 1990s, youth violence, which had reached epidemic levels in the nation's cities beginning in the late 1980s, took an apparently unprecedented form in rural and suburban middle and high schools across the country. Between 1992 and 2001, 35 incidents occurred in which students showed up at their school or at a school-sponsored event and started firing at their schoolmates and teachers. These incidents, represented most starkly by the incident at Columbine High School, in Littleton, Colorado, left 53 dead, and 144 injured.

These incidents shocked the public, partly because so many were killed in single incidents and partly because the targets of the shootings seemed so arbitrarily selected. A third reason is that these incidents occurred in such unexpected places. The previous epidemic of deadly youth violence, which peaked in 1993 and then declined, had occurred among black and Hispanic youth in the nation's most disadvantaged urban neighborhoods and schools. In most of these new cases, communities that had previously thought of themselves as insulated from lethal youth violence discovered that they, too, were vulnerable.

Consequently, Congress requested that the National Research Council study this phenomenon. The Committee to Study Youth Violence in Schools was established in 2001, and its charge mirrored the language in the legislation, which stated:

> The National Academy of Sciences [will] conduct a study regarding antecedents of school violence in urban, suburban, and rural schools,

including the incidents of school violence that occurred in Pearl, Mississippi; Paducah, Kentucky; Jonesboro, Arkansas; Springfield, Oregon; Edinboro, Pennsylvania; Fayetteville, Tennessee; Littleton, Colorado; and Conyers, Georgia.

Congress specifically asked that detailed case studies be developed of the circumstances that led to extreme lethal violence in schools. The goal was to use these cases to learn as much as possible about two important questions. First, what could be said about the important causes and consequences of these unexpected, lethal shootings? Second, what actions could individuals and institutions take either to prevent these events from occurring in the first place or to minimize the damage once they began to unfold?

The most important challenge the committee confronted was to choose the particular cases to be developed, a task with both practical and scientific elements. Congress asked the committee to examine "incidents of lethal school violence in *urban, suburban, and rural schools,*" but all of the specific cases identified in the legislation occurred in suburban and rural schools between 1997 and 1999. From a practical standpoint, cases could be selected from this list. However, the scientific question before the committee was what was the general class of violence of which these eight incidents were exemplars? When the committee examined the data sources on school shootings from this period, we found urban school shootings, but none that appeared similar to the listed cases. It seemed then that the form of lethal school violence that occurred in the late 1990s might represent a distinct form of lethal school violence—different in its causes and in its effective prevention and control. This possibility made the important scientific question of the relationship between the form of lethal school violence that was concentrated in the inner-city schools, and the seemingly newer form of lethal school violence that erupted in suburban and rural schools in the late 1990s, central to the committee's work.

To more fully answer these questions the committee decided to examine the series of school shootings that began in 1997, not by themselves as a separate phenomenon, but instead *against the backdrop of the broader patterns of violence that had recently affected American society*, especially between 1985 and 1995. This also seemed important for policy-making purposes; that is, it seemed important to keep these particular shootings in perspective. An overreaction to events that were so dramatic and so unexpected, overshadowing the importance of other violence problems, seemed likely. It also was important to look at general trends to understand the relationship between the unexpected outburst of shootings in suburban and rural schools and other forms of both youth and adult violence. The committee was particularly interested in understanding the relationship, if any, between the earlier epidemic of inner-city youth violence and the later series of shootings by youth in suburban and rural schools.

To meet its charge, the committee commissioned and analyzed six case studies of schools and communities that had experienced incidents of serious school violence in which more than one person was killed or seriously injured in a single attack. Four of these cases involved schools in suburban and rural communities that were listed in the legislation. Two of the cases involved inner-city schools, one of which had experienced two such incidents earlier in the 1990s.

The committee also reviewed the literature on violence as a context for interpreting the cases, especially in terms of what might be causing the incidents we studied and how they might best be prevented. This included review of a very small literature on incidents that looked similar to the ones we were asked to review; the literature on broad categories of violence, including violence in general, youth violence, school violence, and the relationship between violence and suicide; and an emerging literature on some specialized forms of violence that bore some similarity to the incidents we studied, including mass murders, rampage shootings, and "suicide by cop" (e.g., incidents in which individuals seemed to shoot in order to provoke a response by the police). Finally, because there might be some contagion effects in the events we were examining, we looked into studies that explore the contagiousness of violence. The committee's findings, presented below, are based on our analysis of the cases, the data, and the literature review.

THE FINDINGS

The limitations of the available evidence made it impossible for the committee to reach firm, scientific conclusions about either the causes and consequences of the shootings in rural and suburban schools or the most effective means of preventing and controlling them. However, we did develop some hypotheses that seem strong enough to guide action and research while better information is being developed.

Consequences

The committee found significant and long-lasting harm in each of the communities studied, although the lethal violence took different forms across the urban and the rural and suburban cases. The tragedy and shock of the large numbers killed and injured all at once in the suburban and rural cases still reverberates in those communities. Those closest to the center of these incidents continue to be traumatized; victims' civil suits against the shooters' families and the schools are still pending, and some bitterness remains unresolved. In three cases, business continues to suffer because of the harm to the communities' reputations.

The inner-city school shootings further shocked already traumatized community residents. Most experienced them as an extension of the gun violence in the neighborhoods of the schools, which in the two cities had claimed many lives over time. So many neighborhood youths had themselves been victimized, or had known other victims, that in one of the cases, young men spontaneously pulled up their shirts to show the researchers their scars from violent incidents. The trauma in both cases radiated from those directly affected to involve the entire city, including the mayor and the city council, in the response. Interestingly, the policy responses in all six cases were more homogeneous than the circumstances or causes seemed to be.

Causes

Although the lethal shooting sprees of the 1990s followed closely on and even seemed to emerge from or be influenced by the earlier violence—and may stem from similar underlying factors—the committee also considers it possible that these events represent a separate strain of violence. While the inner-city epidemic of violence was fueled by well-understood causes—poverty, racial segregation, and the dynamics of the illicit drug trade—the violence in the suburban and rural schools more closely resembles "rampage" shootings that occur in places other than schools, such as workplaces, or in other public spaces.

In these six cases, this idea is supported by the notable differences in the motives of the shooters and the circumstances under which the shootings occurred. In the inner-city cases, the shooting incidents involved specific grievances between individuals that were known in the school community. In contrast, the suburban and rural shooting incidents did not involve specific grievances. These shooters felt aggrieved, but their grievances were a more general and abstract sense of feeling attacked rather than a specific threat by an individual. The grievances of these youth were not understood by those around them. As in rampage shootings involving adults, suburban and rural school shooting cases generally seem to involve youth who have these kinds of exaggerated and somewhat abstract grievances.

Evidence from Trends

Whereas events that could be described as rampage violence are only a small component of all violence and seem to move independently of other forms of violence, the committee found a spike for all kinds of rampage killings in the late 1990s. This raises the possibility that there may have been some kind of epidemic of rampage shootings in the late 1990s that cut across all ages, including youth. Consistent with this hypothesis is the evidence from the cases that copycat mechanisms, which clearly were at work in at least one of the shootings, also may have influenced two of the other three suburban and rural school shootings examined.

Trend data on school shootings indicated that the school rampage cases listed in the legislation were not in fact new or unique: there had been similar incidents of school violence as far back as 1974. Remarkably, we could not find similar rampage shootings in the nation's inner cities, a fact that surprised the committee. Only three events in inner-city schools across the country met our formal criteria for inclusion in the study. And when we looked closely at these cases, we found that they looked quite different from the cases in the suburban and rural schools.

The Shooters

Looking across the cases, we found that the eight shooters exhibited a number of similar traits. While these are consistent with risk factors for serious youthful violence identified in the literature, this study can do no more than claim them as tendencies or propensities. All were boys. Five had recently begun hanging out with delinquent or more troubled friends. Five had a relatively recent drop in their grades at school. Five had engaged in previous serious delinquent acts and the other three in minor delinquent behavior. Serious mental health problems, including schizophrenia, clinical depression, and personality disorders, surfaced after the shootings for six of the eight boys in these cases. All had easy access to guns. The rural and suburban boys had experience with guns, and one of the urban teens appears to have practiced with the gun he used.

However, there were also some characteristics that are usually thought of as protective. Half of the shooters came from intact and stable two-parent families, and five of the eight were good students, at least until 8th grade. Only three of the shooters struggled with grades or experienced the early school failure that frequently precedes the development of serious delinquent behavior. Only one of the eight shooters was a loner, and only two were gang members. Most had friends, although the quality of the friendships differed. Most of these shooters were not considered to be at high risk for this kind of behavior by the adults around them.

Community and School Environments

The central differences in these cases can be found in community structure. The two urban neighborhoods were characterized by community social and physical conditions that research has shown create a milieu for the development of youth violence. Most of the rural and suburban communities did not demonstrate these structural conditions, and in fact three of the four were demographically the opposite—thriving economically, having a high degree of social capital, and mostly free of crime and violence. The committee notes that five of the six communities in these cases had experienced rapid social change, which may produce instability even where the changes are seen as positive ones.

A common element across school settings was the presence of numerous informal and exclusive student groups. In the urban schools, these were mostly marginal groups—gangs, including criminal gangs, and "crews." In the rural and suburban schools, they were cliques—some mainstream and some marginal. Membership in these groups determined social status in most of these schools, but there were notable differences in relationships in the different school settings. In the urban cases, the boys' friendships were embedded in these marginal groups; in the rural and suburban cases, the boys were marginal members of both mainstream and marginal groups.

An important similarity across all of the cases was the gulf between the communities' youth culture and that of adults. Parents and most teachers had a poor understanding of the children's exposure to changing community conditions, their experiences in social situations including at school, and their interpretations of those experiences. There was an intense concern among these shooters about their social standing in their school and among their peers. This took different forms in the inner-city and the rural and suburban cases, but for this group of offenders it was similar in that it was almost always about shielding themselves from physical victimization, including bullying or other personal humiliation. Although in most cases the youth had hinted at what was to come, parents and teachers were mostly unaware of the status problems they were experiencing and of their almost universal belief that they had nowhere to turn. In the words of one of the case authors, "the social dynamics of adolescence in these communities were almost entirely hidden from adult view." Whether or not this is characteristic of most communities is a question that remains unanswered.

Community Responses

With only one exception, the cases were treated virtually identically in the criminal justice system. Six of the eight shooters were charged with the highest offense that could be supported by the evidence, usually first- or second-degree murder, and tried in the adult (criminal) courts rather than the juvenile courts. Most were sentenced to long terms of incarceration and correctional supervision, with the upper limit of sentences for most ranging from 20 to 60 years. The exception was where state law required the justice system to treat the offenders, ages 11 and 13, as juveniles. Even though there is little room in the adversarial process of the criminal courts for the special problems these boys had to influence the outcome, most residents of that particular community saw adjudication in the juvenile system as unjustifiably lenient treatment, given the nature of the offense.

Instituting or adding to physical security measures was the most common response of the school communities to these shootings in almost every site. In the urban cases, public officials and residents went well beyond security measures to effect improvements in community climate and communication between youth and adults. The rural and suburban communities also took steps to improve communication but did not focus on community climate, tending to explain the incidents as the act of a troubled youth rather than resulting from community-level or social factors that needed attention.

OBSERVATIONS AND RESEARCH RECOMMENDATIONS

School rampage shootings are rare events that have occurred in middle-class and affluent rural and suburban schools, but they are not found in inner-city schools. They resemble other rampage shootings, especially mass murders, more than other forms of youth violence or urban school shootings. It is virtually impossible to identify the likely offenders in advance; thus, there is no accurate way to develop a profile of students at high risk to commit these kinds of acts.

Little is known about what causes school rampages, so the development of primary prevention mechanisms is difficult. Until more can be learned about causes, case studies such as these can be helpful in identifying some plausible targets of intervention. One approach involves the fact that these young people had such easy access to firearms. Based on these cases—and the fact that all but one of the incidents of lethal school violence involving multiple victims in the United States over the last decade have involved firearms in the hands of children—the committee believes it is necessary to find more effective means than we now have of realizing the nation's long established policy goal of keeping firearms out of the hands of unsupervised children and out of our schools. In addition, there is a need for youth and adults, among themselves and together, to be more sensitive to the often fragile status concerns of young people. Students are often in a position to preempt rampage attacks simply by telling what they know to school authorities, but that requires crossing the gap between the society of youth and that of adults. Specifically, there is a need to develop a strategy for drawing adults and youth closer together in constructing a normative social climate that is committed to keeping the schools safe from lethal incidents.

The committee notes that conducting empirical studies to establish causal processes leading to these rare and heinous outcomes is not the only scientific approach possible in the search for prevention and control. Case studies like those presented here are essential and appropriate scientific tools for use in seeking for causes and effective interventions, especially in the study of important but rare events such as these school shootings. Only by first carefully analyzing the patterns that exist in the unfolding of these occurrences can one gather the information needed to develop studies from which findings can be generalized.

The committee recommends that new research be undertaken to further improve understanding of the factors that might influence school shootings, particularly school rampage shootings, and to develop knowledge on the impact of interventions. Our specific research recommendations cover further exploration of the precursors to these incidents, including nonlethal violence and serious bullying in schools; illegal gun carrying by adolescents; the signs and symptoms of developing mental health problems in youth in grades 6–10; the effects of student attacks on teachers; and the effects of rapid change in increasingly affluent rural and suburban communities on youth development, socialization, and violence. Evaluation studies should include programs targeted at thwarting planned school shootings. Evaluations of security measures and police tactics in responding to school shootings are also needed.

1

Introduction

America is now and has long been a relatively violent nation (Gurr, 1989; NRC, 1993). It has had a particularly notorious history with respect to *lethal* violence (Zimring, 1998). In recent years, the general violence of American society has engulfed the nation's young (Cook and Laub, 1998). From 1985 to 1994, the United States experienced a historically unprecedented epidemic[1] of lethal youth violence that took the lives of young victims, shattered inner-city communities, and ruined the prospects of many young people across the nation (Blumstein, 1995; Moore and Tonry, 1998; Cook and Laub, 1998).

The general violence and the fear it causes have also reached into American schools. Serious violence has always been rare in schools, and it remained so even as society weathered the epidemic of youth violence. Recent figures show that between 1994 and 1999, 220 events of school-associated violent deaths occurred—on average, 36 per year. This is less than 0.3 percent of all violent deaths that occurred during that time period. Young people are far more likely to be killed or seriously injured when they are out of school rather than in it. The school boundary has continued to create a relatively safe haven for the nation's young.

To say that schools have remained relatively safe compared with other social and geographic communities is not to say that current levels of violence and fear in schools are acceptable. Quite the contrary. The prevalence of violent victimizations in schools has more than doubled since 1989: 3 percent of youths reported violent victimizations at school then, compared with 8 percent today. And while it is true that most schools are relatively free of violent crime, some schools experience very high rates, especially those in high-crime areas of U.S. cities (Kaufman et al., 1999).

During the past 10 years, a spate of multiple-victim shooting incidents in school settings has greatly increased and widened public concern about violence in schools. From 1992 to 2001, 35 incidents occurred in which students showed up at their school or at a school-sponsored event and started shooting at their schoolmates and teachers. These incidents, represented most starkly by the incident at Columbine High School in Littleton, Colorado, left 53 people dead and 144 injured (see Table 9-1). These shootings contributed to a significant increase in homicide rates for students killed in multiple-victim incidents on school grounds between 1992 and 1999 (Anderson et al., 2001).

These events shocked the public partly because so many were killed or injured so quickly in a single incident. It was also particularly frightening that in many instances the victims seemed to have been chosen more or less at random. But the intense public concern was also due to the social location of the shootings. Much of the violence occurred in communities that had, until the time these shootings occurred, been spared the kind of lethal youth violence that beset some of their urban neighbors. Moreover, the shootings took place in schools—the place in communities that is supposed to be protective of children. Finally, the fact that these terrible shootings didn't stop—they kept occurring at an apparently increasing rate in a pattern that suggested an emergent epidemic—pushed the level of concern much higher.

THE STUDY CHARGE

Responding to the high level of public concern, the U.S. Congress asked the National Research Council (NRC) to undertake a detailed study of lethal school violence, giving special attention to these particular events. Specifically, the NRC was asked to convene a committee to "conduct a study regarding antecedents of school violence in urban, suburban, and rural schools, including the incidents of school violence that occurred in Pearl, Mississippi; Paducah, Kentucky; Jonesboro, Arkansas; Springfield, Oregon; Edinboro, Pennsylvania; Fayetteville, Tennessee; Littleton, Colorado; and Conyers, Georgia."

Congress also requested that the study should be conducted through the development and analysis of detailed case studies describing the circumstances leading up to the events, what happened in the events, and how the community responded both before and after the event. The goal was to use the case studies to learn, first, about the important causes and consequences of these unexpected lethal shootings and, second, what actions individuals and institutions could take either to prevent such events from occurring in the first place or to minimize the damage once they begin to unfold.

This special mandate from Congress challenged the National Academies' National Research Council (NRC) to focus its attention on a very specific, very urgent national problem. Viewed from one perspective, this is not unusual. The NRC is often called on to offer scientific advice to policymakers as they confront urgent, specific problems. The difference here, however, was that the violent incidents that galvanized congressional concern seemed to many to be *new* and *unique* as well as *urgent* and *specific*.

Of course, the problem represented by multiple, lethal victimizations in school settings could, on close investigation, turn out to be neither new nor unique. Perhaps there had been episodes like this in the past or in other places that had not been noticed as much as these recent incidents. Perhaps the phenomenon could be easily understood against the backdrop of a larger theory of violence, or of school violence more generally, or as some kind of offshoot of the dramatic increase in youth violence that had occurred earlier in the nation's inner cities.

If, however, on close investigation, the incidents that attracted special congressional attention turned out to be both new and different from what society had seen before—producing different kinds of consequences, caused by different processes, and best prevented through different mechanisms than the other more familiar kinds of violence—then the committee charged with characterizing and understanding the problem would find itself in a difficult position. It would be challenged to make sense of a phenomenon for which no established literature existed.

Without a strong science base to rely on, the committee would have to abandon its usual procedures of reviewing an extant literature, arbitrating the disputes contained therein, and synthesizing the results. It would instead have to engage in original research—something that is often thought to be done better by individual scholars pursuing their theories rather than a committee trying to reach a shared understanding. To the extent that original research was less than definitive, the committee would enter a realm in which the standards for making claims about causes, consequences, and effective cures were ambiguous. It would be operating in a realm in which its findings had only a little more weight than the opinions of other thoughtful observers.

THE COMMITTEE'S APPROACH

The task before the committee was to find some satisfactory way to meet these challenges. An important first step was to develop agreement about the operational definition of the phenomenon under study. An abstract concept does not always point unambiguously to a single operational definition. Consider, for example, the problems that arise when developing an operational definition of a concept such as alcoholism. Should the idea include both very high rates of alcohol consumption and frequent periods in which a person is highly intoxicated, or is that definition too narrow? Should it be defined only in terms of consumption or include the idea that individuals have "lost control" of their drinking?

The development of an operational definition requires the exercise of some judgment on the part of the investigators. Much is at stake in that judgment. A particular phenomenon can be made to seem large or small in the world. It can come close to ordinary understandings of the word or concept, thus allowing an informed discussion; or it can depart from ordinary meanings of terms, thus introducing confusion and error into the public discourse. It can harden misconceptions about the nature of a problem, or it can help to illuminate. That is why people fight about operational definitions.

On one hand, the committee's charge invited a very broad definition—one that was almost exactly coterminous with a literature on school violence in general. On the other hand, it pointed to a set of incidents that seemed to have distinctive characteristics. The distinctive characteristics included not only the fact that the incidents occurred in schools and were committed by students, but also that they resulted in multiple deaths and serious injuries in a single incident. Incidents with these characteristics represent only a tiny subset of all instances of school violence: the few cases at the right tail of the distribution of seriousness for such events.

While Congress seemed primarily interested in understanding these very serious incidents of school violence, seeing them in a broader context that could help us explain the size and significance of these very frightening events, as well as what the causes of such events could be, seemed equally important. This led us to adopt an approach that focused attention on four dimensions. Serious school violence involves:

- incidents of lethal violence
- that took place in or were associated with schools
- that were committed by students of the school and
- that resulted in multiple victimizations in a single incident.

We did not have to include "use of a gun" in the definition of the events that interested us. It turned out to be true that all but a few of the cases that fit the definition of multiple victimizations involved the use of guns.[2] We could, therefore, add the concept of "shootings" to the definition without loss of accuracy.

The committee summarized this definition as "lethal school violence including multiple victimizations." This has the virtue of being short and easy to understand, and it focuses on a problem that is obviously of great public importance. It is also neutral and objective. The words have precise meaning. Furthermore, the definition itself does not suggest much about the causes of the events, and therefore it does not bias attempts to explain why the events occur. Moreover, the facts that one needs to assign particular incidents to this category of violence are relatively easy to ascertain for any given incident.

The concept of lethal school violence could include incidents in which only one person was killed or injured—a phenomenon that is much more common and may be somewhat differently motivated and executed than incidents in which many were killed or injured all at once. It could also exclude such incidents as the Rockdale County case (Conyers, Georgia) that Congress had identified in the legislation. That incident included multiple injuries but no fatalities. The committee didn't think it made sense to exclude this incident, since it was certainly more probable than not that someone could have died in this kind of incident.

The committee therefore decided it would be a mistake to apply its operational definition—lethal school violence including multiple victimizations—too rigidly. It seemed clear that any incident in which a student walked into a school and started shooting apparently randomly was of potential interest. This moved us away from defining the violence in terms of its *consequences* (measured in terms of victimization) and focuses instead on the *motivations and behavior of the offender*. This operational definition of lethal school violence seemed best to reflect congressional intent in requesting the study.

The next step was to consider what sources to use to develop an understanding of the incidents. We already knew that we would have cases describing some of these incidents in rich, narrative detail—providing much of what we would rely on.

We also thought it was important to put the cases into a broader perspective. We sought to locate this particular kind of lethal school violence in both the country's overall experience with violence and in our theoretical understanding of violence.

To do so, we needed to construct a new database to identify all the incidents of violence that met the criteria. This would indicate how large this form of violence was compared with other forms, when this form of violence appeared, and how fast it had been growing. It might even allow us to see whether this kind of violence seemed to move independently of or consonant with other forms of violence, thus providing clues about whether its causes were the same as those that caused other forms of violence.

We also needed to identify and read the literatures on violence that seemed related to the particular incidents of violence we had been asked to study. This was important not only in itself, but also as a necessary guide to our efforts to interpret the cases. Our theoretical understanding and review of data would help us select cases and construct the templates for gathering information. Our review and understanding of the relevant literature would help guide our interpretation of the cases.

At the outset, it was not clear which particular theoretical literature would be relevant, since we did not yet know how to locate the form of violence we had been assigned to study. Again, we knew it was at the tail of some distribution because the incidents were extremely serious and therefore quite rare. But it was by no means clear to what theoretical categories these particular instances of violence should be assigned.

In the end, we explored several literatures to help interpret the cases and more generally to help us think about what might be causing the incidents and how they might best be prevented. First, was a very small literature on incidents that looked very much like the ones we had been asked to review (only four studies, and only two of them that met reasonable standards of scientific care and rigor). Second, there are large literatures on broad categories of violence, such as violence in general, youth violence, school violence, and the relationship between violence and suicide. Third, there were also literatures on some specialized forms of violence that bore some similarity to the incidents under study. These included literatures on mass murders, rampage shootings, and "suicide by cop" (e.g., incidents in which individuals seemed to shoot in order to provoke a response by the police). Finally, because there might be some contagion effects in the events we were examining, we looked into the literatures that explored the contagiousness of violence and other social events.

These are the sources the committee used to understand the incidents of lethal school violence involving multiple victims: the cases that describe six specific incidents; a statistical database constructed for the committee's purposes from several existing sources; and several literatures that had something more or less directly relevant to say about this phenomenon.

Perhaps the most valuable resource we had was the commitment and expertise of the case writers who developed the cases, all senior scholars with a significant amount of experience in qualitative methods and most with substantive knowledge in the area of youth violence. The discussions we held were designed to stretch the evidence and knowledge on incidents of lethal school violence as far as we reasonably could, and to make it available to the nation as it sought to deal with this important, urgent problem.

There is almost nothing in this report that meets the usual scientific standard of 95 percent confidence that a statement is true. But there is much that is likely to be true (more probable than not), and some of these things are not obvious. There are also some findings that upset some conventional assumptions about the phenomenon we studied, and there are some pretty clear ideas about where the priorities for future research may lie that could strengthen the science base for understanding and preventing particular kinds of lethal school violence.

The report is organized in the following way. Part I presents the cases. They are preceeded by a short analytic section that explains why they were chosen, lays out the template used to collect the information, describes the sources consulted to obtain the information, and describes the procedures followed by the case authors and the committee for the protection of human subjects. Part I also presents a cross-case analysis of the similarities and differences among the specific cases to develop some plausible hypotheses about the causes, consequences, and effective methods of preventing and controlling these incidents.

Part II puts the cases into the context of the literature reviewed and uses the cross-case analysis to develop some observations about policy implications and research recommendations. In reading Part II, it is important to keep in mind that the work contained therein reflects sources beyond the cases. It presents statistics about trends in violence and findings from the literatures that seemed relevant to the inquiry.

We begin with the cases because, even though these are not all that we relied on, they are in many ways the heart of our understanding and the source of some of our most important ideas. It is valuable for readers to enter this field as we did—in an inductive, exploratory way rather than a deductive, hypothesis-confirming way. The cases, presented here as signed, stand-alone pieces by the case investigators, are full of surprising facts, poignant moments, and rich insights. They put a human face on the tragedies that have beset the nation. They fill in the gaps that empirical studies cannot address. Those who are more inclined toward a deductive approach and would like to see what observations we have drawn first and then to test them against the evidence of the cases should feel free to start with Part II and read the cases afterward.

NOTES

[1]Epidemic means an elevated level; it does not necessarily imply a contagious phenomenon. There could be a contagious mechanism at work, but that requires investigation.

[2]One case in Portland, Oregon, involved a knife, and an international case involved arson.

Part I

Case Studies of Lethal School Violence

The request for original research on what appear to be rare and extreme events necessitated a comparative case study approach. However, the aim of the case studies was not to generate certain, scientific knowledge about the causes, consequences, and effective methods of preventing and controlling these events. It was obvious from the start that these few cases could not support such an ambitious goal. As a scientific matter, there were too few data points to allow us to decide which of many possible explanations were true and which of many plausibly effective responses would actually work. The aim instead was to use the limited experience available to develop some plausible hypotheses about causes and effective interventions and to check commonly held assumptions for their plausibility.

DEVELOPING THE CASE STUDIES

In developing and comparing these cases, the committee was committed to using the discipline and methods of science to ensure that the information gathered was accurate and could be usefully interpreted by others. The desire to be disciplined in the development of the cases forced us to take up three important study design issues: (1) how the cases would be selected for study, (2) what information would be sought across all the cases, and (3) what sources of information would be used. We sought to answer these questions in a way that would maximize the evidentiary and inferential power of the cases—again, recognizing in advance that the cases would fall short of providing definitive scientific answers.

CASE SELECTION

The first challenge was to choose the particular cases to be developed. This task had both practical and scientific elements. Congress had provided some guidance to the committee on this matter by specifying eight incidents that were illustrative of the problem they considered important to take up and that might be suitable for detailed study. The practical part of the problem, given that resources were available for only six cases, was to select from among those listed.

The scientific question before the committee was to determine the general class of violence of which these eight incidents were exemplars. As stated in Chapter 1, Congress asked the committee to examine "incidents of lethal school violence in *urban, suburban, and rural schools,*" yet all of the cases identified by Congress occurred in suburban and rural schools.

It seemed important that the committee address both the spirit and the letter of the congressional request. We therefore decided initially to study incidents that had the characteristics listed in our operational definition of lethal violence regardless of the nature of the community in which they occurred.

Moreover, as part of our work, we developed a dataset of all such incidents, using it as a kind of sampling frame for the set of cases and later as a way of indicating both levels and trends in this form of violence. Since much of the lethal violence among young people had occurred in inner-city schools, we assumed we would find examples of this kind of lethal violence in inner-city schools as well.

To our surprise, we could find *no* cases in urban inner-city schools that met these requirements in the time period we were examining, 1997–1999. There were incidents of lethal violence in urban schools, and there were a few schools that had experienced more than one fatality in a given year. But no incidents had what seemed to be the key characteristic of multiple victimizations including fatalities occurring in the same incident.

That preliminary finding was a very important one to the committee. It seemed that the form of lethal school violence that occurred in the late 1990s might represent a distinct form of lethal school violence—one that might be similar in its consequences for the victims and communities in which it occurred, but different in its causes and in its effective prevention and control.

This possibility encouraged us to take up the important scientific question of what the relationship might actually be between the form of lethal school violence that was concentrated in inner-city schools and the seemingly newer form of lethal school violence that erupted in suburban schools in the late 1990s. Several such relationships were possible. One was that the different forms of violence were the products of similar causes that played out differently in the different community contexts. A second was that the inner-city violence had created the conditions that shaped the later suburban violence. A third possibility was that there was, in fact, little relationship between the urban violence and these new cases.

The decision to take up the scientific issue of whether this was a new and unique form of lethal school violence had important implications for case selection. It would be important to look closely at examples of lethal school violence in inner cities to determine whether the causes of such violence were similar to the causes of the newer forms of violence. In effect, we could choose the cases to get some variation on the dependent variable: within the class of incidents neutrally described as lethal school violence, we could look at the form that this violence took in different kinds of communities—urban, suburban, and rural. If it turned out that the antecedents to lethal school violence in inner-city schools were different from those in suburban and rural schools, then we would have some evidence pointing toward a firmer conclusion that this was a separate strain of violence.

But there was another reason to look at lethal violence in inner-city schools. A preliminary look at the data indicated that levels of overall lethal violence in inner-city schools were much higher than in suburban-rural schools and had been that way for a long period of time. By developing cases on lethal violence in inner-city schools and comparing them with lethal violence in suburban-rural schools, we could put these incidents under a microscope and describe the structure of the similarities and differences in the character and antecedent causes of lethal violence in different settings. This would help us understand whether there was something about inner-city communities that made them immune to the forms of violence that hit suburban-rural schools in the late 1990s, and whether there was something about the suburban and rural communities that seemed to protect them from the violence that struck the inner cities in the decade from 1985 to 1995.

These considerations were sufficient to persuade the committee that a portion of our limited resources should be focused on developing cases of lethal violence in inner-city schools. To find such cases, we simply had to relax the time frame under consideration. In the period 1990–1992, we found two inner-city schools that had experienced incidents in which multiple individuals were killed and injured. A school in Chicago experienced a shooting in which one person was killed and two were wounded in 1992. And a school in New York experienced two incidents involving multiple victimizations. In an incident in late 1991, one person was killed and one was injured. In a second incident shortly thereafter in February 1992, two people were killed. We decided to develop cases on these events in addition to four from among the cases Congress identified.

With regard to the suburban and rural cases, the committee was aware that in three of the sites listed in the legislation, the shooters had been interviewed for case studies conducted by others, including the Secret Service, the Department of Education, and the Federal Bureau of Investigation. We assumed that the previously studied sites would not be particularly amenable to yet another team of researchers conducting an in-depth examination of their community and decided to exclude them from our sample. We decided that we could better add to the knowledge base by selecting sites that had not been previously studied. We had hoped to be able to examine data from the Secret Service and Department of Education case studies of other rural and suburban sites for our analysis, but confidentiality agreements between the researchers and their subjects precluded this possibility. In the end, we chose the following six cases for close examination:

- Heritage High School, Rockdale County, Georgia
- Parker Middle School, Edinboro, Pennsylvania
- Heath High School, Paducah, Kentucky
- Westside Middle School, Jonesboro, Arkansas
- Tilden High School, Chicago, Illinois
- Thomas Jefferson High School, East New York, New York

THE CASE TEMPLATE

Once the cases were chosen, the next decision was what information to gather for each case. The aim was to be sure that there was enough information to be able to consider a variety of possible explanations for the violence that was at the center of the cases, and to gather the information as consistently as possible from one case to another in order to be able to check for the presence or absence of each particular variable. The committee developed a template that each case-writing team could use to guide their data collection efforts. A shortened form of that template follows:

1. **Situational factors**: Narrative description of the events immediately surrounding the incident (leading up to and immediate response to

the incident at different levels of analysis as seen by different participants and witnesses):

- Description of the shooting itself: preparations, precipitating events (including possible provocation by victims), location, targets, relationship of shooter to victims, immediate responses that ended the incident
- Motivations/state of mind of shooter at time of shooting
- Recent trends in objective and subjective life of the shooter
- Witness accounts of shooting (including their interpretations of the offender and motivations)
- Warning signs for incident (e.g., shooter's threats, widely known festering grievances and disputes)
- Immediate conditions in the school affecting motivation for incident
- Immediate conditions in the school affecting response to incident
- Immediate conditions in the community affecting motivation for incident
- Immediate conditions in the community affecting response to incident

2. **Individual factors**: Individual traits and family background of offenders:

- Prior criminal activity of the offender
- School record (both achievement and disciplinary)
- Peer standing/affiliations at school (What groups? What standing in individual groups? Relationships with opposite sex?)
- Important adults in offender's life/quality of communication and connection
- Family relationships (parents/siblings) (strength/quality) (parental knowledge/supervision of kids)
- History of mental illness
- Interest/consumption of violent media materials
- Experience with firearms

3. **Community-level factors** affecting incident and response:

- Economic status of community (mean and variance)
- Stability of community (transience)
- Social coherence/divisions in community (ethnic, racial, religious, political)
- Stock of "social capital" in community
- Engagement of community with teenagers and with schools
- Teen culture in community
- Police strategy/organization/connection to community
- Justice system organization/connection to community

4. **School-level factors** affecting incident and response:
   - Size of the school
   - Organization of the school
   - Teacher characteristics
   - Parental involvement in the school
   - Educational policies of the school (tracking/class size/extracurricular activities)
   - Governance and disciplinary policies and practices of the school
   - Security arrangements for the school
   - Extent and quality of teacher connection to students beyond curriculum

5. **Description of response and consequences** for community of both incident and responses made:
   - Outcomes of court cases (criminal, civil)
   - Consequences for offenders
   - Consequences for victims
   - Consequences for families of offenders
   - Consequences for families of victims
   - Grief counseling/activity following events
   - Policy changes (and apparent consequences) initiated with schools:
     - ♦ New hardware at schools (magnetometers, fences)
     - ♦ Heightened surveillance and control of students
     - ♦ Police officers in schools
     - ♦ Use of transfers of students to other schools
     - ♦ Increased efforts to deal with festering disputes and grievances
   - Policy changes (and apparent consequences) initiated in wider community

SOURCES OF INFORMATION

With the template developed to define the information to collect for each case, it became important to describe the sources to consult to obtain the needed facts. The sources of information relied on in developing facts and observations to fill out the case template include:

1. Journalistic accounts:
   - local newspapers
   - local radio and television
   - national newspapers and magazines
   - national radio and television coverage
   - special documentary reports

2. Official records pertaining to the incident or the offender:
   - court records (criminal and civil)
   - police records
   - school records

3. Governmental statistics
4. Interviews:
   - the offender
   - the offender family
   - friends/acquaintances of offenders
   - the victims
   - the victims' families
   - witnesses to shootings
   - responders to the incident
   - those involved in handling legal cases
   - school officials
   - teens in the community
   - adults in the community
   - political leaders in the community
   - civic leaders in the community

5. Direct/participant observation
6. Surveys

PROTECTION OF HUMAN SUBJECTS

All of the case authors submitted their study designs, including multiple consent forms, to their university's or organization's institutional review board (IRB) for the protection of human subjects, and in one case directly to the National Research Council's (NRC) institutional review board. The NRC's IRB provided a second layer of human subjects review for five of the six case studies once they had been approved by university or organizational IRBs. The NRC then sought and obtained a certificate of confidentiality from the Department of Health and Human Services for the entire project—that is, all six case studies. All subjects interviewed in these cases, no matter what their occupation or public role, signed a consent form that cautioned that every effort would be made to keep responses confidential and anonymous. However, those already in the public eye and/or very close to the events who would be readily identifiable could not be promised such protections. Finally, only the individual case authors have access to the interview data on which these cases are based.[1]

VALUE OF THE CASE STUDY APPROACH

The strengths and weaknesses of the case study approach are discussed in detail in Appendix A. For this study, the committee sees them as valuable in helping people to understand and respond to instances of lethal violence in schools and school rampages. The case descriptions reveal important possible causes and points of intervention that might never have been considered by social scientists working with general models of violence and relying on statistical information to guide their understanding of causes and solutions. To some degree, thick description of events allows for a different kind of causal analysis than is possible by using large samples of superficially described events. In sum, the cases present a different method for developing ideas about causes and potential interventions. Indeed, in looking at phenomena that are very rare and cannot be studied in laboratories, it may be that thick description is the only viable way of learning much about the likely causes or potentially important interventions.

The value of the cases goes beyond their value as evidence in the scientific process of finding causes and effective interventions. The cases are valuable as stories to be used by communities as they make their own judgments about the nature of this threat, what they ought to do to prevent it, and how they ought to react to it if a school shooting should occur in their midst. Despite the enormous advances of scientific knowledge, this probably remains the principal way that most people facing real problems continue to try to learn. We present these cases with the hope that they will support learning in the nation's communities when they are used in formal and informal discussions about the problems of lethal violence in schools and school rampages, as well as when they are used as part of a more elaborate and formalized method of scientific inquiry into these matters.

NOTE

[1]Within the broad outline of the template developed by the committee, the field work for the case studies was conducted and the studies were signed by the case authors as independent researchers. The case authors determined who to interview and what records to review, and each team independently arrived at the findings and conclusions in the individual case studies. Moreover, only the case authors had access to interview transcripts on which these cases are based. The authors were responsible for compliance with the protection for human subjects consistent with the institutional review boards approvals as described.

2

The Copycat Factor: Mental Illness, Guns, and the Shooting Incident at Heritage High School, Rockdale County, Georgia

Mercer L. Sullivan and Rob T. Guerette

On May 20, 1999, one month to the day following the school shootings at Columbine High School in Littleton, Colorado, Anthony B. Solomon, Jr., known as T.J., entered the commons area of Heritage High School in Rockdale County, Georgia, and opened fire with a .22 caliber rifle. He discharged 12 shots, emptying the rifle, and ran from the building. While doing so, he pulled out a .357 magnum handgun and fired three more shots. He then knelt, put the handgun in his mouth, and hesitated. Shortly thereafter, he surrendered the gun to a school official and was taken into custody by law enforcement officials. He had not killed anyone, but he had wounded six students, one of them seriously. He was subsequently convicted as an adult and received a long prison sentence, with a minimum of 18 years before possible parole.

Although the incident at Columbine had provoked a great deal of discussion among youth and adults in Rockdale County, T.J. Solomon's actions came as a complete surprise to everyone in the area, both because of where it happened and because of who committed the act. Rockdale County is an affluent suburb of the city of Atlanta, known as an area of expensive homes, high-quality schools, and low rates of crime and delinquency. Heritage High School had been rated one of the best public high schools in the state and had never experienced significant problems with youth gangs or other patterns of serious violence. T.J. Solomon had never been arrested, had no reputation for getting into fights or being aggressive, and came from a close, upper-middle-class, churchgoing family in which the parents closely monitored every aspect of their children's lives.

This case study examines in more detail this community, school, family, and individual in order to trace the antecedents and consequences of the May 20, 1999, incident. As we show, there were antecedent conditions that can be connected, at least in retrospect, to the eventual occurrence of the incident.

METHODOLOGY

The authors of this case study gathered data during two field visits to Rockdale County over the course of one month, some subsidiary field trips to interview knowledgeable people not present in Rockdale County at the time, many telephone calls, and through collecting a large body of archival material. All interviews were conducted according to procedures for the protection of human subjects approved by Rutgers University and the National Academy of Sciences. Some subjects with uniquely identifiable roles in the event agreed to be interviewed for public attribution, but most subjects spoke under pledges of confidentiality.

Data collection yielded an extensive and diverse record. A total of 42 people participated in interviews, including law enforcement, local government, and school officials; some of the victims and their parents; journalists who had worked in the community; and community members, including adult residents and also young people, a number of whom had known T.J. Solomon and his family. Interviews were also conducted with people who had known T.J. and his family before they moved to Rockdale County.

Besides the interview data, researchers also had access to an extensive archival record. Census data, Chamber of Commerce reports, and school, police, and health records provided background data on the area. Newspaper and other media accounts provided initial glimpses of the incident and those involved, although many of the facts in the early accounts proved to be erroneous.

More directly relevant to the main concerns of the case study, however, were the extensive files provided by law enforcement agencies, including the sheriff's office and the district attorney's office of Rockdale County, which had direct jurisdiction over the case. Since T.J. Solomon was eventually transferred out of juvenile court and convicted as an adult in superior court, all of these records are publicly available under Georgia law. They include several hours of videotapes of the young offender, both immediately following the incident and from a psychological interview conducted three months later, investigative reports, psychological assessments, evidence inventories, crime scene photographs, depositions with family members during subsequent civil lawsuits, and voluminous miscellaneous supporting materials. The investigative reports alone provided transcripts of interviews by law enforcement officials with dozens of people who had knowledge of the offender, his family, the victims, and the school.

While the detail and diversity of this record are notable, it is also true that the largest share of the record, at least in terms of bulk, was obtained from law enforcement sources. While we are fortunate to have this information and personally beholden to the individuals who facilitated this access, we are also aware of the potential pitfalls of relying too extensively on law enforcement sources.

Despite overtures to T.J. Solomon, his family, their attorneys, and the Georgia Department of Corrections, we were unable to interview any of them directly. The only response to any of these requests came from a spokesperson for the Department of Corrections, who replied that our request had received serious consideration but could not be granted at the time because of concern over the mental health of T.J. Solomon, who had attempted suicide in prison in December 2000, six months prior to the data collection period.

Despite possible concern about the extensive reliance on law enforcement sources, it turns out that there is very little disagreement across multiple sources of information, either within the law enforcement data archives or between them and our many other data sources, about the basic facts of the case. The main disputes in the adversarial process that pitted the interests of the offender against those of criminal justice officials had to do with the assessment of T.J. Solomon's mental health. Even in this matter, the facts bearing on that assessment were little disputed, only their interpretation, and the opposing interpretations are well documented.

Extensive psychological assessments were conducted by highly qualified examiners working separately for the district attorney, the juvenile and superior courts, and T.J. Solomon. His family was able to secure prominent legal counsel and reputable psychological consultation. The contending interpretations of his psychological condition, as presented by prosecutors and defense lawyers and decided on by the juvenile court and superior court judges, are matters of public record, available in the transcripts of the transfer hearing from juvenile to superior court and the sentencing hearing in superior court. We turn now to descriptions of the community, the school, the offender and his family, the incident, and the aftermath.

THE COMMUNITY

Rockdale County lies just southwest of Atlanta, bisected by Interstate Highway 20 and State Highway 138. The opening of Interstate 20 in 1963 led to a major transformation of Rockdale from a rural, exurban area surrounding the small town of Conyers, to a rapidly growing suburb of the rapidly expanding city of Atlanta. Although Rockdale ranks as the second smallest of the many counties of the state of Georgia in geographical area, its population tripled in size in a mere three decades. Of the more than 70,000 residents in 2000, only about 10,000 lived in the city of Conyers. Most of the rest lived in a series of recently developed housing subdivisions, ranging from moderate to expensive in price, throughout the rest of the county. Many of these middle-class and upper-middle-class families have moved to Rockdale fairly recently, drawn by the vibrant economy of metropolitan Atlanta and the prospect of a relatively short and direct commute, at least as compared with their competing options for dealing with the notorious traffic difficulties that beset Atlanta.

Table 2-1 compares the demographic characteristics of Rockdale County with those of the nation, the state, and its neighboring counties in the metropolitan Atlanta area. Fulton County contains most of the city of Atlanta. DeKalb County lies between Fulton and Rockdale and contains a small portion of the city of Atlanta as well as another large city, Decatur. Rockdale is the outer ring suburb. As the table shows, levels of family income and home ownership rise progressively from Fulton out to Rockdale, while levels of poverty and minority population fall.

The old town of Conyers serves as an administrative center for the courts and the county government and is also a kind of inner city where the county's small population of blacks and lower-income white families with direct ties to the modest circumstances of a rural past are concentrated. A few signs of gentrification are visible in the form of gourmet restaurants and fancy shops developed in the charming old brick buildings beside the railroad track, but many of these old buildings are still untouched.

TABLE 2-1 Comparative Demographics—Income, Race, Home Ownership, and Poverty by Area

Area	Median Household Income	% Black	% Hispanic	% Owning Home	% Below Poverty
United States	$37,005	12	13	66	13
Georgia	$36,372	29	5	68	15
Fulton	$39,047	45	6	52	18
DeKalb	$42,767	54	8	59	13
Rockdale	$48,632	18	6	75	9

SOURCE: U.S. Census Bureau Figures, 2000. 1997 model-based estimates.

Out in the county, a few country roads with modest old houses remain. A trailer park on the edge of the county called Lakeview Estates houses the largest concentration of poor people in the county. The older residents there are white and of rural origins, while the younger families are Mexican immigrants, most very recent arrivals in the area who work in low-wage service jobs in hotels or temporary construction work. An equestrian center on the northeastern side of the county was built for the 1996 Olympics and now functions as a recreational and sporting destination for horseback riders from throughout the metropolitan region and beyond.

The commercial as well as geographical spine of Rockdale is the Interstate 20/Georgia 138 corridor, lined on either side with strip malls. One resident noted that the area has grown so rapidly and recently that it is just now making plans to build its first regular mall. The recent transformation of the area is noticeable even to teenagers, like the focus group participant who responded to an open-ended question about what the community is like ("What would you tell an email pen-pal in another place about your community?") as follows: "Our community right now . . . it's big . . . it's growing. . . . It went from a small country [town], where somebody can sit on their front porch on the strip and watch cars go by, to Wal-Mart, K-Marts left and right, to hotel buildings everywhere to kids getting worse, the atmosphere, the population is growing. . . . people are getting more money."

The connection of community growth and prosperity to increasing youth problems in the above statement is a contentious issue within the community. While many people interviewed agreed that the connection exists to some degree, Rockdale residents also feel that it is overblown and that they have been unfairly singled out as examples of something that is happening all over the United States.

Their anxiety stems in part from the notoriety conferred on them by T.J. Solomon's deed, but their discomfort stems mainly from a public television documentary entitled The Lost Children of Rockdale County that aired on the series Frontline shortly after the Heritage High School shooting incident (Goodman and Goodman, 1999). The documentary is framed, somewhat gratuitously, by film clips of the Heritage High shooting, but its main subject is a syphilis outbreak among teenagers that occurred in 1996.

The outbreak, which affected over 200 teenagers, grew out of a pattern of extreme sexual experimentation among one clique of local youth. It centered around one wealthy young man who had the run of his parents' large house without supervision. He also liked experimenting with violence and had recruited an associate for his violent exploits who was poorer than he, black, and interested in sharing his wealthy lifestyle. The combination of kinky sex, interracial sex, very young participants, and wealth created a sensational impact nationally, but especially in Georgia and Rockdale County. The film also showed other aspects of the varied youth lifestyles in Rockdale, notably the revivalism of a local grass-roots church and the charismatic leadership of its pastors and their son, the lead singer of a popular Christian rock band.

There was no direct connection between T.J. Solomon and the events portrayed in the rest of the film. The syphilis outbreak had actually occurred three years before the shooting and had been excavated after the fact by the filmmakers. Local health officials had responded quickly, identifying and effectively medicating those affected. By the time of the Heritage High shooting, the disease was gone, but the local stigma remained.

The combination of these two sensational events, both exceedingly rare within Rockdale as well as outside it, one on top of the other, has given Rockdale residents a sense of being under scrutiny for living in a strange and evil place. Most people we interviewed considered this image wildly inaccurate and unjust. In contrast, they point, with much justification, to the excellent quality of their schools, the high level of church membership and participation, and the low crime rates, especially in comparison to the more urbanized counties of the Atlanta area. A frequent point of comparison is to neighboring DeKalb County.

The teenage focus group participants also corroborated the importance of churches and religion in the community: "Our town has grown into, there is a church on every corner. It's Bible, it's kind of like the Bible Belt, we are in it. It is like you are in the heart of it here. . . . There are churches everywhere in Rockdale. It's a real religious based county."

Of course, it is entirely possible for religiosity and deviance to coexist in the same community. In fact, there is something distinctly and traditionally Southern about that combination. The existence of high rates of violence in the Bible Belt, for example, is one of the better known facts in criminological scholarship (Butterfield, 1995; Curtis, 1975). What makes Rockdale County really distinctive is not that its residents are so violent. There is clear evidence that they are not.

Rather, the sensational but rare acts of extreme youth deviance that have brought so much unwanted attention as well as the high rates of church membership and investment in high-quality public schools seem to be related in plausible ways to the underlying structural characteristics of the community, namely, rapid change and prosperity. Rockdale County is a place where people with money have been moving in very quickly. They spend their money on building family and community, yet the community is not stable. Institutions have perhaps not had time to take root. Attachment to the community is recent and far from exclusive.

Adults work outside the community for long hours and negotiate a lot of the worst traffic in the country. People have lots of cars, including young people, with the result that they can drive to neighboring counties or to entertainment in Atlanta.

Under these circumstances, there may be more money than personal time available for building community in this rapidly expanding community. It is clear that there are many young people with time on their hands, money, cars, and little supervision. T.J. Solomon, as will be seen, fit some parts of this profile and not others. He had lots of supervision, for example, much more than others. Still, he lived in this environment and he had to cope with the challenges of adolescent development in this particular context.

CRIME AND CRIMINAL JUSTICE IN ROCKDALE COUNTY

The Heritage High School shooting incident occurred in a community with low and stable rates of crime, among both juveniles and adults. Figure 2-1 shows that crime rates in Rockdale County bear an inverse relationship to its prosperity, in comparison to its neighbors, the state, and the nation. The statistical evidence of this presented in this figure is further confirmed by interviews with law enforcement and judicial officials and community members. Inquiries about other incidents of serious violence repeatedly elicited descriptions of the same small handful

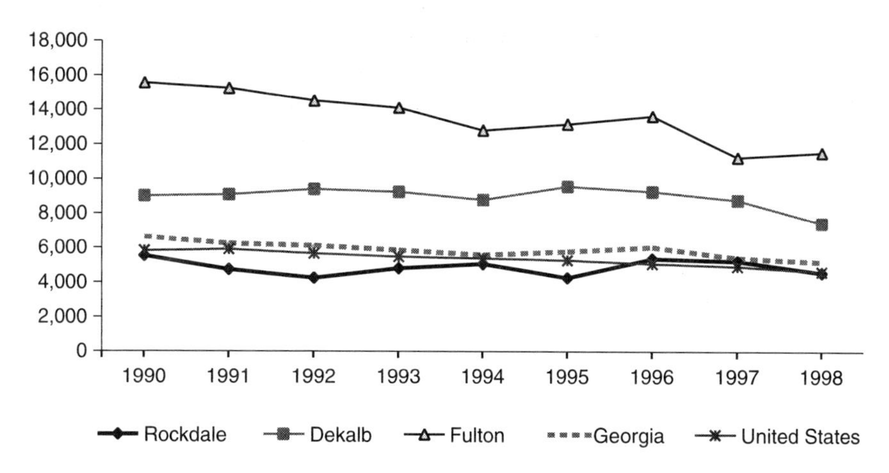

FIGURE 2-1 Overall index crime rate per 100,000 population, 1990–1998.
SOURCE: Data compiled from Georgia's Criminal Justice Coordinating Council—Crime Statistics and FBI Uniform Crime reports.

of incidents, well known to all in the community by virtue of their rarity.

Murders in Rockdale Country occur once or twice a year, almost always involving adults. There had been only two youth homicides in anyone's memory. One was a sensational thrill-killing incident that occurred in 1992, in which a youth ordered a pizza and killed the delivery woman. The other occurred about two months after the Heritage High School incident and was described as a freak accident that occurred in the midst of an ordinary scuffle. Two males arguing over a female began wrestling. One of them grabbed the other in a headlock and ruptured a blood vessel.

To the extent that there is a higher crime area in this low-crime county, that area is the trailer park, Lakeview Estates, which accounts for the most frequent calls for service to the sheriff's office, the law enforcement agency responsible for the areas where over three-quarters of county residents live. Many of these calls are related to domestic violence incidents and drunken brawls. As is true in most communities, there is thus a connection between geographical concentrations of poverty and violent crime. The residents of this area include white people native to the area and a more recent population of immigrants from Mexico. While both groups are involved in these calls for service, one law enforcement official reported that the area had actually gotten more peaceful since the recent influx of immigrants. The problems had been at their worst in the late 1970s and early 1980s and were caused by the "rednecks." This "redneck" population was now aging.

Young people in the county also occupied the attention of law enforcement, but more in terms of order maintenance than of crime control. Law enforcement officials reported that the main places where they had to deal with juvenile problems were the strip malls along the I-20/Georgia 138 corridor, where crowds of teens often gathered. These areas constitute the bulk of public space in the county. Besides being the location of commercial entertainment, such as movies and the bowling alley, their parking lots were frequently filled with crowds of young people just hanging out. The participants in the youth focus groups confirmed that the law enforcement kept a close eye on their public activities:

> SPEAKER 2: Yeah, like a couple of years ago we all used to hang out in front of the Kroger's parking lot and it was like everybody just split and went their separate ways.
>
> SPEAKER 3: Yeah, cops put a stop to that. . . .
>
> SPEAKER 1: Now they're hanging out at Salem Gate. . . .
>
> SPEAKER 3: Yeah, it's finding drugs, finding alcohol, and finding girls.

While not all the young people who participated in the focus groups would have agreed with the statement about finding drugs and alcohol being the center of teen life in the area, clearly these patterns of activity were present and kept law enforcement busy. The same could probably be said for many if not most communities, but it is noteworthy that here both young people and the police agree that the police keep a close watch.

Youth gang activity, in contrast to the partying just described, has not been prominent in this area, despite the fact that widespread increases in youth gang prevalence were reported during the 1990s, particularly in suburban areas in which such activity had not been seen previously (Miller, 2001). Everyone mentioned one particular group of young people who had been criminally active in the mid-1990s and at one point styled themselves as a gang in the manner of the Los Angeles-based groups so prominent in the movies and popular music at that time. That group, interracial in composition, had centered around one particular individual and dispersed when he was arrested and incarcerated after the group set fire to a church. This same individual was also implicated in the syphilis epidemic. There were no reports, however, of anything resembling a gang war, presumably because it takes more than one gang to make a war.

According to official statistics and interviews, Rockdale County is a low-crime area in which much of the crime that does occur consists of relatively low-level offenses, with very little serious violence. T.J. Solomon's offenses were utterly unlike the normal patterns of crime in the area, in stark contrast to other patterns of fighting among youth and the few other cases of lethal or potentially lethal violence.

Two law enforcement jurisdictions served the area, the Rockdale County sheriff's department and the city of Conyers police department. During our brief field visits, the representatives of these agencies conveyed an impression of progressive professionalism. They were highly cooperative with us, over and above providing the access to their records required under Georgia law. There have been no scandals in this county, unlike neighboring DeKalb, where the local sheriff was murdered just prior to our visits, allegedly in connection with kickbacks and extortion involving vendors to the county's correctional system.

The juvenile court judge described a breakdown of cases before him that is probably typical of this kind of middle-class area: 35 to 40 percent involving family concerns such as ungovernable youth; 15 to 20 percent involving acting out at school; and about 40 percent involving crimes related to illegal drugs and theft. Rockdale County does not have its own juvenile detention facility and sends those in need of secure confinement to neighboring Lawrenceville, as long as they are under the jurisdiction of the juvenile court.

The most significant changes in the local juvenile justice system in recent years occurred as the result of changes at the state level. As in most states, the Georgia legislature has passed legislation making it easier to transfer juveniles accused of serious crimes out of the juvenile court so that they can be tried as adults. The Juvenile Justice Reform Act of 1994 identifies seven categories of criminal charges, known as the "seven deadly sins," for which juveniles over the age of 13 are automatically transferred to superior court to be tried as adults. The seven charges notably do not include aggravated assault, the most serious charge that could be brought against T.J. Solomon, since he did not kill any of his victims. For that reason, T.J. Solomon was subject to a transfer hearing in juvenile court, which became the primary judicial forum in which evidence related to his deeds, intentions, and mental health was presented. Following the decision to transfer him to superior court, that evidentiary record then became the basis for the sentencing hearing that followed his guilty plea in superior court. This was the only transfer hearing held in Rockdale County in the memory of anyone we interviewed.

In 1998, the state parole board also adopted guidelines mandating that those convicted as adults of 20 specific offenses serve 90 percent of their sentences before becoming eligible for parole. This change in policy eventually affected the disposition of T.J.'s case.

Another significant change in juvenile justice in this state was the 1998 signing of a memorandum of agreement between the State of Georgia and the U.S. Department of Justice, following a federal investigation of conditions of confinement in juvenile facilities in the state. The results of that investigation revealed severe deficiencies in educational and health services and the monitoring of abuse.

FIREARMS

As is true throughout most of the South, hunting is a common activity in Rockdale County. Consequently, gun ownership is pervasive. Under Georgia law, the age at which a person becomes an adult with respect to most criminal code violations is 17. In order to possess or have control of a handgun, however, a person must be at least 18, unless the minor is using the handgun at a safety training course, for target shooting, for hunting or fishing, or is in transit to or from any of these activities. A minor is also allowed to have possession of a handgun on his or her own or parents' own property if there is parental consent to maintain such possession.

Anyone in Georgia, regardless of age, is allowed to have possession of a rifle or shotgun and is allowed to carry such weapons as long as they are openly visible and not concealed. Concealed carrying of a handgun is permitted for those 21 years of age or older who have undergone certified firearms training and have obtained a concealed carry permit from the municipality in which they reside. Possession of a firearm within a school safety zone, school property, school functions, or on school-sponsored transportation, however, is a felony.

Interviews conducted for this study consistently indicated widespread ownership of firearms across the county. Most of the guns owned were reported to be shotguns and rifles; however, ownership of handguns was also stated to be common. Most people agreed that hunting was the most prominent reason for owning firearms, but also noted that people kept them for secondary purposes of self-protection and conveying authority.

Owning firearms was also common among youth. Youths participating in focus group discussions stated that a large portion of the people they knew own some sort of a firearm. Their estimates of the percentage of people they knew who owned firearms ranged from 50 to 75 percent.

Under these conditions, obtaining a firearm was seen to be easy, even if one did not own one. Unanimously, all focus group participants, male and female, stated that they could easily obtain a firearm if they wanted, most indicating that it would take between 15 minutes and an hour. One said it would take "two seconds, if I had driven." Another stated: "Most guys in our community have a gun. It is not like a handgun, like I am going to go shoot up somebody, a lot of guys just have them." Another said: "Now that it's summer, a lot of people carry guns in their trucks. Friends carry guns in their cars. In the school year you can't bring a gun on to school property. Now that it is summer, they have got gun racks behind their heads and show off your guns to people. I mean, I have one right behind my head in my truck. It's not like a big thing."

Firearms are thus accepted as an everyday part of life for people in this area, including young people. Guns are ubiquitous.

YOUTH

Of Rockdale County's estimated 70,111 residents, about 28 percent are under the age of 18. Public education data for Rockdale County indicate that 71 percent of students are white, 22 percent are black, 3.8 percent Hispanic, and the remainder Asian, American Indian, and multiracial. The gender ratio is equal.

As in most communities, youth in Rockdale County frequently classified one another according to a set of local categories used to identify particular peer groupings. Some common distinctions made by and of local youth were among "jocks," "preps," "Christian kids," "rednecks," "blacks," "Hispanics," "wiggers," "drama kids," "band kids," the "straightedge mafia," and "loners."

Rednecks were identified as those who like to hunt and fish and listen to country music. Sometimes redneck also means lower class, but other times it does not. Some rednecks come from affluent families and drive new pickup trucks with Confederate flags hanging from the back. Jocks are those who are members of the school athletic teams. Preps were described as kids who dress nicely, wearing button-down shirts. In the words of one youth, "They wear Abercrombie all the time and want to be all nice and look like they are 30." Christian kids are self-consciously religious and abstain from sex, drinking, and using illegal drugs. Drama kids and band kids were identified with those activities. Wiggers were white kids who "hang with, act like, or like to date blacks." The straight-edge mafia, also referred to as "vegans," are a group of youth who are vegetarians. They listen to rock and roll and like to have a good time without smoking, drinking, or using illegal drugs; they refer to these as "the three x's," but their ascetism is not defined in religious terms. The ethnic labels "blacks" and "Hispanics" or "Mexicans" were applied to the small number of racial and ethnic minorities, although associations did take place across racial and ethnic lines. During the fieldwork for this study, for example, researchers observed a racially mixed group of young teenage girls bowling, with no apparent notice taken by other patrons.

Despite the common acknowledgment of these categories, researchers consistently heard that there is considerable overlap in group identity and mobility of individuals across identities and groups. For example, it was noted that many of the rednecks were also members of the athletic teams, making them jocks as well. Furthermore, some youth associated with the drama group, the jocks, and the rednecks interchangeably.

Focus group youth also reported that it is easy for new students to make friends with existing students. In the words of one participant, "You are taken in real well. But it's left up to them who to drift off towards and who to get closer to." In spite of the reported ease of acceptance among Rockdale area youth, several people interviewed talked about loners. In some estimates, the number of loners has grown in recent years as the area has increased in size and diversity. School officials reported that 15 years ago there were relatively few loners, but that the number has increased.

In a 14-month period in 1996 and 1997, two years prior to the shooting, Rockdale County experienced a series of three suspected suicides. A local official reported that while they were all unrelated to one another, the three victims were similar in that they were all loners and came from middle-class to upper-middle-class families.

The various cliques in Heritage High School gathered in distinctive places, both inside and outside school. In the commons area before class and in the cafeteria at lunch, the black students congregated in one area, the rednecks in another, the preps another, and so on.

Similarly, there were also descriptions of various groups gathering in different and predictable places outside the school. After school and on weekends, different groups would gather in the parking lots of retail stores, such as Kroger or TJ Maxx. The rednecks, for example, were said to hang out in the Kroger parking lot, while black youth had begun to congregate in the parking lot in front of Wal-Mart. Other focal points of youth activity were the local bowling alley, movie theatres, and church facilities. During football season, Friday night home games are attended by most of the student body and serve as a focal gathering point for the entire community.

As is true in many communities, some Rockdale County youth experiment with sex, alcohol, and drugs, albeit at varying levels. In focus group discussions, several area youth estimated that anywhere from 85 to 90 percent of kids in Rockdale County have used alcohol. Estimates of marijuana use ranged from 60 to 75 percent of area teens. They also reported high levels of sexual activity, estimating that 50 to 60 percent of area adolescents are sexually active. Several reported that they knew one or more teenage girls in the community who have babies.

In terms of intergroup relations, several area teens reported that in general, interactions between peer groups tended to be amicable, with no discernable rivalries at the group level. While this was the case for the majority of peer groups, there were some exceptions. For instance, more than one source identified tension between the rednecks and the straight edge mafia. Knowing the contempt the straight edge mafia had for meat, rednecks were reported to occasionally leave the corpse of a dead deer on the front lawn of one of their member's houses. Another reported group conflict was between some rednecks and black youths.

In contrast to the general lack of group conflict, there were many reports of fights between individuals, over the usual issues of status, respect, and reputation. Very few of these disputes involved the discharge of firearms, but most of the youths we interviewed knew of incidents in which guns were brandished. Some had seen such events. A few could recount a dozen or more instances, others knew of four or five. One stated that he had someone pull a gun on him.

Still others indicated that displaying guns during confrontations is very common. A couple of the focus group participants stated that, when they see a fight, they expect to see a gun pulled. Another diverged slightly from this, saying that he expected to see a gun only when certain people are involved in the fight, indicating that some are more likely to use firearms than others. While it is hard to pin down an estimate of how often these incidents occur, it is clear that guns are available to local youth and are also not uncommonly brandished during interpersonal disputes.

THE SCHOOL

Heritage High School is one of three high schools serving residents of Rockdale County. The oldest of the three schools, Rockdale County High School, is located in the old town of Conyers and serves a population that has higher percentages of minorities and working-class whites than the others. The newest, Salem High School, was built in 1991 as part of a national progressive education movement that emphasizes team teaching across a variety of subjects.

Heritage, built in 1976, is touted as the flagship high school in Rockdale County. It boasts high SAT scores, high graduation rates, top-ranking soccer and baseball teams, and a good band program. Administrators claim that many families move into the Rockdale area so their children may become beneficiaries of what the Heritage program has to offer. The school enrolls about 1,400 students in grades 9 through 12. Of the three county high schools, Heritage maintains the highest average SAT scores, the highest grades, and the lowest dropout rate. SAT scores run about 100 points above the state average. Over 97 percent of students graduate and 70 percent go on to college.

Situated just off Highway 138 in the southwest portion of the county, Heritage was built and designed on a traditional model of educational philosophy that has continued. The school building forms a T, the center of which is the school's common area where students gather before and after school and during lunch periods. Two wings consist mainly of classrooms and offices. The third houses the school auditorium/gymnasium and auxiliary areas. Parking lots are located in the front and rear of the school. A circular drive in front allows for drop-offs, and buses load and unload in the rear, near doors opening into the commons area.

School officials reported that both before and since the shooting incident, day-to-day problems have been minor and infrequent, the most common being smoking in the restrooms, thefts from lockers, fender benders in the parking lots, and an occasional confrontation or fight between students. On a few occasions, students have been caught bringing drugs or knives into the school.

While overall school problems have been and continue to be minor, the school system has nonetheless maintained an active security program. Prior to the shooting, each of the area high schools was assigned one school resource officer, a sworn law enforcement official responsible for both the high school and its linked feeder middle school. Heritage also maintained four video surveillance cameras prior to the shooting. The TV monitors and recording devices were located in the main office and were primarily used for determining the identity of a student subsequent to a reported problem.

THE OFFENDER

Prior to the day of the shooting incident, T.J. Solomon (hereafter T.J.) had never hurt anyone. Other than a few oblique remarks to peers in the weeks prior to the shooting, he had never threatened or bullied anyone. He had no record of arrest and had apparently committed only a few delinquent acts, all but one quite minor, and none violent. He was well-mannered, neat, and respectful of adults. The subsequent investigations, criminal and psychological, revealed that he suffered from depression. He was found guilty but mentally ill. The depression probably began in early childhood and became much worse after his family moved to Rockdale County from Kernersville, North Carolina, when he was in the eighth grade.

The task of reconstructing his state of mind in the period leading up to the shootings is one that has consumed the efforts of many people since the incident: family members, victims, neighbors, mental health and criminal justice professionals, journalists, and members of the public. The present study, conducted two years after the incident, thus comes in the wake of many previous efforts. The present study suffers from the limits of the increasing distance in time. It also suffers from lack of direct contact with T.J. or his family, although, if contacted at this point, they would not be the same people. At the same time, this study has the benefit of an extensive and diverse record compiled by previous investigators with a wide variety of personal and professional motives and challenges.

The present study differs from past efforts by those involved in the adjudication of the legal charges. Their responsibility was to decide what to do with the offender. In contrast, the goal here is to present an objective, scientific case study. A key difficulty for the present study is how to reconcile the differing viewpoints of the defense and prosecution in the legal deliberations on the question of T.J.'s mental health. The prosecution contended that he was mildly depressed, the defense that he was severely mentally ill.

In assessing the available evidence, we conclude that he was severely mentally ill prior to the incident and that the shooting at Columbine High School, combined with this preexisting illness, triggered the incident. Most of the parties who dealt with him agree that Columbine played a key role in stimulating his behavior. The disagreement was about the previously existing illness.

The importance of assessing T.J.'s mental health at this point no longer has anything to do with his legal status, which has been decided through due process of law. Rather, the issue at present is to understand what led to his heinous act of violence. In our view, it is necessary to recognize the salient role of mental illness in order to understand what occurred. An extraordinary national event coincided with a developmentally vulnerable period in his life in a chance and tragic manner. Had they not coincided, he might not have done what he did—but he still would have been mentally ill. Contrary to the prevailing arguments in his court proceedings, our study concludes that his illness was severe (see Table 2-2).

No one recognized that illness, because of his age and circumstances, the gradual developmental progress of mental illness during adolescence, and the nature of the particular form of depression that he suffered. This study is therefore relevant not only to efforts to understand, prevent, and control the sensational school shootings that have terrified the public, but also to broader concerns about mental illness and its prevention and treat-

TABLE 2-2 Timeline of T.J.'s Significant Life Events

September 6, 1983	T.J. is born in Baton Rouge, Louisiana.
June 1988	T.J.'s father and mother separate. All contact between T.J. and his biological father ceases. T.J. is 4 years old.
1990	T.J. moves to Denim Springs, Louisiana, with his mother. and sister. His mother remarries. At age 7, T.J. begins use of firearms.
1993	T.J. moves with family to Penbrook Pines, Florida. At the beginning of 4th grade T.J. begins taking Ritalin after diagnosis of attention deficit disorder.
1994	T.J. moves with family to Kernersville, North Carolina.
November 1996	T.J. moves to Conyers, Georgia, with family. T.J. loses interest in organized sports and fails to make meaningful friendships. Grades begin to decline.
May 1998	T.J. tells two fellow students that he doesn't want to live. Counselor notifies T.J.'s parents.
December 1998	T.J. steals .22-caliber handgun from stepfather's boat and sells it to a teenaged neighbor.
February 11, 1999	T.J. steals a CD from a teacher's desk at school.
February 26, 1999	T.J. allegedly takes gun to school showing it to a fellow student. Skips school in afternoon returning home at midnight after drinking all day. Parents reported T.J. as missing.
April 20, 1999	Columbine, Colorado, massacre takes place.
April/May 1999	T.J. makes statement to others in reference to Columbine about doing it differently and saying how cool it was.
May 19, 1999	T.J. is caught with cigarettes and receives lecture from parents.
May 20, 1999	T.J. enters Heritage High School and opens fire. Begins incarceration.

ment. Although a study of one event cannot provide definitive answers, it does attempt to raise awareness and pose useful questions.

A particular form of psychological investigation is employed here to organize the analysis of T.J.'s state of mind leading up to the shootings—that of ecological psychology. This approach is introduced here, with a minimum of jargon, primarily because of its value in helping to make sense of an extremely puzzling situation, the commission of a heinous violent act in a nonthreatening situation by an individual with no previous aggressive tendencies.

Ecological psychology is an approach to the study of human development that assumes that development is profoundly influenced by environment and sees the continuous interaction of person and environment in terms of nested levels of environmental context. These levels can be designated here as family, community, and society.

In our view, T.J.'s illness had its roots in his own family, stemming from the event of his biological parents' divorce in his early childhood and exacerbated by emotional distance from his mother from that point forward. His illness continued at a low level through middle childhood and the onset of adolescence and then became rapidly much worse as a result of a change at the community level, when his family made one more in a long series of residential moves, this time from North Carolina to Georgia. His illness then erupted in unprecedented and irrational violent behavior as the result of an extraordinary event at the national level, the Columbine High School holocaust. It was the unfortunate and unpredictable confluence of processes across these three levels of environmental experience that led to the events of May 20, 1999.

T.J.'s biological parents, Anthony Solomon and Mae Dean Blundell, were married in 1974, nine years before starting a family. T.J., born Anthony Solomon, Junior, and called T.J. for "Tony Junior," was born in Baton Rouge, Louisiana, in 1983. His younger sister was born three years later.

His parents separated when he was 4 years old, and their divorce became final two years later. In psychological interviews after the shooting incident, T.J. immediately spoke of the separation from his father at age 4 when asked to describe the worst thing that had ever happened to him. This event appears to have had a permanent effect on his development.

Excavating the circumstances of his biological parents' marital breakup from the available record is difficult, but the breakup was clearly not amicable. Two versions of the story are available. One, contained in the transcripts of criminal and civil court proceedings, is that of his mother, who repeatedly testified that T.J.'s father abandoned the family suddenly, for no reason, and terminated contact with his children thereafter. The other account, from his father, is brief, provided to investigators after the incident. He reported that she had left him for his best friend, Robert Daniele, whom she subsequently married.

The divorce was uncontested. T.J.'s father was granted visitation rights, which he never exercised. His mother received custody of the children. Although the couple agreed to divide much of their joint property, T.J.'s mother also took sole possession of their jointly owned stock of firearms, which she estimated in her testimony as numbering between 10 and 15 guns. Hunting and target shooting were central recreational activities for the family. She participated directly in target shooting and also went on hunting trips, although she said that she did not shoot during the hunting. The guns, however, had been purchased jointly.

The record strongly indicates that T.J. was suicidal before and during the Heritage High School shooting incident, despite the contention of prosecutors to the contrary. It is possible, though uncertain, that he may have been influenced by biological predispositions or examples in his own family. His mother's brother killed himself; his suicide occurred after he discovered that he had an incurable illness. And T.J.'s father had been institutionalized for suspected suicidal behavior; whether he had actually been suicidal is unclear, but he had been committed for observation after the breakup with T.J.'s mother. T.J. himself never learned of the incident involving his father prior to the incident at Heritage High School. Other experiences much later in his life are far stronger indications of T.J.'s own suicidal intentions.

After T.J.'s mother married Robert Daniele, the family went through a series of residential moves, including three moves in the six years prior to the shooting incident. His stepfather achieved success in business during this period, and the family became affluent. Mr. Daniele traveled often on business, and T.J.'s mother was the primary caretaker for the children, including an older stepbrother from Robert Daniele's previous marriage.

While T.J.'s interviews indicate that he bore permanent psychological scars from the abrupt separation from his biological father in early childhood, all the evidence points to an outwardly normal and untroubled childhood prior to the family's eventual move to Rockdale County, Georgia, from Kernersville, North Carolina, when T.J. was in the eighth grade. T.J. appears to have suffered only mild depression prior to that, manifested to others, if at all, only in a tendency to shyness and conformity.

T.J.'s parents separated when he was four and the family was living in Tennessee. A period of two years ensued that are not well documented but that led to his mother's remarriage to his stepfather, in Louisiana, where the newly constituted family continued to live for the next three years. T.J.'s stepfather was then transferred to Florida, where T.J. attended the fourth grade. Immediately following this move, he experienced trouble with his grades. At this point he was diagnosed with attention deficit disorder and began taking the medication Ritalin. His grades reportedly improved afterward. The following year, the family moved once again, this time to Kernersville, North Carolina, following another change in his stepfather's business career.

Research for this study included interviews with a number of people who had known T.J. in North Carolina, all of whom had been shocked when they learned what had happened at Heritage High School. They uniformly volunteered that he seemed "normal" at that period of his life. In Kernersville, T.J. had been involved in many activities, including friendships and play in his neighborhood and organized baseball. He had won fishing contests and been a successful athlete, skilled with a baseball bat. There was some indication that his older stepbrother was "macho" and bullied and harangued him occasionally, but other reports indicated that this was "normal sibling stuff" and that T.J. also respected and looked up to his stepbrother, who "stuck up for him" with others.

While the record of T.J.'s childhood before Kernersville is less detailed, some aspects of the quality of family life are consistently evident. The testimony of T.J. himself, of his mother and his stepfather, and neighbors and relatives all indicate that family life following his mother's remarriage was very close and somewhat turned in on itself, probably because of the frequent changes in residence. T.J.'s mother took primary responsibility for the children and monitored them very closely. Even though his stepfather traveled frequently on business, he was also very involved with the children when he was at home.

The family engaged in many activities together. Over the years these included hunting and fishing, golf, and frequent weekend camping trips. To the extent that they socialized outside their own immediate family, it was usually with extended family members. They were religious, Roman Catholic, and attended church services regularly.

The household rules and routines were detailed and supervised. The children had regular chores. The house was always extremely neat and the children neatly attired. T.J.'s mother closely monitored his Ritalin medication from the time it began in fourth grade up through the morning he opened fire at Heritage High School. She paid close attention to medical advice about possible adverse effects of long-term use and made sure that T.J. took Ritalin "holidays" whenever he did not have to attend school. She never entrusted the medication regime to him, however, and personally gave him the medicine to take each day that he went to school.

Corporal punishment was a feature of childrearing in this household, administered by T.J.'s stepfather. While some experts classify any use of corporal punishment as abusive, there is no evidence that it was frequent or severe in this case or at odds with the standards of the various communities in which they lived, all of them culturally similar despite their number and geographic dispersion across the South. It did not happen frequently and was meted out according to a consistent set of standards. There is no indication that T.J. was an abused child according to the standards of his culture. In light of the existing literature on childrearing and delinquency, the extreme consistency of childrearing practices here is particularly notable. There was none of the unpredictable fluctuation between inattention and sudden imposition of harsh discipline that is known to be associated with behavior problems, particularly aggression (Baumrind, 1978).

All the orderliness and close supervision of the household, however, appears to have to been related to another kind of problem for T.J.: a lack of emotional connection to others, particularly to his mother.

There is some indication in his mother's own descriptions of him in court testimony. One community member who observed the hearings judged her to be "cold" and noted that she referred to him throughout as "this child" rather than "my son." Inspection of the transcripts tends to bear this out. In response to multiple lines of questioning, she spoke repeatedly of her determined efforts to make sure that he performed well in school and behaved appropriately. She testified defensively to his normality in earlier childhood and voiced her fears about his well-being, especially in relation to taking Ritalin. She stressed her own careful monitoring and supervision as her primary response to these fears. She never expressed warmth or enjoyment of his company.

Whatever the state of T.J.'s mental health and his relationship with his mother in the earlier part of his life, the indications are substantial that both were in serious and worsening condition from the time of the family's move from North Carolina to Georgia.

After the move, when he was in the eighth grade, T.J. became increasingly passive and withdrawn from others. This tendency affected every aspect of his social relationships: at home, in school, in organized recreational activities, and in informal relationships with peers.

After the move, T.J. enrolled in organized activities, with his parents' assistance, but resisted active participation. Formerly a home run hitter in baseball, he would stand at the plate and let three strikes go by without swinging, while onlookers laughed at him. He also resisted joining a Boy Scout troop, even though he had previously enjoyed scouting. He did eventually join and appeared to continue enjoying camping activities. Being outdoors and alone or indoors and listening to music by himself were increasingly the main activities he enjoyed.

Around this time, T.J. did succeed in one achievement that earned him the approbation of others. At the age of 13, he killed his first deer. His stepfather spoke of it approvingly in court, and peers reported that talking about guns and hunting was one thing that T.J. always did with enthusiasm. In retrospect, however, there appears to have been a dark side to this. The psychologist who interviewed him on retainer from his defense counsel later testified that "guns were the love of his life. He enjoyed watching animals die and looking into their eyes and trying to figure out what it's like. He's trying to get, to go to the other side. He enjoyed that."

His grades also began to suffer again, a process that had already begun prior to the move and worsened afterward. His parents tried to motivate him by saying that he could not get a drivers' license until they improved. Obtaining a learner's permit at the age of 15 is social milestone for Georgia teens. He replied that he did not care about driving.

At the time he was arrested, T.J. had been in Georgia for over two years. He had formed no close friendships and had never been on an individual date. He did associate with young people his own age, but only in the context of family, with his cousins, or in organized activities, principally those related to school and scouting.

The press widely reported in the immediate wake of the incident that it had been precipitated by a romantic breakup with a particular girl, but that turned out to be false. The young person in question was merely one of a group of people he spoke with in school. There may have been an element of flirtation, but the closest he had come to dating was going to movies in the same group. There had not been a breakup incident, in her mind or his. Girls frequently called the house, but T.J. never accepted the calls. His mother's expressed irritation about the calls probably did not encourage him, but he himself stated in interviews that he did not like to have friends and did not want people coming over to his house.

The strongest evidence that a great emotional distance had opened between T.J. and his mother comes from the videotape made by law enforcement officials immediately after the arrest. It is a remarkable tape, obviously not a directly representative record of daily patterns of interaction under normal circumstances, but difficult to ignore nonetheless. Although discussing it at this point is out of chronological sequence with the presentation of the course of his development, it is crucial to the analysis of his state of mind in the months leading up to the incident.

The tape begins at 8:42 a.m., less than an hour after he discharged the firearms at others. He is alone in a room in the sheriff's office, with the tape running in real time. He is sobbing and breathing heavily. The first person who comes in is the chief of detectives, Warren Summers. Summers tells him he only needs to ascertain his name and birthdate and will hold other questions about what has happened until T.J. has a chance to seek an attorney. Summers is then solicitous of T.J.'s welfare. He tells him repeatedly to calm down. He offers to get him food. T.J. is frightened and respectful, addressing the detective as "Sir."

T.J. asks Summers if he has hurt anyone. Summers says "Well, there's some people not in too good shape. I tell you this, I don't think anybody is critical. I don't think anybody is gonna die." Summers leaves the room. T.J. breaks into sobs, says "oh, God" once and then "fuck" twice. He then gets down on his knees and puts his hands in praying position. Someone brings him food and coffee. Summers comes in and out, talking to him about baseball and trying to calm him.

At 9:53 a.m. his stepfather arrives. Robert Daniele immediately puts his arms around T.J. and strokes his head. They converse inaudibly. At this point, Summers reads them a Miranda warning. Daniele says he wants to get a lawyer, and Summers agrees but expresses his concern that they do not know what else T.J. may have left at the school. Daniele agrees that he is concerned about that also; he then asks what happened and why and T.J. repeatedly says that he does not know. At one point, Daniele tells T.J. to give him his hands and look him in the eye. T.J. does so and appears to relax somewhat, but he still cannot articulate an explanation.

At 11:35 a.m. his mother arrives. She does not touch him. She asks if that is his food and has it removed. She begins asking him what happened and scolds him. During the subsequent conversation she expresses her bewilderment, her anger, her shame, her concern for his sister, and the victims. She asks if he has prayed for the victims. He says he has.

She also asks him three separate times why he did not follow through killing himself. The first time, she says, "And you were going to kill yourself, I understand? How did that not happen?" T.J. replies "I decided not to." She then repeats, in a mocking tone, "You decided not to, after you shot these five [sic, there were six] innocent people, you decided not to hurt yourself?"

She then expresses her ongoing frustration in communicating with him, saying, "Why didn't you talk to me?" T.J. replies, "I was scared." His mother says, "You were scared? You were too scared? But not scared enough to pull a gun on people? T.J., people are tryin' to help you. I been tryin' to help you. I couldn't reach you. You pushed me away."

After some more failed attempts to get him to explain his actions, she says, "This is the best place for you to be. I mean really. If you don't know why you did it, how they can help somebody that don't why they hurt human beings?" T.J. then raises his voice for the only time in the conversation and says, emphatically, "They can." His mother replies, "No, they can't. Those were children." Several minutes later, she challenges him for the third time about his failure to kill himself: "I don't know how you took innocent children but you were afraid to do anything to you. That really has me puzzled. You didn't think twice about doing it to them."

Almost immediately after this last statement, investigators enter and announce that they are going to have to search the house. T.J.'s mother becomes concerned about damage to her home and leaves shortly thereafter. At 11:55 a.m., she says to her husband, "I don't know what else to say, he obviously does not want to talk to us." She turns to T.J. and says, "I don't know what to say to you, T.J., you've made your decision." She and her husband leave. She never touched T.J.

As difficult as it is to base a psychological assessment on data of this extraordinary nature, collected so invasively under such stressful circumstances, these data starkly illustrate a profound lack of emotional connection between T.J. and his mother for some time prior to this. Other evidence is completely in accord as to her attentiveness to his outward needs, her diligence in managing their household, her bewilderment over his passivity and increasing withdrawal, and the fact that she repeatedly sought professional help and responded actively to the more visible aspects of his psychological abnormality.

His needs and her responses, however, were caught in a vicious cycle. The more she tried to manage and control his life, the more he withdrew, and the more frustrated with trying to reach him she became. And his withdrawal was successful. He maintained enough of the appearance of normality, to his family and to others, that he seemed just a little immature, not mentally ill. The profound and deepening nature of his depression was concealed under the polite exterior of a boy who always addressed adults as "Sir" and "Ma'am," who generally did what he was told, and who rarely talked back.

After the initial signs of T.J.'s worsening depression following the move to Georgia, an appearance of normality reestablished itself. His withdrawal was continuing, however, and it was accompanied by changes in his habits. He began to listen to music at his computer for hours at a time, and his musical tastes shifted from country rock to rap and rock music. He copied lyrics from these songs for hours at a time, to the distress of his parents, as it appeared to detract from his attention to schoolwork. They removed the Internet connection from the home computers. At some point, T.J. downloaded bomb instructions from the Internet, which were found in his room after the incident; he said he did it long before the incident, although it is not clear when.

T.J. also shifted his tastes in music, away from country music to more transgressive and rebellious styles. He copied and tried his hand at writing violent rap lyrics, in the manner of Tupac Shakur and other "gangsta' rappers." At one point, he told a female classmate that he wanted to be a rap star and write songs about sex and violence, but then added that he did not want to actually hurt anybody because he was not that kind of person.

His favorite group, however, was a rock group, Korn. Among the myriad genres of contemporary music, Korn is classified by one music rating service in the categories "post-grunge, alternative metal, heavy metal" and described as being "ominous, gloomy, nihilistic, aggressive, detached, visceral, bleak, angry, hostile" (Guide, 2001). A review of Korn's lyrics discloses a few aggressive threats but a much higher proportion of statements of self-loathing, disgust with others, and suicidal longing.

After T.J. entered Heritage High School in the ninth grade, an unambiguous warning sign appeared. There had been local concern about teenage suicide since the string of three deaths two years before, all involving youth who seemed to be loners, like T.J. Although one of these was subsequently reclassified as an accident and a countywide agency report was issued saying there had not been a suicide epidemic, fears were aroused, and Heritage High School began teaching suicide awareness classes. Two of his classmates reported that T.J. had expressed suicidal thoughts. The school and his family reacted appropriately: the school notified the family, and T.J.'s mother took him for a psychological examination. The examiners concluded, however, that he was not at risk.

During his tenth grade year, leading up to his shooting rampage at the end of that year, T.J. grew more and more remote from others. His grades continued to decline, and he engaged in some delinquent acts. One of these was serious. In December, he stole a handgun from his father's boat and sold it to another youth, who subsequently claimed that he had acquired it for self-protection. T.J. subsequently claimed that the other youth had badgered him to get him a gun because T.J. frequently talked about guns, but that he had disabled the gun before selling it. This act is the only substantiated instance of serious delinquency on his part prior to the shooting. There was also a report, in February, that he had brought a gun to school and showed it to another student, but the report was investigated and never substantiated.

Other instances of delinquency were quite minor. In February, T.J. stole a CD from the desk of a teacher he disliked. On the same day in February when he was reported to have brought a gun to school, he left school early with another boy, got drunk with him, and returned home late. His stepfather punished him by spanking him with a belt, reportedly the only time he ever applied corporal punishment that severe, but also rather striking for the age at which it was administered.

Aside from these incidents, the only form of delinquency T.J. is known to have engaged in consisted of some experimentation with alcohol and marijuana, but this does not appear to have been extensive.

There is some indication here of association with delinquent peers, a known correlate of the development of delinquent behavior. There were three boys he was said to have spent time with who had reputations for illegal behavior, including those involved in the gun and drinking incidents just described. It does not appear, however, that T.J.'s associations with them were much closer than those with any other youth. They never came to his house, and they seem to have sought him out because of his knowledge of and access to firearms.

T.J.'s own position in the configuration of peer groups and group identities in the area was not well defined. Although there were recognized labels in the area, they were not hard and fast but rather fluid and overlapping. Although a handwritten note introduced in court proceedings contained language about getting revenge on "jocks and preps," there is no indication that T.J. was strongly identified by others or himself as a member or opponent of these or other categories. One boy who knew him in school said that the group T.J. sat with at lunch could be considered "nerds, preps, or jocks." In subsequent psychological interviews, T.J. identified categories of "skaters, jocks, freaks, people that were and weren't cool" and said that he fit none of them.

The record does contain some reports that T.J. had been taunted on some occasions by other boys in school, but there does not appear to have been a pattern of persistent bullying, and T.J. was not seen as different from others in having to put up with more of this kind of behavior than others.

Despite T.J.'s rather amorphous social identity, there was one source of identity that was salient for him and reported subsequently by many others. He was proud of his knowledge of guns, shooting, and hunting. He talked avidly of his excitement about getting a new gun and his hunting trips with his family. He also invoked this aspect of his identity inappropriately at times. A teacher later reported that he had been socially inappropriate in class, making sexual remarks that fell flat and also saying at one point, "Don't mess with me. I have access to guns."

The other significant event indicated in the record as occurring in the months just prior to the shooting was a suicide attempt, or at least a strong ideation. This came to light only in psychological interviews after the incident, but he had apparently gone down to the basement one night and put the barrel of a gun into his mouth. He was not able to articulate his intentions at that time, just as he has never been able to articulate his intentions behind the shooting since it occurred.

THE INFLUENCE OF COLUMBINE AND OTHER CIRCUMSTANCES IMMEDIATELY PRECEDING THE INCIDENT

It is our contention that, at the time of shootings at Columbine High School, T.J. was suffering from mental illness that had grown steadily more serious since the family's move to Georgia two years before and had already manifested itself in withdrawal from others and suicidal thoughts. In this condition, the events at Columbine High School made an enormous impression on him. Withdrawn and lonely, immersed progressively in constructing meaning out of the materials of transgressive popular music, he became obsessed with Columbine. This obsession and its suicidal implications are evident in writings T.J. made before he began shooting, reports by others of remarks he made to them, and writings and statements to psychologists that occurred after the incident at Heritage High School.

Writings that were found in his room in the immediate aftermath of the shooting contained the following statement:

> No one could ever know how I feel. No one will ever know. Even the smallest of scars, can run the deepest. I've overcome a lot in my few years, but I understand I'm still leaving a lot behind me that I haven't even experienced, *yet*! There aren't many words that I can say, to describe how I feel. One big Question everybody's probably wondering about now is *WHY*?! Well, for the sake of my brothers and sisters related to the trench coat mafia, that will have to remain a mystery to the public eye. I have been planning this for years, but finally got pissed off enough to really do it.

The same statement also contained references to bombs that would be discovered after this fantasized incident. Printouts of bomb-making instructions downloaded from the Internet were also found.

Other fragments of writing found in his room appeared to be attempts at writing song lyrics. One of them read, "Laughin' at my victims as they drop to their knees. Beggin' for their life, screamin', 'please, Dear God, don't let this crooked motherfucker murder me'" and was signed "Me."

While there is no question that these documents establish that T.J. had previous thoughts about enacting his own version of Columbine, their timing is uncertain. The prosecution in his court proceedings maintained that the former document had been written the morning of the incident. In his psychological interviews, however, T.J. said that that it had been written some time before. It is unclear when or how he had downloaded bomb instructions from the Internet, since his parents had removed the household Internet connection quite a while before, but there was never any evidence that T.J. made or placed bombs.

After the incident, T.J. wrote explicit statements while in detention prior to court hearings that explicitly attributed his actions to the spur of the Columbine example. He wrote:

> I was feeling anger, rage, envy, and fear all together. It wasn't the first time that I have had that feeling, but I wanted it to be the last. I felt that

> expressing myself through explicit and violent lyrics, and poems wasn't enough anymore. I was tired of fighting off emotions that I had to let out. Though I did not feel at ease to discuss my thoughts with anyone but myself. I felt the next thing left to release anger would be through violence. I had just gotten the idea from the shooting at Columbine High School on April 20. So the Monday of the May 20 shooting, I decided to open fire May 20, one month after the Colorado shooting. But I didn't really assure myself that I would.

Investigations after the incident disclosed multiple instances in which T.J. said things to other people before the incident that indicated his feelings of identification with the Columbine killers. He began to talk about Columbine in odd ways to other people. His remarks did not make sense to them and were disregarded at the time. They were not direct threats but rather oblique speculations. During this same period, another boy in this same area had made direct threats to enact a Columbine scenario, and this boy was identified and committed to a mental institution. Those hearing T.J.'s remarks discounted them.

A member of his scout troop reported that during a discussion of Columbine, T.J. had said, "I should do something like that." T.J. also made a number of remarks in the presence of a small group of students he ate lunch with. Those students later stated to investigators or in court that, with explicit to reference to Columbine, "He said it should have happened to our school a long time ago," that he "could understand" the Columbine killers "wanting to shoot the jocks and preps," and "that the kids at Columbine were aiming at certain people and that slowed them down, and if he ever shot at Heritage, that he wouldn't take any time to aim, that he would shoot at everybody."

The record contains two different versions of reports by one student, a boy named Trey Carver, that T.J. had made statements about shifting blame to the band Korn if he did something heinous. The difference between the two statements is that the heinous act in question in the previous version, recorded by investigators immediately following the incident, is suicide. In the latter version, from testimony at T.J.'s transfer hearing, the act in question is shooting up the school.

In subsequent legal proceedings, the defense and prosecution attorneys agreed that T.J. had been obsessed with Columbine but disagreed about whether his identification with the Columbine killers was or was not bound up with suicidal ideations. Evidence on this point was tied to two different reports by a classmate. The investigator's notes from the interview on May 20, the afternoon of the shooting, record the classmate as saying "T.J. told me one day that he thought it would be real cool if he could put some song lyric on his calculator about suicide and if he committed suicide, people would blame the band that wrote the song lyrics for his suicide." Appearing as a government witness at the transfer hearing on August 8, several weeks later, the classmate changed the story, saying "He said that if he ever shot up a school or did something like that . . . he would put song lyrics to a band in his book bag so people would blame it on the band." The classmate further identified the band in question as Korn. During his testimony, the classmate denied his earlier report that T.J. had been talking about suicide but admitted that the Korn lyrics in question were in fact about suicide.

It is not clear why the classmate changed his story, but, by conventional standards of social science research, the earlier investigation report has the greater credibility, on the basis of its immediate proximity to the events as well as by virtue of the way in which the account emerged during the natural flow of the conversation. Although the classmate's alteration of his story later played a significant role in the outcomes of T.J.'s court cases, there is ample evidence from multiple sources both that T.J. was suicidal and that he was obsessed with Columbine. Columbine was, after all, a suicidal undertaking.

On May 19, 1999, the day before he opened fire at Heritage High School, T.J. had gone to school as usual. That afternoon around 4 p.m., T.J.'s sister discovered a pack of cigarettes in T.J.'s pocket and informed their mother. Later that evening, T.J.'s parents sat down with him for a discussion. A lecture regarding the health dangers of using tobacco turned more broadly to one about accepting responsibility for his actions and understanding the consequences of his behavior. Aside from this conversation, no punishment was given. At 10:00 p.m. the talk was finished and T.J. was sent to bed.

He put his favorite Korn CD on the stereo underneath his bed and, as he often did, set the player to "repeat." He fell asleep listening to the suicidal lyrics that continued to play throughout the night.

THE INCIDENT

The following morning, Thursday, May 20, 1999, precisely one month to the day after the Columbine killings, T.J. woke at 7:00 a.m. He sat up in bed and listened for a moment to the music that was still playing. As he usually did, he went out to get the newspaper and returned to the kitchen where his mother gave him his Ritalin medication. He then proceeded downstairs to get ready for school. While T.J. was downstairs, his sister and a friend were upstairs, getting ready to receive awards for academic achievement at a ceremony that day. T.J.'s mother was taking pictures of the girls as they prepared for their triumphant moment.

Meanwhile, downstairs at approximately 7:10 a.m., T.J. broke into his stepfather's locked gun cabinet and selected a .22 caliber rifle and a .357

magnum handgun. He proceeded to saw the stock off of the .22 rifle and hide the discarded butt under a nearby couch. T.J. showered and dressed in baggy blue jeans and a loose fitting white t-shirt. He concealed the rifle inside his pants by securing it to his leg with a leather strap and placed the handgun in his book bag. At 7:45 a.m., T.J. left the house for school and boarded the school bus.

Remarkably, several of the other students who were on the bus that day reported to law enforcement officials that they had not seen anything unusual about T.J. One, however, later reported he had noticed T.J. limp a little as he boarded the bus. No one realized that he had two firearms in his possession. T.J. sat at the back of the bus and kept to himself.

At a little before 8:00 a.m., T.J. arrived at Heritage High School and proceeded toward the rear door of the school's commons area (see Figure 2-2). On his approach, T.J. left his book bag near some woods just behind the school and continued toward the rear entrance. One student witnessed T.J. kneeling by his book bag. In an interview after the incident, T.J. recounts, "When I got to school I was walking up there. I didn't even

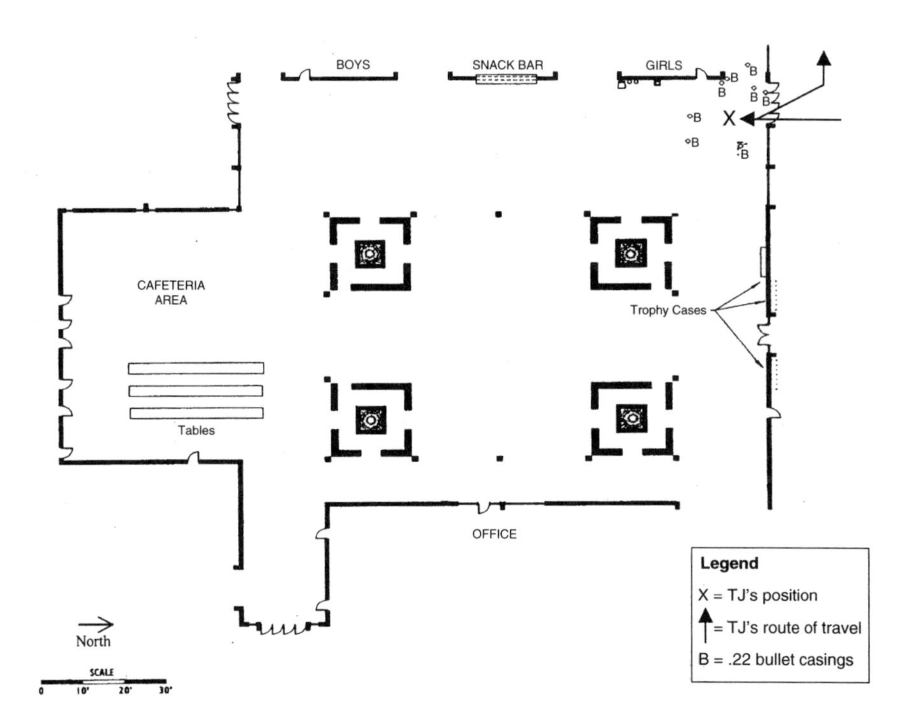

FIGURE 2-2 Heritage High School commons area.
SOURCE: Diagram adapted from Rockdale County district attorney's files.

know what I was going to do yet. . . . I just know I had them there. So I was walking, and the thing came out for the gun . . . and the bullets started dropping out. So I put them back in there . . . and then after that it was pretty much—I pretty much felt I had to do it because, you know, there was somebody that had already seen me with it at this point."

With the .22 rifle in hand, T.J. continued to walk toward the rear commons door. Two other students observed him as they also approached the rear door. T.J. said to them, "Y'all stay here. Y'all are cool. You'll be okay." T.J. entered the commons area and opened fire.

Witnesses reported that T.J. held the .22 rifle at his hip as he fired 12 shots throughout the room, emptying the firearm. Eight students were hit. Two of them escaped injury because the bullets lodged in objects they were wearing (a backpack) or holding (a book). Six were wounded, one seriously, none fatally. Two male students then chased after T.J. as he went back outside through the same doorway he had entered.

After dropping the rifle, T.J. pulled the .357 magnum revolver from his pants and turned to fire at his pursuers. They dove for cover, escaping injury. T.J. continued down a sidewalk behind the school and fired two more shots to his rear at the school building but did not strike anyone else.

By this time, two of the school's assistant principals had made their way through the gym bringing them outside in front of T.J.'s path. T.J. stopped, dropped to his knees, and placed the revolver in his mouth. In a note written by T.J. sometime after the incident he recalls, "When I walked out I had, somewhere between the door and the sidewalk, turned suicidal. I felt like death was the only way out of the situation. I was scared and mad at myself. I thought that no one really cared what happened to me from then on."

After several minutes of coaxing from the assistant principals to put the gun down, T.J. complied and handed over the firearm. As he relinquished control of the .357, T.J. hugged one of the assistant principals. He was crying hysterically and said repeatedly, "I don't know why I did this!" He later wrote, "Then the assistant principal walked out, and I saw he cared so much as to walk into a dangerous situation to help me. So I gave my weapon up and it felt good to know that everything was over." The incident had lasted 12 minutes.

In spite of the sudden and rapid unfolding of events, local officials responded quickly. While several people had called 911 to report the event, it was a student in the school using a cellular phone who first contacted law enforcement. Within four minutes, sheriff's deputies were on the scene and immediately took T.J. into custody. Table 2-3 details the incident's timeline of events. By 8:40 a.m., T.J. was sitting in a Rockdale County sheriff's office interrogation room in preparation for questioning.

TABLE 2-3 Incident Timeline of Events

May 19, 1999	**10 p.m.**	T.J. goes to bed following a lecture on accepting responsibility for his actions after being caught with cigarettes.
May 20, 1999	**7 a.m.**	T.J. is awakened by his mother for school.
	7:05–7:40 a.m.	After getting paper for his mother, T.J. goes downstairs breaks into gun cabinet and takes possession of a .22 caliber rifle and a .357 handgun. T.J. saws off the stock of the .22, showers and dresses in baggy jeans and a tee shirt.
	7:45 a.m.	With .22 rifle concealed inside his pants and a .357 in his book bag, T.J. catches the bus for school.
	8 a.m.	T.J. arrives at school, enters into common area and opens fire. Six students receive non-fatal injuries.
	8:15 a.m.	Within four minutes Sheriff deputies arrive on scene taking T.J. into custody. EMS arrive within 11 minutes treating victims. The school is evacuated.

After 11 minutes emergency medical personnel had arrived and began treating injured students. Several of the students fled from the school in panic, running across major roadways into neighboring fields. Others were evacuated from the school by bus.

The school resource officer, who staggered his hours at the school campus from day to day, was en route when he heard the call on his radio that a shooting had taken place. He arrived minutes later and began assisting in evacuating the school and helping to tend to the injured. Some in the Rockdale community expressed contempt that the officer was not present to prevent or respond to the shooting incident. Yet one school official conveyed relief that the officer was not there, citing the possibility of fatal consequences for T.J. or others.

Authorities estimated that there were anywhere from 125 to 200 students in the common area that morning. Many sat on the floor talking to friends before class. Others worked on homework assignments from the previous day. Remarkably, the spraying bullets struck only eight. Of the six students who were wounded, five were taken by ambulance to an area hospital. The sixth, sustaining the most serious injury, was taken to an Atlanta medical facility by helicopter. Of the injuries sustained, one of the victims was shot behind her right knee. Another was shot twice in his left thigh. Two were shot in the buttocks, one sustaining more serious injury than the other. One was shot in his left foot, and the final injured victim was shot in the back of her right ankle.

Two other students were shot but did not receive physical injury. One of the students was shielded from harm by her book bag, which she wore on her back; a single book held at chest level saved the other.

Soon after word of the shooting incident hit the airwaves, the media swarmed. By one estimate, there were 67 news agencies that rushed to the scene. Their equipment and trucks lined the roadways in the area surrounding the high school. Helicopters circled in the air. Several students provided interviews and information to the press.

In an effort to control the dissemination of information, school officials designated the athletic field house behind the school as the locale for press releases and briefings. Portable toilets were brought in to accommodate reporters and their crews. The phone lines were flooded by an enormous volume of calls to school and government offices by concerned community members. There was panic and increased awareness across all schools in the county. Several officials echoed the sentiment of one government leader who stated, "The shooting used every resource this community had."

As the community grappled with the immediate aftermath of the incident, there emerged varying accounts of what T.J.'s supposed intentions had been. Early media portrayals of the incident indicated that he was distraught over the break up with a girlfriend. Others pointed to the idea that he had been bullied by classmates and had sought out revenge. Neither of these stories was found to be true.

In an interview with a psychologist while in custody, T.J. reported, "I just wanted to cause a big panic. I didn't want to kill or hurt anyone. I wasn't really aiming at anybody. I aimed at objects, not people—the wall, the floor, a cinder block. I was scared and was thinking I'm going to let it out and afterwards it's going to be gone." In another such interview, he said, "I didn't really want to hit anybody. . . . It was as an accident as far as anybody getting hurt. That wasn't supposed to happen."

Prosecutors later discounted these subsequent statements as self-serving and contended that, while T.J. may not have been aiming at anyone in particular, the intent to inflict harm was still present. In support of this, one law enforcement official pointed out that while T.J. was shooting from the hip and aiming low, many if not most of the students in the commons area that morning were seated on the floor or on benches and as such were directly in harm's way. Apart from the issue of how much harm T.J. intended, it is clear that he did not go there to single out any one person but rather shot randomly throughout the room.

Another statement by T.J. that later figured in his court proceedings emphasized how much he envied the attention that the Columbine killers got as a result of their deeds. He said that he was thinking of that when he acted.

All of those who subsequently interviewed T.J. were in agreement that his act was prompted by the events at Columbine. The assistant prosecuting attorney on the case stated in an interview for this study, "Columbine was the trigger that gave T.J. the permission to do it. It showed a way that T.J. could gain power; he could be in control. He envisioned he could be someone; that he could be infamous."

And dead.

THE OFFENDER AND THE JUSTICE SYSTEM

T.J. was immediately apprehended by the sheriff's department of Rockdale County. A massive law enforcement investigation began at once, involving local and state agencies. T.J.'s family hired a local attorney within about three hours and later retained the services of one of the most prominent criminal defense law firms in the state.

The principal criminal justice proceedings that ensued were a transfer hearing in juvenile court on August 2, 1999; a sentencing hearing in superior court on October 2, 2000; a decision by the state parole board in May 2001; and, most recently, a court order in superior court amending the original sentence in response to the action of the parole board.

At the conclusion of the transfer hearing, the juvenile court waived the offender to superior court, to be tried as an adult. T.J. entered a plea of guilty but mentally ill before the judge, the Honorable Sidney Nation, and received a complex sentence that appeared to make him eligible for parole after a minimum of 18 years. In May 2001, however, the state parole board reinterpreted Judge Nation's sentence under changes in parole board policy that had been instituted the previous year to mean that T.J. would not be eligible for parole for 36 years. In August 2001, Judge Nation issued an ordering reducing the sentence so that, even under the parole board's newly announced guidelines, T.J. would still be eligible for parole in 18 years. The sentence reduction also imposed additional conditions, including successful completion of psychiatric treatment and a specific admonition that evidence of his being a danger to himself or others would be considered a violation of probation.

At each of these four critical junctures in the judicial response to T.J.'s actions on May 20, 1999, issues of the state of T.J.'s mental health, the standards by which his mental health should be assessed, and the relationship of his mental health to state law were central and controversial.

In 1994, Georgia joined the list of 47 states and the District of Columbia that adopted legislation between 1992 and 1995 making it easier to try juveniles charged with serious crimes as adults. The Georgia legislation specified seven violent offenses, soon known as the "seven deadly sins," for which charges would become the exclusive jurisdiction of the adult criminal justice system. Aggravated assault, however, and the other charges brought against T.J. were not among the seven. Had any of his victims died, he would have gone immediately into the adult system. Since none did, a transfer hearing in juvenile court was necessary before T.J. could be handled as an adult in superior court. The Rockdale County district attorney proceeded in this manner.

The mental health issue quickly came to the fore, but in a manner quite different from insanity pleas in adult cases. Georgia law states a number of conditions under which a person age 17 or younger may not be transferred of out the juvenile system. Among them is the requirement that the juvenile not be committable to an institution for mental illness. Other provisions of state law define the conditions that would require such involuntary commitment to a mental institution to include "disorder of thought or mood which significantly impairs judgment or behavior" and presenting "a substantial risk of imminent harm to that person, himself, or others based on recent behavior."

These standards thus involve much less pronounced forms of mental illness than the severe psychotic disorders that typically must be demonstrated in cases of pleas of not guilty by reason of insanity in adult courts. This situation set the terms for most presentations and interpretations of evidence during the transfer hearing. T.J. and his attorneys did not deny what he had done. Rather, they argued, and the prosecution disputed, that his actions were associated with mental illness sufficiently severe to require that he be involuntarily committed to a mental institution. Since state law recognizes substantial risk of suicide as evidence that a person needs to be so committed, the transfer hearing essentially turned on whether T.J. was seriously suicidal, for which the current study finds there to be powerful evidence in the affirmative.

The arguments back and forth between defense counsel and prosecution were thorough, informed, and cordial. There was ample evidentiary support provided by both sides and sharp questioning by the judge, Honorable William Schneider. All parties agreed that T.J. was clinically depressed, but the court-appointed psychologist asserted that the depression should be categorized as "mild disthymic disorder," while the psychologist retained by the defense characterized it as "major depression, recurrent, severe with psychotic features."

A number of issues left room for interpretation and, ultimately, judicial discretion in assessing opposing arguments. The question of whether T.J. should be considered suicidal was parsed, often implicitly, by the time period in question: before, during, or after the shooting incident. Competing assessments of the extent and seriousness of his suicidal intentions before the incident turned on whether the previous reports by fellow students who reported him after the suicide prevention class were significant, or whether, as those persons later said during postincident investigations, they had been "joking"; on whether T.J. had really been close to suicide when he put the gun in his mouth in the basement the winter before the incident; and whether the classmate got it right the first time, when he talked to investigators on the day of the incident, or later, during court testimony, when he changed his story.

Then there was the question of whether the shooting incident itself was either suicidal behavior, as a full emulation of Columbine would have been, or indicative of sufficiently severe "disorder of thought or mood" to warrant involuntary commitment to a mental institution. On one hand, T.J. did a pretty bad job of copying Columbine. He did not successfully kill any of his victims or himself. On the other hand, he might well have. The concept of "secondary gain" was important here, the issue of whether there was anything positive that T.J. was getting out of the act. The defense argued that the obvious lack of any possible positive consequences showed disordered thought and suicidal intentions. The prosecution argued that this was a "me" crime, in which T.J. wanted to get a lot of attention so badly that he was willing to do something that he knew was very wrong. T.J. had made statements indicating that he mainly wanted attention, not from the national media but from those in his immediate environment.

The arguments over whether he was committable for mental illness also had to deal with events that occurred in between the shooting and the transfer hearing. While in detention in Lawrenceville, T.J. displayed symptoms that might be classified as psychotic. He banged his head on the wall, scratched an X across his chest, and began to hear screams and experience flashbacks to the incident. He had initially been put on antidepressant medication, Zoloft.

After these symptoms and behaviors began to appear, he was switched to a different medication, Depakote, which is also an antidepressant but one with more sedative effects. After the change in medication, T.J. stopped hearing the screams and acting out. The court-appointed psychologist testified that T.J. would have been committable after the incident without medication but that he should not be considered committable as long as he was securely confined and given appropriate medication. In other words, he could be tried as an adult as long as he was kept incarcerated and sedated. The prosecution also argued that T.J. could get adequate treatment in the adult corrections system.

Judge Schneider returned a decision explicitly agreeing that T.J. was not committable to a mental institution under the standards of state law and waived the case to superior court. He stated: "The severity and the viciousness of these offenses makes the public's interest in treating the juvenile as an adult paramount to any other interests before the Court including the juvenile's interest in remaining in the juvenile system."

Waiving T.J. to superior court exposed him to far more severe penalties than he could have received in juvenile court. Had the case remained in juvenile court, T.J. could have received a maximum sentence of 5 years and would have been free at the age of 20. This issue was never argued directly by the prosecution and defense at any point in the hearings, as it was formally irrelevant to the questions of mental health and court jurisdiction that had to be decided under state law.

During the 14 months following the waiver, T.J. was evaluated by psychologists and indicted in superior court, where he entered, and the superior court accepted, a plea of guilty but mentally ill for 29 criminal charges, including aggravated assault (for the gunshots), cruelty to children (for the injuries to the victims), and illegal possession and use of firearms. Only one charge was tied to his age, that of possession of a pistol by a person under the age of 18.

The sentencing hearing was much shorter than the transfer hearing. The judge, Honorable Sidney Nation, relied heavily on the transcript of the transfer hearing as an evidentiary base. Four of the six victims and the father of a fifth made statements, along with members of T.J.'s family and psychological experts for the court, the state, and the defense.

The issue of mental illness played quite a different role in the sentencing hearing than it had in the waiver hearing. Under state law, a verdict of guilty but mentally ill has no bearing on length of sentence. It merely entitles a prisoner to treatment, which is not mandated for adults as it is for juveniles.

Under the charges to which T.J. pleaded guilty, he could have been sentenced to over 300 years in prison. The defense asked for 10 years, the prosecution for 60. The judge asked some pointed questions of the prosecution. One was about T.J.'s intentions. Both witnesses' accounts and T.J.'s written statements afterward suggested that he might not really have wanted to kill anybody. His bullets had gone low, and most of the victims were shot in the foot or the leg. The judge wanted to know why someone with as much experience with firearms as T.J. had not in fact killed anyone. The prosecution pointed to the bullet that was stopped by a book held over a girl's chest.

Another question was whether T.J. deserved any discount for pleading guilty, which the judge referred to as a not uncommon practice in the state's courts. The prosecution declined to ask for a discount. The judge did not ask such sharp questions of the defense, instead responding rather dismissively. At two points in the colloquy with the defense, he posed the rhetorical question "Do we have the guts to protect ourselves from our own children?"

Judge Nation handed down a sentence of 60 years of custody by the Department of Corrections along with a concurrent 40 years on probation, and a prohibition against ever again setting foot on school property in the State of Georgia. The way in which the judge set the consecutive and concurrent relationships of the sentences made it appear to most observers that T.J. would have been eligible for parole after a minimum of 18 years.

In reaching and issuing his decision, Judge Nation said that he considered T.J. mentally ill. He said, "I understand mental illness. I know it's real. I know it exists," but, that, in reaching his decision, "What I kept searching for was for somebody to tell me when, if down the road it would be safe. Nobody could give me that answer," as a result of which he chose to err on the side of public safety.

T.J. was remanded to the custody of the Georgia Department of Corrections immediately after the sentence was handed down, in October 2000. He was sent to Arrendale State Prison, in the mountainous northeastern part of the state, a facility that provides secure confinement and a variety of diagnostic, educational, and treatment services for adjudicated juvenile and convicted adult male offenders. Two months later, in December, he attempted suicide, overdosing on pills that he had managed to obtain from another inmate. He came close to death.

In May 2001, the state parole board issued its finding that, under their guidelines, Judge Nation's sentence should be interpreted to mean that T.J. should serve a minimum of 36 rather than 18 years. The parole board's decision was presented as conforming to its 1998 guidelines that 90 percent of sentences for 20 specific crimes, including aggravated assault, be served in full. Fieldwork for the current study during June 2001, however, disclosed speculation from a number of knowledgeable parties that the parole board's action was in fact an angry response to T.J.'s suicide attempt.

During fieldwork for the current study, researchers contacted the Georgia Department of Corrections with a request to interview T.J. A spokesperson replied that the request would be given serious consideration but mentioned that T.J. had recently tried to kill himself and was considered likely to be a problem for the agency for many years to come. Shortly afterward, a second spokesperson informed us that the request could not be granted because of concern about his mental health.

In August 2001, Judge Nation reduced the sentence so that it would conform to the originally understood minimum of 18 years. His order contained an explicit and stinging rebuke to the parole board, affirming his original intentions and pointing to eight other cases in which the board had not conformed to the guidelines it cited as justification for its action in T.J.'s case the previous May.

The judge imposed further conditions of the sentence, requiring that T.J. receive and comply with psychiatric treatment or risk violating probation. These additional conditions are consistent with his original statements at the sentencing hearing that he believed T.J. to be mentally ill and in need of treatment. His specification that any further evidence that T.J. is a danger to himself and others be considered a violation of probation can be read as a statement to the state parole board not to tamper with his express intentions, inasmuch as the probation department falls within the jurisdiction of Rockdale County.

The import of the judge's new sentence for T.J. himself is less clear now. If T.J. tries to kill himself again and fails, he risks having his probation revoked. If he succeeds, the issue is moot. The structure of incentives provided by the new order thus penalizes only a failed attempt. In light of his previous actions and his most recent attempt, which was reported to have come close to succeeding, it is uncertain whether or how the judge's order might affect T.J.'s future actions. This uncertainty illustrates some of the difficulties of dealing with mental illness through the criminal justice system.

AFTERMATH: THE VICTIMS

For all but one of those injured, physical recovery has been virtually complete, with only the presence of scars where the bullet had once entered their flesh. The exception is the case of the most seriously wounded student, a female. The bullet fragmented after entering her abdominal area, piercing her colon multiple times with part of it lodging in her hipbone. Splinters of bullet and bone also hit her right ovary. She has since undergone two rounds of surgery and months of physical therapy. It is not clear whether she will ever be able to bear children.

Aside from physical afflictions from the shooting incident, several of the victims and their families have experienced psychological and emotional distress as a result of their experiences. All of the victims underwent psychological counseling in the months succeeding the shooting. Their parents also suffered emotional anxiety, and several family members experienced depression requiring prolonged treatment. Some continued psychological treatment over two years later.

The event also placed strains on marriages and relationships between children and parents. More than one report indicated parents becoming, in their words, "overprotective" of their children, causing tension. Another family went in the opposite direction, giving virtually unfettered freedom to their child, with some negative results.

There were also reports of several of the victims acquiring indifferent dispositions following the shooting. We collected reports of some victims who "just didn't seem to care about much after the shooting" or "seemed to live for the moment."

In addition to the direct physical, psychological, and emotional impacts sustained by the victims of the Heritage High shooting, some families also felt additionally victimized by the mass media. They felt used by reporters and felt that their privacy had been invaded in unethical ways.

In spite of these difficulties, however, for the most part all of the victims of the Heritage High School shooting have moved on with their lives, following the same succession of life course events that many others not affected by the shooting have pursued. At the time of our fieldwork, some had gone off to college while others were still at Heritage.

One victim has had some difficulty adjusting, having initially gone to college, only to drop out after a year and a half. After experimentation with drug use and a carefree lifestyle, he now holds two jobs and is considering going back to school. These difficulties may have been related to his victimization.

Throughout the judicial proceedings, all of the victims had at least access to an active role and were given the opportunity to voice their opinions. In the time prior to hearings, prosecutors met with most of the victims concerning their views of what punishment T.J. should be given. Two of the victims testified in the juvenile court transfer hearing and others provided testimony in the sentencing phase.

Victims and their families differed considerably in their attitudes toward T.J. According to one recollection, three of the victims wanted the longest possible sentence to be handed down in response to T.J.'s actions. In the words of one, "No, I don't forgive him and I don't like him. Him doing the shooting makes him less of a person. The DA asked the victims what we wanted the sentence to be and I told him 'I want him to be in jail as long as he possibly can be.' . . . You see things differently when it happens to you."

Conversely, two other victims expressed attitudes of forgiveness and concern over T.J.'s fate. Another wanted T.J. to be assigned to some type of service so as to give something back to the community.

A family member of one of the victims' felt that T.J.'s incarceration is the appropriate recourse and believes that T.J. exhibited signs of trouble that should have been picked up on by parents and school officials. "He robbed a lot of people of a lot of things. . . . I don't have any anger against him; he's just a mental case."

Three of the six injured victims filed civil suits against T.J. and his family. One was settled. Two were still pending.

While no one died, T.J.'s bullets impacted the lives of the victims and family members in multifaceted ways. Victims were clearly affected physically, psychologically, attitudinally, and within their interpersonal relations.

AFTERMATH: THE COMMUNITY

During our fieldwork, conducted two years after the shooting, we were able to assess several aspects of the incident's aftermath for the wider community. One common theme was the necessity of having to deal with the external reputation of Rockdale County and its residents. Virtually everyone resented the portrayal of the area in the mass media. Besides resenting the content of these portrayals, many people who had direct contact with reporters, especially the victims but many others as well, felt violated by the way journalistic practices in the wake of a hot story invaded their privacy.

This media resentment, however, was shaped more by the public television documentary on the teenage syphilis outbreak than by the shooting incident. It was easier to rationalize T.J.'s actions as those of a lone "mental case" than to deal with external perceptions of local youth related to the syphilis outbreak, but the combination of the two created a special burden.

The young people interviewed for this study said they were embarrassed to say where they came from when they met peers from outside the county. One reported having players on opposing football teams refuse to shake his hand. Others said they planned to keep a low profile until they got to college.

Local adults were more diverse and measured in their reactions. One person felt the area had been "raped" by the media. Another reported that when he talked to people outside the county, he actually got supportive feedback, to the effect that these incidents could have happened in "Anywhere, U.S.A."

A few people, including some who had worked with the documentary filmmakers and also some public officials responsible for services in the county, thought the spotlight on these problems was at the same time painful but important. From their perspective, real problems did exist and needed to be confronted. They shared the "Anywhere, U.S.A." perspective and the feeling that Rockdale County was being unduly targeted for attention about problems found in many other places. At the same time, they expressed hope that the attention could lead to needed changes.

Another reaction was avoidance. One person referred to a cultural norm among middle-class Southerners that discourages discussing unpleasant things in polite company. Others expressed milder versions of this perspective, saying that there was not so much avoidance as a desire to move on.

Across the community, as among the victims, there was a wide range of attitudes towards T.J., both in terms of basic emotions and in terms of how he should be and was being dealt with by the criminal justice system.

A wide range of people stated that they were more sorry for him than angry at him. This was especially true of young people in the community. One adult who worked with some Heritage students immediately after the shooting said "kids were compassionate with T.J. They understood his problems. They described it as 'Yeah, he had problems with X, Y, Z and he just lost it.' They were not angry with him."

Two years later, in two separate youth focus groups, we heard similar statements. There was a lot of doubt about whether he had really wanted to kill anybody. One person said, "I don't think T.J. knew who he shot." Another said, "He just started pulling the trigger, I don't think he intended to seriously harm people," and another "Yeah, he was shooting towards the ground, the whole time. I don't think—I really truly don't think he was trying to kill anybody." Another pointed out how much T.J. knew about guns and said, "if he wanted to kill somebody, they'd be dead."

These youths also felt a good deal of sympathy for the emotional stress that they perceived as driving his actions. They said he was giving a "cry for help" and a "cry for attention . . . look at me." One said, "I think he got tired of everybody's bullshit and said 'Screw it,'" to which another added his construction of T.J.'s state of mind as "Nobody else is going to help me, I'll do it myself. . . . I think he wanted to scare the hell out of everybody. And he accomplished that." When asked what kind of help they thought he wanted, they said "Friends, somebody to care about him" and "Somebody to stop judging him; judgment is a huge issue, right now."

Not everyone held compassionate attitudes. Some were simply afraid and wanted him put him away for as long as possible. Others vigorously disputed the idea that he had not really wanted to hurt anyone, echoing the prosecution's emphasis on the girl who missed a bullet through the chest only because she was holding a book in front of her and pointing out that 9 hits and 6 wounded victims out of 12 initial shots was a high percentage. One of the victims was aware of and impatient with the sympathy for T.J., saying, "People feel sorry for him. Shouldn't it be the other way around?"

It is of course quite likely that if there had been deaths or more serious injuries, this mix of views would have been angrier and more punitive. The feelings of empathy among the young people, however, are striking. The tone of their remarks suggests fairly prevalent feelings of emotional identification.

Public officials and community members described a wide range of responses by community institutions in the aftermath of the shooting. There was clearly a flurry of activity right after the event. Extensive efforts were made to provide counseling to victims, students at Heritage, and young people throughout the county. County officials worked to bring in counseling resources from outside the county and were successful in doing so. Churches became involved and held community meetings. One church organized a retreat at which some students from Columbine High School came to Rockdale to talk with Heritage students.

On a more long-term basis, the school system has increased the number of psychologists and social workers in the system, and a new parent education program has been introduced in the county. These increases, however, come in an area that already has a high level of public services.

Despite this, public health officials and youth workers that we interviewed expressed ongoing concern that too many local youth still are in need of adult attention. One of them offered the opinion that "It's not kids. It's the parents." The circumstances of rapid community change and of many young people on their own while their hard-working, affluent parents are making long commutes are ongoing structural problems in the area that social service providers continue to confront.

One of the youth focus group participants also reported some changes in community attitudes toward young people with guns since the incident:

"A year ago guns are bad and you are not allowed to touch them. Now it has gotten more loose. . . . parents are real strict about it a lot, I know my parents are about me having one. . . . I don't think it will ever get as loose as what they were. But they are getting looser about you carrying a gun."

The primary area in which there is evidence of decisive, long-term change in institutional policy is that of relationships between school officials and law enforcement with respect to weapons in schools. Prior to the incident, school officials retained some discretion about whether to report weapons discovered in school to the police. By all accounts, that is no longer the case.

After the incident, there was a series of meetings at which school officials and parents discussed what changes were needed in school security. There was general agreement that bringing in metal detectors was not the answer and that the community "did not want the schools to become like prisons." There was some pressure also from some parents for systems to detect potential offenders in advance through some kind of psychological screening. School officials successfully resisted that also, without much controversy.

There were a number of behavioral code changes in the schools. Heritage students to whom we spoke grumbled about new dress codes, including such nonsafety-related requirements as that shirts must have collars, along with a rule directly related to T.J.'s shooting strategy that strictly forbids baggy pants of the type in which he concealed the .22 rifle.

The major change, however, appears to be the new policy of immediately referring any incident involving weapons in school to law enforcement. There was an incident following T.J.'s shooting in which a student was discovered to have brought a pistol to school. Since that student was already 17 years of age and therefore an adult under state law, he went directly before the superior court. Judge Nation gave a sentence of 10 years, to serve a minimum of 7, and had the sheriff post a copy of the sentence in every school in the county.

CONCLUSIONS

This case study, although focused on but one of the recent incidents of extremely serious school violence that have troubled the United States, raises a number of issues of potentially more general interest. These include public perceptions of the role of bullying in generating these incidents, the possible existence of a copycat wave of behavior, the role of mental illness, and the handling of mental illness among youth both by the criminal justice system and by society.

Even one case is enough to refute theories that oversimplify by attributing universal causal linkages. There has been a tendency in much of the commentary about school violence to see it as a response to bullying behavior. While revenge against bullies has been a significant factor in some cases, it was not for T.J. Solomon. His problem was not rooted in his direct interactions with peers, except in the negative sense that he was disconnected from them and from every other form of social interaction with others. Bullying is a serious problem, and one that has proved amenable to systematic intervention (Olweus, 1991). While that intervention is worthwhile for many reasons, it may not be the only or even the best way to think about preventing these kinds of incidents.

On the other hand, this case study clearly demonstrates the existence of copycat behavior. T.J. Solomon was stimulated to do what he did by the sensational media coverage of the events at Columbine High School. The next logical question is why it was T.J. that responded this way, rather than one of the legions of other young people exposed to this media coverage. Here, the study provides some potentially helpful answers. First, although the record contains conflicting points of view on the issue, it appears that T.J. was seriously mentally ill and suicidal. The contentions to the contrary advanced during his court proceedings have been contradicted by the subsequent event of his nearly successful suicide attempt in prison. That he had never been diagnosed so is not surprising. It would be surprising if he had, since serious mental illness can be difficult to diagnose in middle adolescence. The course and timing of his own developing psychopathy made him extremely vulnerable to the effects of Columbine.

One other factor was crucial in this case, the role of firearms in his family history and their inadequately secured presence in his home. Mental illness, the example of Columbine, and emotional attachment and access to guns were the crucial factors that came together in the Heritage High School shooting incident.

These things are clear from this case. Could there be other cases in which the same three factors did not produce the same result? Undoubtedly there could. The fact that there is at present no way to assess how many such cases there are is not reassuring.

Still, there is one potentially hopeful element here. To the extent that there is a copycat thread connecting the recent school shootings, this presents the possibility that it could run its course, as the infamy of being just one more suicidal loser dims. At a minimum, the increasing passage of time without further recurrence would seem to be favorable, to the extent that a copycat process has occurred.

If, however, the trend of mass school shootings does subside and, in so doing, diverts attention from the plights of other youths as seriously mentally ill as T.J. Solomon, then that diversion of attention would be unfortunate. The problems of family communication, rapid community change, and lonely young people sinking into despair in the midst of an affluent, heavily armed society are widespread. Those like T.J. still need help, even if most of them never hurt anyone but themselves.

Finally, this case raises a wide range of questions, most of which cannot be answered here, about youth, mental illness, and justice. The paradoxes of T.J.'s course through the juvenile and adult justice systems are several. In order for him to be tried as adult, it was necessary to deny his mental illness, a denial that has been definitively mocked by subsequent events. In an era in which the long-standing assumptions of the juvenile justice system are under sustained attack, this case reminds us that youth are not adults, and it points to a very particular aspect of this difference. Mental illness does not arrive full-blown. It cannot be assessed in a 15-year-old as well as in a 25-year-old.

This realization leads directly to another paradox. Even though in this case state law formally tied transfer to adult court to an assessment of mental health, the ability of state institutions to provide adequate treatment to mentally ill juveniles may have been severely compromised. Just two years before the incident at Heritage High School, the State of Georgia resolved an investigation by the U.S. Department of Justice into conditions of juvenile confinement by signing a memorandum of agreement designed to improve education and mental health services and improve monitoring of physical abuse. The judicial processes that sent T.J. to a long prison sentence as an adult did not happen in a vacuum. They occurred in the context of a broader structure of criminal justice and public health systems.

REFERENCES

Baumrind, D.
1978 Parental disciplinary patterns and social competence in children. *Youth and Society* 9:239–276.
Butterfield, F.
1995 *All God's Children: The Bosket Family and the American Tradition of Violence.* New York: Alfred A. Knopf.
Curtis, L.
1975 *Violence, Race, and Culture.* Lexington, MA: Lexington Books.
Goodman, R.D. and B. Goodman
1999 The lost children of Rockdale County. Boston: WGBH Educational Foundation, PBS Frontline.
Guide, AMG All Music
2001 Available: http://www.allmusic.com. [Accessed July 15, 2002].
Miller, W.
2001 The growth of youth gang problems in the United States: 1970–98. Washington, DC: Office of Juvenile Justice and Delinquency Prevention.
Olweus, D.
1991 Bully/victim problems among school children: Basic facts and effects of a school-based intervention program. In: *The Development and Treatment of Childhood Aggression Among Children,* D.J. Pepler and K.H. Rubin, eds. Hillsdale, NJ: Lawrence Erlbaum.

3

Bad Things Happen in Good Communities: The Rampage Shooting in Edinboro, Pennsylvania, and Its Aftermath

William DeJong, Joel C. Epstein, and Thomas E. Hart

At approximately 9:40 p.m. on Friday, April 24, 1998, Andrew Jerome Wurst, age 14, shot and killed science teacher John J. Gillette, at an eighth grade school dance held at Nick's Place, a banquet hall near the Parker Middle School just north of Edinboro, Pennsylvania. Armed with his father's .25 caliber semiautomatic pistol, Andrew also wounded another teacher and two classmates. Adjudicated as an adult, Andrew eventually accepted a plea on third-degree murder. Judge Michael M. Palmisano sentenced Andrew to serve 30 to 60 years in prison. He will not be eligible for parole until age 45.

* * *

April 24, 1998: There is a photograph of Andrew Wurst taken less than an hour before he shot and killed science teacher John J. Gillette. The occasion is the end-of-the-year dinner dance for the graduating eighth grade class at Parker Middle School in Washington Township, just north of Edinboro, Pennsylvania. Andrew sits awkwardly at a banquet table, dressed in a dark blue sports coat, white shirt, and blue and tan tie. He looks like an unexceptional 14-year-old boy, with blue eyes, glasses, and a mop of thick, dark brown hair covering his forehead. His arms are tightly crossed, his face expressionless. Looking at the photograph, it is hard to imagine that this slightly built and callow middle school student would soon be pulling out a semiautomatic handgun and killing John Gillette at close range.

June 7, 2001: Parker Middle School is a large, sprawling complex, in many ways typical of modern, suburban-style middle schools found across the United States. From a distance, there is nothing remarkable about the school, but near the front entrance that impression changes. The flagpole is surrounded by a large square wall made of brick, with a cement slab on top for seating. Mounted on one side is a small brass plaque with a three-line inscription: "John's Bench/Friend, Teacher, Builder/John Gillette." A few yards away is a well-tended garden with several stone walkways, each made of decorative stones with messages and designs made by Parker students. One corner of the garden, which was dedicated a year after the shooting, is dominated by a large gray stone engraved with these words:; "In memory of John Gillette, April 24, 1999."

* * *

After John Gillette's death, people who live in this small and picturesque lakeside town have repeatedly asked themselves, "How could this horrible tragedy happen in such a good community?" "That's the wrong question," a school administrator told us. "The real question is, Why not here?" By that he meant that a school-related shooting like this, involving a troubled and impulsive boy with access to a gun, could happen anywhere in the country. What he also meant is that if we are to prevent future shootings, we need to reflect on how we care for our children, manage our schools and communities, and live our lives.

In this case study, we examine the events that led to the shooting death of John Gillette, with the aim of understanding the mix of individual, peer, school, and community factors that combined to create this tragedy. We also explore the aftermath of the shooting, focusing in particular on how people have come to explain the event and what they appear to have learned from it.

Our report is based on extensive interviews we conducted during a site visit to the Edinboro area during June 2001, which involved individual and group sessions with 32 students, teachers, school administrators, community residents, town officials, police investigators, lawyers, court officials, and journalists. We also interviewed Andrew Wurst's parents, Catherine and Jerome J. Wurst. We learned that Andrew's mother does not permit interviews with Andrew, and we did not ask for one. Based on advice from friends of the Gillette family, we also did not seek interviews with them.

We conducted all of the interviews according to procedures for the protection of human subjects approved by the institutional review board of the National Academy of Sciences. Everyone we interviewed, including public officials, spoke under conditions of confidentiality. Several minors participated in a focus group with informed consent from their parents or guardians, in addition to their own assent. We have not identified sources by name unless their statements were part of the court record.

We also received official court documents, including procedural transcripts, psychiatric reports, and other materials entered into evidence. It was clear that, three years after the shooting, people's memories of what occurred were clouded, making it difficult to rely on the interviews to piece together a clear narrative. We benefited enormously from having access to the complete police investigative file, whose several thousand pages included summaries of witness interviews, signed depositions, Andrew's writings, evidence inventories, and other useful material. This file was provided to us by a source outside the Pennsylvania State Police. We also reviewed media reports, but after discovering several inaccuracies in those accounts, we relied on other sources for basic facts about the case. To provide background on Edinboro and the General McLane School District, we consulted a mix of census data, town records, school and police reports, and town-related web sites.

THE INCIDENT

The basic facts of the case are beyond dispute (Box 3-1). On Friday evening, at approximately 9:40 p.m., April 24, 1998, Andrew Jerome Wurst, age 14, shot and killed science teacher John J. Gillette, age 48, at Nick's Place, a banquet hall located just south of Parker Middle School on Route 99. The event, which Gillette had helped plan, had drawn about 240 students. The theme fit the occasion: "Had the Time of My Life."

BOX 3-1 Incident Timeline of Events: April 24, 1998

2:50 p.m.	The school day ends, and Andrew takes the bus home.
5:00 p.m.	Dressing for the eighth grade dinner dance, Andrew takes his father's handgun and puts it into a holster belt under his shirt.
5:10 p.m.	Andrew leaves a suicide note on his pillow.
5:30 p.m.	Andrew's parents drop him off at Nick's Place.
6:00 p.m.	The dinner dance begins.
9:25 p.m.	Andrew goes to the bathroom to take the handgun out of the holster belt and put it into his pocket.
9:40 p.m.	Andrew shoots and kills teacher John J. Gillette. He then shoots and wounds teacher Edrye May Boraten and students Jacob Tury and Robert Zemcik.
9:50 p.m.	Andrew is apprehended at gunpoint by Nick's Place owner James A. Strand.
10:00 p.m.	Andrew's mother arrives at Nick's Place to pick up her son, instead finding him under arrest.

NOTE: All times, except the start of the dinner dance, are approximate.

About 20 minutes before the dance was to end, Gillette went out on the back patio to ask a group of kids, including Andrew Wurst, to come back inside. As Gillette turned to leave, Andrew pulled out a .25 caliber semiautomatic pistol belonging to his father and shot him twice, once in the face and once in the back. A friend standing nearby recoiled in shock. "Don't worry," Andrew reassured him, "I'm not going to shoot you."

Andrew moved back inside, carrying the gun, and called out for Eric Wozniak, another student. Patricia M. Crist, Parker's principal, was cowering nearby, trying to take cover. "He's not here," she told Andrew. Andrew turned and pointed the gun at Crist's head. "That won't save you," he said sternly, looking at the principal's makeshift barricade. Andrew didn't fire but continued walking toward the dance floor area. As Andrew came through the double-door entry, another friend pleadingly asked why he was doing this. Andrew responded by putting his free hand to his head, twirling his extended finger, and yelling, "I'm crazy, man! I'm crazy, man!"

Andrew scanned the crowd of terrified teachers and classmates who were scrambling to hide or escape. "Shut up," he yelled, "or someone else is going to die." He fired two more times, grazing a second teacher, Edrye May Boraten, 51, and wounding a classmate, Jacob Tury, in the back. Andrew's eyes locked onto Justin Fletcher, one of the toughest boys in the eighth grade, who stood defiantly and stared at him. "If you're going to shoot, shoot me," Justin said firmly, "just don't shoot anybody else." Andrew fired. The bullet passed through Justin's shirt sleeve and hit classmate Robert Zemcik in the foot.

Seeming confused, Andrew left the banquet hall and went to a grassy area in back. Nick's Place owner James A. Strand, who lives next door, heard the gunshots, grabbed his 12-gauge shotgun, and ran over to confront the shooter. He spotted Andrew about 40 yards in back of the building. In his statement to police, Strand said that Andrew pointed the pistol at him. He drew a bead on Andrew with his shotgun and twice yelled for the boy to drop the gun. Andrew hesitated but did not drop his weapon immediately. Strand then heard someone else yell for Andrew to drop the gun, and he did.

Strand got Andrew onto the ground and began to search him for other weapons, assisted by science teacher David A. Masters and James J. Washok, a student teacher who worked with Gillette. Strand found a dinner fork hidden inside the top of Andrew's sock. In his statement to police, Washok stated: "Wurst was rambling and crying and said something to the effect, 'I died four years ago. I've already been dead and I've come back. It doesn't matter anymore. None of this is real.'" Strand and the two teachers walked Andrew back to the building and held him until state police arrived a few minutes later. Strand later said to police, "If it hadn't been a kid, I'd have killed him."

Parents began to arrive at Nick's Place just a few minutes later to pick up their children from the dance, finding instead a scene of complete pandemonium, with police cars, ambulances, television news crews, and clusters of sobbing children. Panicked parents rushed in to find their children, making it difficult for police to maintain the crime scene. One of those parents was Catherine Wurst. As she frantically looked for Andrew, she was pulled over to police investigators by Patricia Crist, the principal. "Here's the shooter's mother," Crist announced.

Andrew Wurst was arraigned before District Justice Denise Stuck-Lewis on April 25, 1998, on a charge of criminal homicide. Andrew was very quiet and showed no emotion in court. Were Andrew to be convicted in juvenile court, he would be released automatically when he turned 21 years old. Were he to be convicted as an adult of first-degree murder, he would face life in prison with no chance of parole. (According to Pennsylvania law, the death penalty cannot be imposed on offenders younger than age 16.) Erie County district attorney Joseph P. Conti made clear from the beginning that he would seek to have Andrew tried as an adult.

THE PERPETRATOR: ANDREW JEROME WURST

Rumors circulated immediately after the shooting that there was a list of targets and that two of Andrew's friends had vowed to "finish" the job he had started. A girl told police that one of those friends had said this to her directly. Early news accounts revealed that Andrew had tipped off several classmates about his plans, which fed concerns about a possible conspiracy. The police later received reports of an overheard conversation, with one boy going up to Andrew to say, "I'm out." As the police got to know more about Andrew's friends, they even began to wonder if one of them was a "puppet master" who may have manipulated Andrew into using the gun or at least encouraged him. Police were quite right to pursue these theories, but presently there is no credible evidence of any conspiracy. Andrew appears to have acted alone.

What was Andrew's state of mind when he opened fire? His very actions show that he was troubled, but was he mentally ill? This claim was the crux of Andrew's legal defense. Robert L. Sadoff, a nationally reputed forensic psychiatrist and a sought-after expert witness, examined Andrew during four interviews totaling over eight hours. Based on those interviews, plus meetings with Andrew's parents and various other records, Sadoff concluded in his official report that Andrew suffers from "a major mental illness, with psychotic thinking and delusions of persecution and grandeur" and is in need of long-term inpatient treatment with medication.

Andrew was average in size at 5′8″ and 125 lbs. According to Sadoff, while Andrew said he enjoyed football and soccer, he was not particularly athletic. He had worn glasses since second grade and talked of getting contacts to change his image. He also wanted to get stronger by lifting weights, noting that, when trying to work for his father in the landscaping business, he was unable to do some of the physical work due to his lack of strength. Andrew was not under a doctor's treatment and did not take any medications.

Andrew struggled academically, with his grades slipping year by year until he was getting mostly Ds and Fs in eighth grade. In a local newspaper article, one of Andrew's long-time friends said that Andrew's parents had grounded him a few weeks before because his grades had fallen. Andrew liked to read, and Stephen King was one of his favorite authors. Like other boys, he watched television and played computer games.

Andrew's taste in music ran to heavy metal, with Marilyn Manson and Nine Inch Nails being among his favorite bands. Indeed, a friend had given Andrew the nickname Satan because he was such an avid fan of Marilyn Manson's dark and angry music. Andrew told Sadoff that he didn't like the nickname. The Wurst family is Catholic. Andrew attended religious classes but didn't go to church, explaining to Sadoff that he questioned the existence of God because of all the suffering in the world. He said he didn't believe in Satan either, noting that without God there can be no Satan.

Andrew informed Sadoff that he frequently drank whiskey or vodka with orange juice, getting a "buzz," but not drunk. In eighth grade, he began to use marijuana occasionally, which he said made his body go numb. Several classmates told police that Andrew had bragged constantly about his drug use in recent months. According to the police complaint, Andrew was in possession of a "small amount of marijuana" at the time of the shooting, which he later told Sadoff he had bought for $80 to split up later with a friend. According to Andrew, he did not drink or use marijuana the night of the shooting. Urine and blood tests conducted by police investigators found no evidence of cannabinoids.

Andrew had briefly dated a classmate until about three months before the shooting, when she called him on the telephone to end their relationship. About this, Sadoff wrote, "He said he really didn't care, but maybe just a little bit." Clearly, the relationship meant something to Andrew. After school, when he was alone, Andrew would call the girl and talk at length with her, racking up expensive long-distance telephone bills. Andrew told Sadoff that he had never had sex with her and seemed uncomfortable talking about the subject.

By his own report, Andrew had difficulty falling asleep, often lying awake in bed for an hour or two, and he sometimes had nightmares of monsters chasing him. In his last interview with Sadoff, Andrew said that he was afraid of spiders and heights. He also spoke of monsters in his closet. Andrew's mother, Catherine Wurst, confirmed with Sadoff that her son had these fears about his closet, as well as the space under his bed. Each night she had to make sure there was nobody under the bed or in the closet and to leave a light on in his room. Often, she would lie on the covers with him for awhile, talking affectionately with him, trying to help him settle down to sleep.

Sadoff wrote that, according to Andrew, he was a frequent bed wetter until about age 9, which his mother confirmed. Andrew said that his two older brothers teased him about this a lot. Sadoff reported that Andrew had no history of fire setting or cruelty to animals, which are often warning signs of future violence. His mother told the psychiatrist that when at play Andrew would sometimes wear camouflage clothing and carry a toy gun and then crawl on the ground as if he were stalking someone.

Andrew told Sadoff that he began having suicidal ideas when he was 10, but could not say why. One time he had put a plastic bag over his head to see what it would be like. Other times he thought about hanging himself or shooting himself with one of his father's guns, the semiautomatic handgun he eventually used at Nick's Place or a long rifle, but he had never taken steps to act out those thoughts. He denied ever getting angry and, in fact, could not even describe the last time he was.

Andrew recounted to Sadoff that, on the evening of the shooting, he took his father's handgun and put it in a holster belt under his shirt. He loaded the gun with nine bullets in the magazine clip and one in the chamber. Before leaving for the dance, Andrew left a suicide note on his pillow, Andrew's mother told Sadoff. Andrew told the psychiatrist that a friend at the dance had seen the gun under his shirt. The friend, in his own statement to police, said that Andrew had asked him to feel something under his shirt, which he had surmised to be a gun.

Later, according to Andrew, he went to the bathroom to take the gun out of the holster belt and put it into his pocket, intending to kill himself but not hurt anyone else. About 15 minutes later, he was out on the patio with several friends, when Mr. Gillette came out to ask them to come inside from the cold. Andrew pulled out the gun and shot him in the head. The teacher fell to the ground, and Andrew shot him a second time. Why did he do it? Andrew doesn't know, he told Sadoff, saying he had no reason to kill the teacher.

Andrew remembered going inside then and shooting, but not at anyone in particular. One boy, Justin Fletcher, was staring at him, Andrew recalled for Sadoff, and that is why he shot at him. The psychiatrist asked

Andrew why he had called for Eric Wozniak. Having wanted to die, Andrew replied that he didn't think one bullet would kill him, and he wanted Eric to take the gun and shoot him several times to make sure he'd die.

Andrew stated that he went outside and started putting extra bullets from his pocket into the gun clip. Then James Strand, the banquet hall owner, came out with a shotgun, and Andrew put his gun down. Why did he do that, rather than incite Strand to shoot him? Andrew told Sadoff that if Strand had shot him from a distance, it would not have killed him, and he did not want to be in a coma.

Later press accounts stated that, as Andrew sat handcuffed in the back of a police car, television cameras recorded him laughing, which angered many people. (We were told that this video clip was often reused whenever a new development in the case was reported on television news.) When Sadoff asked about this, Andrew explained that he had heard a friend call out his name and had turned to smile at him.

In the second interview, Sadoff asked Andrew about hearing voices, which he denied at first. The psychiatrist persisted, knowing that Andrew had mentioned in a letter to a friend that "the voices are coming again." Andrew then opened up. In his report, Sadoff stated that Andrew thinks he is real, but everyone else is unreal. Other people are "programmed to act and say what the government, mad scientists, or a psycho want them to say." Andrew elaborated later, saying that people are like robots, programmed with "time tablets" that give them differing levels of intelligence and different personalities. Had Andrew found any real people in the world? No, he answered, but he is still looking.

Sadoff asked Andrew several questions designed to get Andrew to elaborate on his belief system. In response, Andrew explained that people can be real in his presence, but once they leave, they no longer exist. Everyone has a body, but what is different in people is the mind. Andrew himself can think, but everyone else is programmed. According to the psychiatrist, Andrew's thoughts are vital to his sense of self. As Andrew put it, "If I can think, I am free—the last freedom are my thoughts."

There was nothing wrong with killing Mr. Gillette, said Andrew, because "he was already dead or unreal." At another point, Andrew said that the unreal world has rules that are not his rules, including the rule that killing is wrong. He is in jail, he explained, because he killed an unreal person.

According to Sadoff, Andrew stated that "when he wakes up, he will wake up and be real and be back in his own real world when he dies. Then he will be going home again." He explained that this real world would be "the ideal world, where there are no wars, no crime, no viruses—it is a utopia."

When did Andrew begin to have these thoughts? Andrew told Sadoff that, when he was 8, he got caught between two swings and lost consciousness. He thought he might have died but "didn't know if he had gone to hell or to heaven or a different world." By the time he was 10, he was certain that he was the only real person living in an unreal world.

At another point, Andrew stated he had returned from the future and has "a mission to prevent something terrible that has happened or that will happen in the future." He wasn't certain what that mission was, but he did know that he had an "arch enemy" who would try to prevent him from accomplishing it. He had never seen this arch enemy but knew that everyone has one, which meant that he had to be on guard.

In the third interview, Andrew explained that he also had to watch the unreal people, stating, "They are going to screw me over." As Sadoff pressed him to reveal more about his thoughts, Andrew replied, "If I tell you my thoughts, that's not good. All I've got left are my thoughts, and I would lose my purpose or mission in this unreal world if I told you my thoughts."

When asked about important relationships in his life, Andrew stated that he loved his dog, Tasha. Sadoff asked Andrew about his parents. The psychiatrist later wrote, "He wasn't sure that he loved his mother. He said he liked her, and he even liked his father, but he wasn't sure he loved him." Andrew next said "the only thing that could save him was killing himself." Then he could go back to his real world. Sadoff noted that Andrew questioned whether the Wursts were his biological parents, stating that he was "brought here when he was about 4 years of age from his world, and his parents were an appropriate candidate to have him."

Also in the third interview, Sadoff asked Andrew how he felt the evening of the shooting. His answer: "Miserable." He said he was planning to kill himself, explaining that the worst thing was the hopelessness he felt. What was he feeling hopeless about? "Everything." What was making him so miserable? "Nuclear wars, viruses, murders, robberies, school." Sadoff asked Andrew if there were specific things about himself that made him miserable, but he couldn't recall anything specific.

In his report, Sadoff concluded that Andrew is "not old enough to be labeled a schizophrenic, but he clearly has pre-schizophrenic ideation. . . . I can diagnose him as having a psychotic condition, not otherwise specified, with paranoid ideation" (diagnosis 298.9 in the *Diagnostic and Statistical Manual of Mental Disorders—Fourth Edition*). It should be noted that this is a condition that is often accompanied by severe depression. In Sadoff's medical opinion, Andrew was psychotic at the time of the shooting and "did not have a realistic appreciation of the nature and quality of his behavior or the wrongfulness of it because of his delusional belief that Mr. Gillette, along with all the others, were unreal and programmed by others. It was not in his thinking that it was wrong to kill a person who is already dead or who is unreal."

A psychiatrist who examined Andrew for the Erie County prosecutor reached a different opinion about the boy's mental health. John S. O'Brien, a forensic psychiatrist and staff psychiatrist at Pennsylvania Hospital, conducted a nearly three-hour evaluation of Andrew in August, just over two months after Sadoff completed the last of four interviews with Andrew. O'Brien's opinion was that Andrew did not suffer from a delusional disorder or manifest symptoms of any major psychiatric illness. Andrew had a history of emotional upset, he said, which was characterized by "depressed moods associated with aggressive and suicidal ideation." He also declared that Andrew was competent to stand trial.

Hard-nosed police investigators are often skeptical about claims of legal insanity or mental illness, and that was certainly the case for the Pennsylvania State Police. Their theory was that Andrew wanted notoriety. He wasn't a star pupil. He wasn't an athlete. The shooting, they assert, was Andrew's way of gaining the attention he craved. In their view, claims that Andrew was mentally ill were overblown. Feeding their doubts was a mistake made by the defense at the competency hearing held on September 25, 1998. Sadoff characterized some of Andrew's handwritten notes as "showing evidence for hallucinations, delusional thinking, loose associations of thought." It turns out, however, that what Andrew had written down were song lyrics by Marilyn Manson. When confronted about this at the decertification hearing on March 9, 1999, Sadoff said that learning this did not change his diagnosis, and that while these turned out not to be Andrew's original writings, it was still important that he had selected them to copy.

From a lay perspective, what makes Sadoff's diagnosis difficult to accept is that people with delusions will often function well and appear to be normal in their everyday lives. The illness can become evident when they discuss their delusional belief system, but they are generally reluctant to do that if they anticipate disapproval. For this reason, Andrew's mental illness could easily have escaped his parents' notice. This also explains why Andrew would reveal his thoughts to Sadoff but not to O'Brien, the psychiatrist brought in by the Erie County prosecutor.

Sometimes this type of illness is revealed more tragically, as with Andrew's rambling declarations after the shooting spree that he had died four years ago and come back and that none of this was real. Based on his interviews, O'Brien asserted that what Andrew had talked about with Sadoff were not fixed beliefs or convictions, but just ideas he was thinking about. In our view, that argument cannot adequately account for what Andrew said after James Strand and the two teachers had him pinned to the ground.

Police investigators' doubts about Sadoff's diagnosis were also fueled by suspicions that Andrew was trying to con the psychiatrist into thinking he was mentally ill. The most crucial evidence against fakery is that, shortly after they started going out back in February, Andrew told his former girlfriend his fantastical thoughts about "real" and "unreal" people. According to Sadoff's report, Andrew said to her, "We are all in reality in hospital beds being monitored and programmed by these mad scientists, and this world is not real for them. . . . The scientists watch over us to see what we're doing." She found the conversation disturbing and asked him to stop. He brought up the subject again about two weeks later, just before they broke up.

THE VICTIMS

Mr. Gillette was known to Andrew as the eighth grade class adviser, but he had never been Andrew's teacher. No school administrators, faculty, or students could remember any kind of previous disagreement between the two. Police questioned eyewitnesses: Did Gillette seem angry at Andrew? Did he say anything to Andrew when he came out on the patio? Did he say anything to Andrew earlier in the evening? Some students recounted that Gillette had called out Andrew's ticket number for a small door prize, and that Andrew seemed reluctant at first to come forward. His friends coaxed him to retrieve the prize, but it was clear to the girl who handed it to Andrew that he didn't want it. In fact, he immediately passed the prize to one of his friends, indicating that he wouldn't need it. On reflection, no one who was at Nick's Place that evening could recall anything that might explain what happened. As noted, Andrew himself had no explanation. A former football coach, Gillette didn't brook any nonsense, but he was also among the more popular teachers. Police investigators firmly concluded that there was no reason why Andrew targeted the science teacher.

When Andrew entered Nick's Place, having just shot and killed Gillette, he asked for Eric Wozniak by name. Was he Andrew's intended victim, or was there another reason he wanted him? It's not clear. Police investigators believe that, when Andrew was scanning the crowd of teachers and classmates, he was looking for specific kids to shoot. Investigators did not find a written "hit list" of targets, but one friend whom Andrew told of his plans had asked about a list, and Andrew did mention one girl by name. During the shooting, however, Andrew did not call out for her. As mentioned earlier, Andrew told Sadoff that he had called out Eric Wozniak's name because he wanted Eric to shoot him to ensure his own death.

Justin Fletcher, the boy who stood up to Andrew and defied him to shoot him, was well known to area police. According to police investigative reports, there was a girl that Andrew and one of his friends had been harassing that spring, "one minute saying she was beautiful, next minute saying she was a whore." Justin had confronted Andrew and told him to knock it off, which he did. We don't know if Andrew had planned ahead to shoot Justin. In describing to Sadoff what had happened that night, Andrew indicated that Justin had been staring at him and challenged him to shoot him. Clearly, Robert Zemcik, the boy hit by the bullet intended for Justin Fletcher, was not an intended victim, nor do police think that classmate Jacob Tury was targeted. Regarding the teacher who was wounded, Edrye May Boraten, police simply have no idea whether Andrew shot at her in particular.

According to police investigators, some kids at the dance thought that Andrew's former girlfriend was an intended target, and there were reports that several of them shielded her from Andrew's sight to protect her. Even though they had stopped dating, it was clear that Andrew was still infatuated by her. He had asked her to the dance, but she declined and came instead with one of Andrew's friends. After being apprehended by James Strand and the two teachers, Andrew asked for this girl by name, hoping to talk with her. Stories also circulated that Andrew had targeted a second girl who had laughed at him when he asked her to go to the dance. While there is no firm evidence that Andrew had wanted to target either girl, it is revealing that their classmates thought that he did.

THE BYSTANDERS

Andrew had signaled to friends for a few weeks his intentions to do something "memorable" at the dinner dance. According to these students, they did not take Andrew's statements seriously, knowing he had a "sick sense of humor," and they failed to notify either their parents or school officials. Even the night of the dinner dance, a group of Andrew's friends thought he might have brought a gun to kill himself, and while they checked in with him to make sure he was okay, no one alerted a school chaperone about their fears.

During her interview with Sadoff, one girl said that Andrew had told her he was not going to be around after the dance because he was going to shoot nine people and then himself. She asked him why, and he answered that he hated his life. In other conversations with her, Andrew said he wanted to shoot his parents. Why didn't she take him seriously? Andrew was always saying odd things, she said, such as calling himself "Your God, Satan." He wanted people to think he was dangerous, she explained, but she didn't really think he was. Similarly, a friend of Andrew's visited the Wurst home on April 11, almost two weeks prior to the shooting. While at the house, Andrew showed the boy a handgun and the gun clip, stating that "he was going to kill nine people he hated and then use the last bullet for himself."

After hearing Andrew talk about his plans to kill nine people and then commit suicide, another friend asked him whom he would shoot if he were to bring a gun to the dinner dance. He named a specific girl, but still the friend dismissed it all as a joke. In a subsequent conversation, Andrew appeared to be trying to recruit him into his scheme, asking him vaguely, "Are you in?" His friend said "No." Then, two days later Andrew asked him again. "So, are you in with me?" His friend again declined, to which Andrew said, "Well, you have so much going for you," citing his friend's girlfriend.

One girl was sure Andrew was joking the day he called her at home and talked about the boys in Jonesboro, Arkansas, who brought guns to school and opened fire on classmates and a teacher, killing five. "He said he was going to do something like that someday," the girl told reporters. "I said, 'Remind me not to come to school that day.'" Another friend remembered Andrew saying, "That Jonesboro thing, that would be like me bringing a gun to the dinner dance," and then Andrew laughed, as if it were a joke. Another time, this friend remembered, Andrew looked out the classroom window and told everyone to duck because the kids from Jonesboro were outside. In his report, Sadoff noted that Andrew didn't recall saying anything about Jonesboro, but that if he did, he was probably joking.

Another critical incident occurred a few weeks before the shooting. At the Wurst home, Andrew showed his father's handgun to several friends who were visiting. Later, the stepfather of one of the boys came by in his truck and drove the boys over to their house. Andrew furtively brought the gun along, and the boys went to a field behind the house for target practice with the gun. One boy said that they took turns shooting the gun, trying to hit a tree about 25 feet away. Andrew, he reported, was the only one who missed. Andrew described the same incident when police investigators asked if he had ever taken the gun out before that evening at Nick's Place.

In the bathroom at Nick's Place, a friend reported later to police, Andrew asked him to feel something under his shirt, which he thought might be a gun, although he wasn't sure. Andrew said his friend saw the handgun. In any case, the friend immediately remembered Andrew's talk of suicide, and he grew alarmed. He shared his concerns with several friends. He told them to keep it quiet, but he also recruited them to keep an eye on Andrew, which they did. Andrew's former girlfriend told police that her date, who was one of Andrew's friends, asked her, "Would you dance with a friend if it would prevent him from committing suicide?" She knew he meant Andrew. She wouldn't dance with Andrew, but she did confront him about the rumor she'd heard. Andrew denied it, she told police, and walked away. Several other classmates also asked Andrew about his intentions, and his response was similar. It was out of concern for Andrew, according to the alarmed friend, that a group of them had gathered around Andrew on the back patio when Mr. Gillette came over and asked them to come inside.

THE WURST FAMILY

Andrew is the youngest of three sons born to Jerome J. Wurst and Catherine Wurst. Andrew's oldest brother is Peter, who was 19 in April 1998. His other brother is Patrick, who was 16. Mr. Wurst was married previously and had two sons by his first wife who are several years older than Andrew and not a real presence in his life.

Jerome Wurst, who was 61 at the time, is the owner of a nursery and landscaping business in McKean, north of Edinboro. For Jerome Wurst, his business has been a seven-day-a-week proposition, with each work day ending long after sundown. He is what could be called "old school"—a hard-working breadwinner, a strict disciplinarian, and a reserved and emotionally guarded man who is sometimes gruff and irritable. He expected all three of his sons to work hard, be self-reliant, and keep out of trouble. Importantly, there was no evidence of any of the three boys receiving abusive or neglectful treatment in the home.

It is striking that the characterizations of John Gillette that we heard echoed what we had been told about Andrew's father. Gillette moonlighted as a bricklayer and builder, trades similar in some ways to Wurst's landscaping work. Like Andrew's father, Gillette worked long hours that took him away from his family. Remembering him, a school colleague told us, "[John] liked to joke that he worked as a teacher part time and worked as a builder the other 16 hours a day." Both men also had a reputation for strictness and occasional irritability. Another person who knew Gillette said, "He was a big, strong guy—an authority figure. He was kind, but in terms of discipline, he ran a tight ship."

Catherine Wurst, 20 years younger than her husband, met him when she went to work at the nursery. Where he is quiet and reserved, she is extroverted and outspoken. Other parents thought of her as a normal mother. She read to her sons. She worried about their grades. She made sure they dressed nicely for school, even as they got older. She gave them responsibility for chores around the house. One parent who knew Andrew said that he always seemed very happy to see his mother when she came to pick him up. Almost daily, Andrew's mother told Sadoff, she and her son told each other that they loved one another. More recently, Andrew's mother was working at the nursery in the afternoons, which meant that Andrew sometimes spent time at home alone. Apparently, it was on one of those afternoons that he went through his parents' belongings and discovered his father's semiautomatic pistol and a box of bullets, which were kept in his dresser drawer.

At the time of the shooting, the Wursts had been having trouble in their marriage, which created great tension in the household. In his report, Sadoff noted that the Wursts had several fights, some of them physical, which the boys sometimes saw. As a result, the Wursts had kept separate bedrooms in the home for the previous three years. How did this affect Andrew? Several of his friends and classmates knew about the tension between Andrew's parents and his unhappiness about it. Sometime after the shooting, Andrew's mother moved out of the family home, and the couple has divorced.

One source of conflict was Andrew himself. Compared with his other sons, Mr. Wurst found Andrew to be lazy and immature, and he wanted him to work harder to help the family business. Andrew resented his father's expectations, telling one friend that he felt trapped into going into the landscaping business. Andrew's mother was inclined to cut Andrew some slack, allowing him to stay home instead of working. She realized Andrew was less mature than Peter and Patrick, but simply thought that he was different from the other boys—less physical, more sensitive, the "baby" of the family. Many times the parents disagreed over how Andrew should be disciplined or what kinds of privileges he should have. Andrew may have found this inconsistency to be confusing, Sadoff wrote in his report.

After the shooting, this conflict between the parents made it difficult for them to agree on a defense strategy, with Andrew's mother more inclined to pursue every legal avenue open to them, and his father more resigned to Andrew having to face severe punishment. The Wursts had to remortgage their family home to raise the funds for his defense.

Andrew's mother told Sadoff that Andrew had been caught drinking with friends at the house. In fact, she had suspected that he'd been sneaking whiskey for several months. After the shooting, the Wursts learned for the first time that one of Andrew's older brothers had caught him smoking marijuana in the bathroom and beat him up for it.

Andrew's mother also told Sadoff that she and her husband had not been aware of Andrew's depression or his suicidal thinking until they found his suicide note after the shooting, which devastated them. Later, police investigators searching Andrew's school locker found a note labeled "Andy's Will." In a childish scrawl, Andrew had written that one of his friends should receive his music CDs. Another student told police that his friend had been surprised to find the note in his notebook, and that a teacher had confiscated the note and sent him to the school office with it. The Wursts do not recall learning about this from school officials.

According to Sadoff's report, Andrew's mother said that her son had seemed emotionally flat, especially for the two months prior to the shooting, and that he appeared to have lost his usual enthusiasm. She also remembered that Andrew often had headaches and seemed fatigued, and that he frequently had trouble getting out of the bed in the morning. She didn't think this was anything out of the ordinary. Andrew's mother also reported that she had never seen her son angry. Jerome Wurst also acknowledged to Sadoff that Andrew did not get angry or show much emotion. But Andrew's father did say that he noticed something different about his son, what he called a "dark look," which he also described as a "faraway look or a day-dreaming look."

Andrew may not have told his parents how angry he was with them, but he did reveal it to his former girlfriend in a letter written a few weeks before the shooting. Andrew wrote: "Are you doing ok? I got yelled at again by my dad. Let me think that's—4 times now give or take. He started that 4th commandment on me you know 'Honor thy father and mother.' Gee I feel soo bad. (cry,cry) Not. Fuck them thanks to them I'm in my shit life on the edge of insanedity [sic], murder and suicide." A classmate informed police that, in a wood shop class, Andrew had told everyone his father was an alcoholic and his mother was a prostitute.

The mother of one of Andrew's long-time friends remembers Andrew fondly, recalling how as a younger boy he liked to play flashlight tag and eat pancakes and go on amusement park rides. She loved his wit, she told us, but she and Andrew also had serious conversations about religion and the existence of God. She found Andrew to be bright and intellectually engaging, and she attributes his poor academic performance to boredom. She also began to see changes in Andrew in eighth grade. Although her son and Andrew were no longer in the same classes, Andrew telephoned her son a lot and she would still see him occasionally. Andrew seemed unhappy, she reported. One time she overheard him speaking in a harsh, disturbing voice, making a crude and angry comment about girls, which upset her. Another time, she reported, as Andrew looked out her back window, he asked her plaintively, "Why do you think my mother married such an older man?"

A nonfamily source told us that there is a history of depression on both sides of Andrew's family. In addition, Jerome Wurst has a sister who has been confined in the nearby Warren Psychiatric Hospital for several years.

ANDREW WURST'S PEER RELATIONSHIPS

Early press reports described Andrew as an outsider at school, someone who often seemed quiet, even withdrawn, and whose nickname was

Satan. It was a dark and disturbing portrait. Andrew was not one of the popular kids. He was not a good student. He was not an athlete. If he was involved in any extracurricular activities, no one mentioned them to us. Contrary to some press accounts, however, Andrew was not a loner. He did have a small group of friends that included both boys and girls. Police investigators characterized Andrew as being a fringe member of the group, but none of his friends described him that way in their statements to police.

Two of Andrew's long-time friends objected to what they were reading in the newspapers, explaining in a story in the Edinboro *Independent-Enterprise* that this disturbed loner was not the Andrew Wurst they knew. He was talkative, they said, always laughing and fun to be around. They didn't know Andrew as Satan, but as Brown Bag, a nickname they gave him years ago because of a reusable lunch bag he used. One of the boys, who said he had been friends with Andrew since third grade, reported that he had talked with Andrew every day at school, but he also admitted that since they weren't in any classes together this past year, Andrew had started hanging out with a different group of kids.

One of Andrew's newer friends was known to school officials for problem behavior. A teacher later told police that, following a fire drill shortly after the Jonesboro shooting, this boy made vaguely threatening statements and pretended to have the remote control to a bomb, which turned out to be an electronic guitar tuner. When the teacher responded angrily, he just laughed at her and said he was kidding. The next day she overheard this same boy speaking to Andrew and four other boys about how Andrew could kill his mother: "We'll wait until she gets back and as she's coming in the door, that's when you'll shoot her, and we'll just say we didn't know who she was, that we thought she was an intruder." The teacher reported both events to a school administrator, but she doesn't know what happened after that. Andrew's mother does not recall being notified about the overheard conversation.

Andrew's shift to this newer group of friends coincided with the change in his mood and behavior that his mother, teachers, and classmates all described after the shooting. He didn't have a reputation for being a bully, but that may have begun to change during the last semester of eighth grade. One teacher remembered that he had seemed to become more animated and cockier during the past semester. During that time, Andrew had made clear to other students that he didn't like the "popular students" and "athletes," saying they were stuck up. This was echoed by one of his friends, the boy overheard talking about shooting Andrew's mother, who said that he and Andrew hated these classmates. As mentioned before, police learned that Andrew and this same friend had been picking on one of the girls in his class, which had brought Justin Fletcher into conflict with Andrew. Another student described playing basketball in gym class on the day of the dinner dance. Andrew was supposed to "sub" for him, but refused. When the student insisted, Andrew told him that if he ever saw him alone, he'd shoot him.

Just after Valentine's Day, Andrew had a very brief romance with a girl in his class, which she broke off. As noted, she later told Sadoff that she did this because of the strange things that Andrew had told her. She informed police that Andrew referred to himself as "Your God, Satan." Other times he talked about the anti-Christ and Hitler as being "cool." She also said to police that Andrew seemed obsessed with her. She noted, for example, that she had given Andrew her class schedule, and that he began to show up to see her throughout the day. As noted before, even after they stopped seeing each other, they continued to stay on friendly terms as part of the same circle of friends. Occasionally, Andrew's anger and disappointment would show. On a page in his yearbook, the girl wrote, "Andy, you're sooo mean. To think I actually liked you and you can't even sign my yearbook."

Clearly, Andrew still hoped that his former girlfriend would get back together with him. In a letter he wrote to her a few weeks before the shooting, Andrew mentioned a boy she had dated previously and his eagerness to get back with her: "Not that I blame him, with a body like yours plus your personality—god I want you." The girl was not interested. Andrew asked her to be his date for the dinner dance, but she refused and instead went with one of Andrew's other friends. In his yearbook, another girl wrote of Andrew's frustrations: "Andy, Give it up—[She] will never go back out w/u! Hey, maybe we can get together w/[male student] & smoke up or something this summer?!"

Andy also asked a second girl to the dance, who later told police investigators that she laughed at him. A student who claimed to have witnessed this exchange said that Andrew had responded by saying, "Then I'll have to kill you." Another girl told police that, about two weeks before the dance, Andrew had offered her $10 to beat up the girl who had laughed at him, knowing that the two girls hadn't gotten along for awhile.

Andrew's inability to get and keep a girlfriend tapped into his insecurity, touching him to the core. Earlier that spring, Andrew had also written a letter to a girl who was dating one of his friends. In the letter, which he signed "Satan," Andrew asked if she was going to the dinner dance with his friend, and whether he was getting "any head" from her. At the end, Andrew wrote, "I'm so stupid I never thought to ask you, but know—can I live at your house? You know how parents are and it wouldn't be very long. Please? but that ok I haven't found anybody to say yes but I will. Well til next time."

JAMES W. PARKER MIDDLE SCHOOL

The Parker Middle School is part of the General McLane School District, which also serves the Borough of McKean, Franklin Township, McKean Township, and Washington Township. The district enjoys an excellent reputation. In its October 2000 issue, the parenting magazine *Offspring* rated General McLane as one of the top 100 school districts in the nation. Parker Middle School houses grades 5–8. In 1999, average student assessment scores for grade 8 math and reading put Parker at the high end of the range for socioeconomically similar schools in Pennsylvania. Scores for grade 6 writing exceeded the range in which similar schools tended to score.

Statistics published by the Pennsylvania Department of Education for the 1998–1999 school year provide a portrait of a large, well-appointed school with a stable student body, largely from middle-income and upper-middle-income families. Parental involvement in the school is high, as are parental expectations.

In the 1998–1999 school year, Parker had nearly 900 students. There were 49 teachers and only two counselors, plus the principal, a librarian, and three other staff. Class sizes ran somewhat lower than state averages for public secondary schools. At Parker, 65.8 percent of classes had 23 or fewer students, compared with 52.9 percent statewide. The attendance rate at Parker was 95.8 percent, compared with a state average of 93.0 percent. Enrollment was fairly stable, with 97.0 percent of students having started in the school district prior to the 1998–1999 school year. Only 16.9 percent of Parker's students were classified as low income, compared with a state average of 31.7 percent.

Parker is a modern, resource-rich school with an abundance of academic programs and extracurricular activities. According to the Pennsylvania Department of Education, the school's library had over 17,000 volumes, with 14,300 titles checked out during the 1998–1999 school year. In that same year, the school had 252 computers available for student use; 146 of these had CD-ROM capabilities. Internet access was available in teacher workrooms, the computer lab, the library, and classrooms.

Historically, Parker has not had major student disciplinary problems. Prior to the shooting, violence and students bringing weapons to school seemed like a distant problem, not the sort of thing that would ever be a problem in Edinboro. There were cliques, which is typical of middle school students, but administrators and faculty were not aware of any disruptive divisions among the student body that warranted special attention. Some sources said that, when students moved into the middle school, Edinboro kids looked down on kids from McKean, where Andrew lived, but others denied that was the case.

Prior to the shooting, Parker had begun a peer mediation program to help resolve student conflicts, but there was no violence prevention curriculum. Washington Township, where Parker is located, falls in the jurisdiction of the Pennsylvania State Police. One trooper was assigned to serve as the resource officer for nine school districts, including General McLane, with a presentation series that included grade 5 and 6 lessons on drug and alcohol abuse. The trooper was able to do presentations at schools only up to twice a year and sometimes was unable to reach certain schools. In 1998, for example, he did not visit the General McLane School District.

Parents we talked with complained that, while the district's elementary schools did a good job of monitoring the children's progress and staying in touch with parents, the middle school did not. Many of them blamed the principal, Patricia Crist, but others noted that parents had begun to register this complaint about the school years before, when the new building was erected and the grade 5 classes were transferred to the middle school. A school administrator informed us that Crist's predecessor was similarly criticized. For many years, Parker has had a student support program to which students who may need intervention can be referred. There is no record that Andrew was ever formally referred.

Andrew Wurst's eighth grade class was the largest cohort at the school, with over 50 more students than the other grades. Looking back on Andrew's time at the school, no school official can remember him standing out in any way, either positively or negatively. In his statement to police, David A. Masters, Andrew's science teacher, described him as withdrawn and reserved, but not as a loner. He also noted that Andrew had never been disruptive in class. Indeed, Andrew had no disciplinary record at Parker. There was also no record of his being bullied. In short, Andrew was not a kid that teachers were concerned about.

BOROUGH OF EDINBORO

Both Parker Middle School and Nick's Place are located just outside Edinboro in Washington Township, a mostly rural area that surrounds Edinboro. With a population of just under 7,000, the Borough of Edinboro is the largest of several municipalities in the General McLane School District. As one Edinboro official put it, "We're the pinpoint on the map for the greater Edinboro area."

People speak of Edinboro as a good place to raise a family. Located 18 miles south of Erie, the borough is the home of Edinboro University of Pennsylvania, originally named the Edinboro State Normal School. With nearly 8,000 undergraduate and graduate students, the university gives the town the look and feel of a small but prosperous college community and enriches the town's cultural and social life (Table 3-1).

University students who live in Edinboro are counted in the census, which makes it difficult to develop a meaningful statistical portrait of the borough's permanent residents. For example, according to U.S. Census Bureau data for 2000, 59 percent of Edinboro's residents are between the ages of 15 and 24 years. Fully 27 percent are classified as living in group quarters (in this case, college dormitories), and 6 out of 10 of Edinboro's households (60.2 percent) are nonfamily households. Including university students, Edinboro's population is 92.3 percent white and 4.4 percent black.

Edinboro is generally considered middle to upper middle class and has a relatively large percentage of professionals compared with other small towns in rural Pennsylvania. The cost of living is low by national standards. Housing is reasonably priced. The borough's web site directory lists 17 churches, and many of the people who spoke with us openly revealed the importance of religion in their lives and the life of the community.

In general, Edinboro's business climate is good. The two largest employers in the area are the university and the Northwest Tri-County Intermediate Unit, a regional agency that provides education programs and services to 17 school districts in three counties. Contacts at both places helped bring in several grief counselors immediately after the shooting. Two other major employers include Penn Union, which manufactures electrical connectors, and Morrison Food Service.

The downtown area along Erie Street has a good mix of stores, services, and restaurants. The Crossroads Dinor [sic], a landmark restaurant

TABLE 3-1 Comparative Demographics: Income, Race, Home Ownership, and Poverty by Area

Area	Median Household Income	% Black	% Hispanic	% Owning Home	% Below Poverty
United States	$37,005	12.3	12.5	66.2	13.3
Pennsylvania	$37,267	10.0	3.2	71.3	10.9
Erie County	$35,341	6.1	2.2	69.2	12.7
Edinboro	NA	4.4	1.0	NA	NA

NOTE: University students who live in Edinboro are counted in the census, which makes it difficult to develop a meaningful statistical portrait of the borough's permanent residents. NA = Information not yet available.

SOURCE: U.S. Census Bureau, Census 2000 (http://www.census.gov/census2000).

owned by John Gillette's brother, James, sits at the corner of Erie and Plum Streets. There are a few taverns in town, one of which is owned by Edinboro's mayor. At the south end of downtown is a refurbished fountain, which was installed in 1913 to "serve men, horses, and dogs." Recently, the Edinboro Kiwanis Foundation led an effort to revitalize the downtown area, placing new benches and trash receptacles. Leading north out from Edinboro, there are small suburban-style shopping centers along Route 99. A new Wal-Mart is under construction near Interstate 79.

In many respects Edinboro is vintage small-town America, with a volunteer fire department, historical society, garden clubs, and a long list of community service organizations, including the Independent Order of Odd Fellows, the Kiwanis Club, the Knights of Columbus, the Lions Club, the Rotary Club, and Veterans of Foreign Wars. The local newspaper, the Edinboro *Independent-Enterprise*, is full of typical small-town news.

A calendar of events published by the Greater Edinboro Area Roundtable reveals an active community: free concerts; golf outings and tournaments; sporting events (e.g., Edinboro Triathlon, Annual Duck Race); auto rallies and classic car cruises; historical society lectures (e.g., "History of the Hardware Store"); family events (e.g., "Baby Bison Days" at the Wooden Nickel Buffalo Farm, the Edinboro Kiwanis "Peanut Days"); seniors meetings; and adult band camps. The July 2001 calendar included a fundraising event, the John Gillette Memorial Golf Scramble, which was initiated by the Gillette family to fund a memorial scholarship. The town's web site (www.edinboro.com/Events) lists even more events.

Edinboro Lake, on the borough's north side, brings vacationers and tourists to the area to enjoy swimming, boating, waterskiing, jet-skiing, and trout fishing. Coming off Interstate 79, the approach into town on Route 6N brings drivers past a golf course, one of several in the area. In the fall, there is hunting season. Winters in this part of Pennsylvania can be harsh, but there is skiing just across the border in New York. In addition to standard spectator sports, Edinboro University has several cultural events during the academic year that are open to the public, including plays, music programs, and lectures. Each summer the university runs summer youth soccer tournaments and state-wide band camps.

Given its size, Edinboro offers a surprisingly wide range of recreational options for children. On Route 6N can be found the new offices of the Penn Lakes Girl Scout Council, a new YMCA, and a large outdoor complex for children's sports. Elsewhere in town there is a bowling alley, a skateboard park, a large outdoor swimming pool, and sports fields. Next to Nick's Place is a miniature golf course, a driving range, and batting cages, and across the road is a go-cart track, Fastrax. Notices posted on bulletin boards in grocery stores and other locations advertise dance studios, church groups, and sports leagues.

Reviewing the list of options, one person who met with us cautioned against "the false doctrine of salvation by activity." That said, he noted that there probably were not enough activities in town for middle school students, a point echoed by several other community residents. One school official explained that, as children reach middle school, many parents think they need less supervision and are more willing to let them stay home alone while both parents work. Transportation is another problem. The community is rural, and many residences are several miles from places or activities that students this age would enjoy. When they reach high school, young people have additional opportunities available to them, many of them through the school itself. And, as they get older, many of them are able to get themselves to community-based programs or after-school jobs by driving or getting a ride from an older friend.

Edinboro has a low crime rate, but crime is not unknown. When applying for a grant from the Pennsylvania Department of Community and Economic Development, the police department noted that Edinboro had seen a very large increase in drug violations between 1994 and 1998. The department attributes the increase to college students, stating that several areas along the periphery of the university have experienced increased drug-related activity over the past 12 years. Several renter-occupied dwelling units have developed into drug trafficking bases, which have attracted some gang members from nearby major cities. In June 2000, two parents and their 18-year-old daughter were arrested for growing marijuana in their home and on their property. The drug problem extends into the middle school. LSD, in the form of "liquid acid," is a new worry, according to some sources. The dealer takes an LSD tab, crushes it, mixes it with alcohol and a sweetener, and sells it a few drops at a time. School officials note, however, that Parker's surveys of student drug use show rates slightly below the national average.

THE COMMUNITY RESPONSE

In a commentary published by the Erie *Times-News* just two days after the shooting, an Edinboro University professor wrote, "Is there anything we can learn to do differently after a tragedy like the one we are living through today in Edinboro? Or do we shed a tear and then go back to our usual ways?" Any community that suffers misfortune has the opportunity to learn and be strengthened by the experience. Has that happened in Edinboro?

Virtually everyone who spoke with us offered opinions about what lessons should be learned from this tragedy. The idea that parents need to spend more time with their kids was a recurring theme. A newspaper quote from an Edinboro police officer captured this sentiment: "When your kid comes home from school, listen to what he has to say. Don't blow him off." Many people pointed out to us how important it is for parents to know their children's friends and to discourage friendships that might harm them or lead them astray. Others took a broader view, expressing the need for Americans to "get back to basics," including making religion the cornerstone of family life.

When asked about lessons the community might learn, one public official offered the following list: (1) Recognize that this kind of crime can happen anywhere; (2) If a kid is talking about shooting or killing someone, that must be taken as a serious threat and not put aside as a joke; (3) Parents need to connect with their children and look for signs of anger or resentment; (4) Parents need to be more active in their children's schools; and (5) Responsible gun ownership means locking up the guns so that children do not have access to them and cannot fire them.

Few Edinboro residents appear to have given much thought to the fact that Andrew had such easy access to his father's semiautomatic handgun. When we asked about this, many people reminded us that the Edinboro area has many hunters, and that having guns of all types in the home is accepted as a matter of course. Reminders of the gun culture are everywhere. One school official told us that there is no school on the Monday after Thanksgiving because it is the first day of deer season. We asked for directions at a gas station, and the attendant asked if we were looking for a certain gun club that was down the road. Ironically, there is a "guns and ammo" shop, Uncle Sam's Trading Post, located on Route 99 in between Parker Middle School and Nick's Place.

There are some residents who have called the Edinboro region's gun culture into question, especially after the shooting, but most remain largely silent. Gun ownership is just too much a part of the community's social fabric. But there is another factor: in support of gun ownership, people had only to point at Nick's Place owner James Strand, who had grabbed his shotgun and forced Andrew Wurst to drop his weapon. How many more teachers and children may have died, they asked, if Strand had not owned a gun?

After the shooting, an Erie television station worked briefly with the district attorney's office to offer gun trigger locks at a discount, but the program died out relatively soon. Several people expressed the idea that locking up a handgun would make it useless for home defense. One person said that a trigger lock wouldn't work because children would know where the key was kept. What we heard instead is that it is a family responsibility to teach respect for firearms.

We also heard ideas about what needs to be done in the schools. A source who was involved in the case drew the following lessons: (1) Teachers need to be educated about mental illness and be aware of the possible signs; (2) There needs to be a mechanism in place for teachers to identify and refer students who might need a clinical evaluation; (3) Teachers should be more available to their students and get to know them better; (4) Bullying should not be tolerated at any level; and (5) Students need to be educated about telling their teachers about others students' behavior that they find disturbing, and they need to feel safe in doing so.

Putting most of these ideas into practice would require systems change in the schools. Every community wants teachers who are motivated to know their students well, but whether teachers can sustain their motivation over time will depend on how school administrators have organized the school's operations to build stronger teacher-student bonds and a true sense of community. This is especially important at a large institution such as Parker. One set of options, for example, would be to divide teachers and students into smaller administrative subunits, and then having teachers meet as a group to talk about and develop plans for all of the students in their subunit. No one we spoke with brought up the need to look at these kinds of systems issues.

After the shooting, the General McLane School District, with a grant from the Safe Schools Office of Pennsylvania, developed a character education program for its students. The program is built on six Pillars of Good Character—respect, responsibility, fairness, caring, trustworthiness, and citizenship. Under each is a set of principles. For example, under respect, the principles include "Follow the golden rule. Use manners. Don't threaten, hit, or hurt anyone." Across Route 99 from the high school and middle school is an old barn whose side has been painted with the program's now familiar symbol, a stylized, triangle-shaped pediment supported by six pillars, each labeled with one of the six program elements.

The curriculum is based on several programs that were already in place, including WiseSkills for grades 3–8, plus selected children's books, videotapes, and other new materials and resources. In addition, the program incorporates a curriculum infusion model, with the principles of character education woven into the main academic curriculum. There are also special displays and activities planned throughout the school year.

Posters supporting the character education program appear in several storefronts in town. Opinion about the program is divided. Some parents wish that more time in school were devoted to these subjects, while others believe that character-building should be done at home and in the churches, with school time devoted to academic basics. One source complained to us that the school still needed a stronger antiharassment policy to make clear that bullying would not be tolerated.

In the aftermath of the shooting, the school district's teachers now receive more training on conflict resolution and how to deescalate potentially violent confrontations. Administrators have also exhorted teachers to use the student support program as it was intended. If teachers are concerned about a student, they should tell someone. Likewise, school officials have talked to students about how important it is for them to come forward with information about a student who is acting in ways that disturb or worry them. The greater challenge, of course, is creating a climate in which students will actually feel safe in doing that.

In the immediate aftermath of the shooting, Parker Middle School began to make changes in security around the school, including restricted building access and name badges for staff. Some experiments failed, such as using metal detector wands to check students before they entered the building; this only served to create long queues of students vulnerable to attack. To their credit, school officials tried not to overreact, which was difficult given the pressure placed on them by worried parents. After the shooting, the Catholic diocese in Erie County temporarily cancelled all school dances. In contrast, the General McLane School District avoided that step and still held the high school prom, although metal detector wands were used to screen students before they entered. Prior to the shooting, there was no policy to hire police officers for security, and that has remained the case. One positive development is that Parker's end-of-year celebration for graduating eighth graders was revamped. The event is now held at the school and no longer has the trappings of a junior prom.

Over the next summer, the General McLane School District received recommendations from outside professional organizations, teachers, parents, and students, which were reviewed in a series of meetings and reports. The recommendations ran the gamut, from banning book bags and backpacks to creating a character education program. Ironically, few of the recommendations would have made a material difference in the Andrew Wurst case. Most of the focus was on school security rather than on creating a relationship-centered school environment in which every student is connected to a teacher or counselor, or on identifying students who may be "falling through the cracks" and need extra attention or professional help. Moreover, the security measures that were recommended were better suited for dealing with outside threats, not those from the student body.

There have also been efforts to get a program started to provide school resource officers, although these so far have not worked out. Erie County sheriff Robert Merski, for example, applied for a U.S. Department of Justice grant, for which the General McLane School District signed on as a partner. However, the Justice Department program will fund only the lead law enforcement agency, which in this case is not the sheriff's office but the state police. In Edinboro, the police department applied for a federal grant for a resource officer but, to receive it, town officials had to maintain the same department budget level for the duration of the grant. That commitment was not forthcoming, as two police positions in town were also funded by grants that would end during the time of the new grant, and it was very unlikely that the town would continue to carry them. With that, Edinboro's police chief, Jeffrey Craft, contacted all of the municipalities that send students to the General McLane School District about funding the position, but McKean Township was unwilling to take that step.

In 1998, Patricia Crist had been Parker's principal for 11 years. Like many principals, Crist received her share of criticism from parents and teachers, especially after the shooting. Those who had always disliked her policies or style now found ample reason to call for her ouster. She served for two additional years and then stepped down, moving next door to the high school to be a teacher and mid-level administrator.

The town government has launched no major initiatives since the shooting. In the private sector, beginning in September 1999, the Christian Institute of Human Relations has sponsored The Hangout, a youth center built in a defunct two-screen movie theater. Plans for the center were under way prior to the shooting. Under the direction of high school coprincipal Rick Scaletta, the center provides a coffeehouse and band hall for older teens to listen to live music, play pool or ping pong, and talk. The Hangout does not have programs for middle school kids, although some will occasionally attend a concert. The center's ultimate purpose is to provide a venue for Christian ministry, but all young people, even those uninterested in the religious program, are welcomed.

The regional bands that play at the center are heavy metal and hardcore groups. Most are secular, but some, such as Closer Than Dying, play music with religious themes. Advertising for the concerts is youth-oriented and hard edge (e.g., "You Ain't Cool Unless You Pee Your Pants"). The Hangout's religious educators then try to make the link between this youth culture and Christianity. One brochure, explaining Christian belief about the crucifixion, speaks of Jesus Christ "receiving the ultimate body piercing." After kids have come to the center a few times, the educators approach them one-on-one to talk and to let them know about The Gathering, a Tuesday-evening prayer session held there.

Edinboro's residents are quite aware of The Hangout and view it as a positive force in the community. The Christian Institute of Human Relations, a nonprofit corporation, is dependent on private donations to meet its operating costs of about $2,500 per month, many of which come from local churches. Adult volunteers serve as chaperones, help maintain the youth center, and lead religious programs. So far, the center has been a success.

CONCLUSION

The Edinboro shooting was one of a series of highly publicized school-related shootings in the United States in 1997 and early 1998. In quick succession, there had been shootings in Pearl, Mississippi; West Paducah, Kentucky; and Jonesboro, Arkansas, before Edinboro, Pennsylvania, was added to the list. Andrew Wurst was fully aware of these dramatic incidents and talked frequently with his friends about the Jonesboro shooting, indicating to them that he might do something like that someday. Violence experts have noted that extensive news coverage of these school shootings can provide some troubled youth with a script for how to deal with their anger, hurt, and frustration. Andrew Jerome Wurst appears to be one of those cases.

In the end, Andrew's defense attorney, Leonard G. Ambrose, recommended that Andrew accept a plea on third degree murder rather than be tried on first degree charges and run the risk of a life sentence with no chance of parole, the maximum possible sentence. Juries are unsympathetic to insanity pleas, Ambrose explained to news reporters, and this would be especially true with public fears about school shootings. Andrew and his parents agreed with Ambrose that there was too much at stake to take the risk of going to trial. On September 9, 1999, Judge Michael M. Palmisano sentenced Andrew to serve 30 to 60 years in prison. He will not be eligible for parole until age 45.

The Gillette family is seeking another day in court. Deborah L. Gillette, John Gillette's widow, has filed a civil law suit against the Wursts, which is still working its way through the legal system. Some town residents expressed the wish that the Gillettes could forgive the Wurst family and doubted that any good would come of the lawsuit. One said it was understandable why Mrs. Gillette would want to take this step, especially in light of the plea bargain that prevented Andrew from receiving a life sentence without parole. In their view, no one is in a position to second-guess Mrs. Gillette or any member of the Gillette family; others cannot know the depths of their despair, nor can they say how they themselves would react if a loved one were taken from them this way.

In our view, there is little doubt that Andrew Wurst was mentally ill. Whether he was legally insane and therefore incapable of forming criminal intent is a harder issue to resolve, which is reflected in the contradictory psychiatric testimony heard in this case. Should Andrew's parents have known something was wrong? So far as they knew, their son had no history of violence or bullying, and he never seemed angry. His grades were poor, but he was not a disciplinary problem at school. His parents had caught him drinking, but they had no idea he was smoking marijuana. Andrew also did not reveal to his parents any disturbing thoughts or fantasies. The Wursts are not trained clinicians, and so it is not surprising that they never thought Andrew was mentally ill or even depressed. The fact is that lots of parents see their kids having problems with the awkwardness and pain of growing up without ever thinking they need psychiatric help or might commit a horrible crime. The Wursts were no different.

Every middle school has procedures in place for teachers to refer students who require intervention, but that system appears to have broken down in Andrew Wurst's case. Following the shooting, the General McLane School District launched a high-visibility character education program. The program got mixed reviews, but few people doubted that it was a step in the right direction. In addition, school officials took reasonable steps to tighten security while avoiding a mindless overreaction that would have compromised the quality of life at the schools. In our view, however, school officials so far have avoided many of the harder issues:

- How might the Parker Middle School be operated differently so that there is a stronger sense of community among students?
- What steps can be taken so that every student feels valued and cared about as a member of the school community?
- What can be done to ensure that students do not merely tolerate diversity, but respect and even embrace it?
- How might the inherent divisiveness of cliques be minimized?
- How can the school protect every student from bullying?
- How might the school be reorganized so that every student is well known by several teachers?
- What can be done to make students feel it is safe to notify teachers, counselors, or other adults when another student says or does something they find disturbing or threatening?
- Does a greater investment need to be made in after-school programs so that fewer middle school students are home alone?
- How can the student support program be reinvigorated so that teachers and other school officials will better recognize when students are showing signs of distress, alcohol and other drug problems, or mental illness, and then advocate for their being referred and receiving appropriate help?

In laying out these questions, we do not mean to imply that Parker school officials should be blamed for what Andrew Wurst did. In ordinary times and by normal standards, Parker Middle School would be considered a well-run and progressive school. Times have changed, however, and the standards by which schools are now judged must change with them.

Andrew Wurst's classmates are now in their senior year of high school, anticipating their next graduation celebration and taking another big step toward adulthood. For them, the events of April 1998 seem almost a lifetime ago. Time has moved on, and so have they. In fact, four years after John Gillette's death, there is a general sense in Edinboro that the whole community is ready to move on. Most of the people we met see the shooting as a bizarre and puzzling aberration, largely disconnected from the fabric of their lives. For them, the story of what happened at Nick's Place boils down to this: Andrew Wurst was a troubled boy from a troubled family, and he happened to live here. He committed a horrible crime, and he deserved to be punished. People expressing this viewpoint often spoke of parents needing to provide stricter moral guidance and to stay better connected as their children move into adolescence.

We met a few people, however, who expressed dismay about the prevailing desire for life in Edinboro to return to normal. They believed fervently that there was still much soul-searching for the community to do, that there were still more lessons that needed to be reflected on and absorbed. What does it mean, they seemed to be asking, that Andrew Wurst could not feel he was part of a school community that valued him? What does it mean that so many people could not look beyond this horrible crime and see a boy who was mentally ill and needed treatment and who might be saved? What does it mean that the Wursts felt shunned by a community that they had been part of for so many years? Without for a moment forgetting the sorrow inflicted on the Gillette family, why could people not see that the Wursts were suffering, too?

When Catherine Wurst gazes at Andrew, she does not see evil, or even a criminal, but a boy who became desperately ill and didn't get the help he needed. In her view, Andrew is still not receiving that help. After battling state prison officials to improve Andrew's therapy, she has become a member of the Pennsylvania Prison Society and an advocate for prison reform. She refuses to believe that Andrew is beyond redemption. She visits her son frequently, concerned about how well he will endure his long sentence, but also hoping that with proper psychiatric treatment, coupled with her love and continued attention, Andrew might come out of prison at age 45 with a hopeful future. Whether that can happen, and whether the town of Edinboro can accept that if it does, is the next chapter of the story.

ACKNOWLEDGMENTS

We wish to thank the students, teachers, school administrators, community residents, town officials, police investigators, lawyers, court officials, and journalists who met with us or provided background materials for our research. We especially thank Andrew Wurst's parents, Catherine and Jerome J. Wurst, who invited us into their homes to talk about their son. No one had to help us, but they did, and we are grateful.

We also thank reviewers from the National Research Council for providing comments on an earlier draft of the report.

We were profoundly moved by our visit to Edinboro. We hope our report will help the people of Edinboro heal and learn from the tragedy that befell their community. We also hope that both the Gillette family and Andrew Wurst's parents will eventually find peace.

4

A Deadly Partnership: Lethal Violence in an Arkansas Middle School

Cybelle Fox, Wendy D. Roth, and Katherine Newman

On March 24, 1998, in the third and deadliest in a series of recent school shootings in a Southern community, Mitchell Johnson, 13, and Andrew Golden, 11, both students at Westside Middle School in Northeast Arkansas, opened fire on 96 of their classmates and teachers. As a result, four students and a teacher died and 10 others were wounded.

This chapter is based on interviews completed in June 2001 with 98 students, teachers, administrators, lawyers, judges, court personnel, parents of the victims and one of the shooters, church leaders, and community residents, as well as media reports and official court documents. Gretchen Woodard, Mitchell Johnson's mother, was particularly helpful. Some key individuals were unwilling to be interviewed, particularly members of Andrew Golden's family, hence we were able to capture only how others in the community saw them. The passage of time and the differences in vantage point among the many people touched by this episode has left its traces in multiple, sometimes conflicting perspectives on its causes and consequences. In what follows, we have done our best to adjudicate amongst the competing accounts of this tragedy. Yet often no single truth emerges, and the researchers' job is best served by reporting them all.

THE SETTING

Residents describe the Westside community as the most unlikely setting for any kind of violence, let alone a mass school shooting. Nearby Jonesboro is a small but growing city in the northeast part of Arkansas, whose residents number 55,000 but whose diverse manufacturing, service, and retail sectors attract a trade area population of nearly a half million from the surrounding areas, including southern Missouri. Located 130 miles northeast of Little Rock, Jonesboro is a Bible Belt town, proud of its religious heritage, embodied in the more than 75 churches within its boundaries. While Jonesboro proper sits astride the gentle rolling hills of Crowley's Ridge, many of the surrounding communities are on the flat farmland—producing cotton, rice, and soybeans—that is characteristic of the Delta. With its thriving economy but small town feel and low crime rates, it is hard to find a person in the area who does not think Jonesboro is a great place to live.

Yet the area has also had its share of natural disasters. In 1968, a tornado ripped through the area, killing more than 40 residents. Once again in 1973, tornados caused millions of dollars of property damage. These events stepped up community preparedness for crises—particularly in the area hospitals—which was to prove very valuable when the school shootings took place. The natural disasters also brought the community together in the wake of trouble, teaching the value of interagency cooperation and interpersonal social support.

The Westside School District is one of five public school districts in the immediate area and serves the Bono, Cash, and Egypt communities, which lie adjacent to Jonesboro and are largely rural towns.[1] The community is closely knit; many of its residents, including many of the teachers at Westside, are graduates of the middle school or parallel institutions in Bono, Cash, or Egypt, which were consolidated in the late 1960s to form Westside. The elementary, middle, and high schools in the district share one large property, but each campus maintains its unique character. During the 1997–1998 school year when the shooting occurred, there were approximately 250 students in Westside's middle school, 125 students in each of the sixth and seventh grades. The school district as a whole includes 1,600 students and is largely middle-class, Christian, and almost exclusively white; approximately a third of the students at the middle school qualify for the free or reduced lunch program. Before the shooting, the biggest safety concern was whether the school buses would arrive safely every morning. The school's location, demographic composition, and lack of any significant history of violence made most people in the community think the school impervious to serious violence. Hence, the school had no violence prevention programs beyond anger management counseling in place at the time of the shooting.

THE SHOOTERS

Mitchell Johnson was 13 years old and in the seventh grade at Westside Middle School on the day of the shooting. He had only recently joined the school. Born in Minnesota, he lived for a time in Kentucky before moving to Bono in 1995. His mother, Gretchen Woodard, married three times. Mitchell's father (and Gretchen Woodard's second husband), Scott Johnson, had an explosive temper. According to media reports, by his own admission Scott Johnson was a screamer. While there is no evidence he was physically abusive toward Mitchell, he punched holes into walls and was verbally abusive, but after his tantrums rarely disciplined Mitchell in a way that would teach him what he had done wrong. Mitchell reacted quite strongly to his father's temper; on several occasions, he was found trembling and physically ill in response, and it could take hours to calm him down.

In Minnesota, Gretchen worked as a correctional officer in a federal prison, and Scott worked at a meat packing plant. Since the couple worked long hours, Mitchell and his younger brother spent a considerable amount of time in the care of their grandmother. While Mitchell was staying at his grandmother's home, he was apparently sexually assaulted, repeatedly and violently, by an older boy in the neighborhood. His mother did not find out about this abuse until after the shooting, however; Mitchell did not tell because his attacker had threatened to kill his grandmother if he ever told anyone about it. He was both ashamed and afraid that his father would be angry with him if he knew.

Mitchell's parents split up and went through a difficult divorce. Shortly thereafter, Gretchen Woodward received a job promotion to a federal prison in Kentucky. Mitchell was 10 years old when they moved. His mother soon encountered and later married Terry Woodard, whom she had first met as an inmate, convicted on drugs and weapons charges, in the prison where she worked in Minnesota. Since Terry had grown up in the Jonesboro area, the new family made their way to Bono and enrolled Mitchell and his younger brother at Westside. Even though they didn't have much money, life in Bono was peaceful and things were finally looking up. Mitchell got along well with his stepfather, who worked at a heavy equipment hauling company. For the first time, Gretchen Woodard could be a stay-at-home mom after Mitchell's younger half-sister was born.

Mitchell seemed to be doing well at Westside. His teachers described him as a normal kid and a good student, who generally made As and Bs. Many of the adults at the school commented on how polite, respectful, and charming he was. Indeed, he received commendations for his good behavior. At ease in talking with adults, Mitchell had a reputation for being a real pleaser. His mother was not very involved at the school but there wasn't that much to be involved in. Gretchen came to Mitchell's parent-teacher conferences faithfully and was supportive of the school in its disciplinary decisions. Mitchell was interested in the Bible and attended Central Baptist Church in Jonesboro. He loved music, performed in the school choir, and even sang at a nursing home on the weekends through his church youth group. He was active in sports, including football, basketball, and baseball. He had friends and was especially close to his younger brother, whom he always looked out for.

But there seem to have been many different sides to Mitchell Johnson, some of which are hard to reconcile with the "model child" aspects of this description. By some accounts, he had an explosive temper reminiscent of his father. Peers and teachers occasionally landed on the wrong side of Mitchell. He got into serious enough trouble to land himself on in-school suspension at least three times while he was a student at Westside Middle School. The first incident came when he was in sixth grade: he got mad and hit a thermostat in the hallway, breaking its glass case. Shannon Wright, the English teacher whom he would later shoot and kill, placed him on in-school suspension for that incident. The following year, Mitchell was on in-school suspension for cursing at a teacher.

The third occasion happened just a few weeks prior to the shooting. Mitchell was wearing a baseball cap even though school policy requires that no hats be worn in school. The teacher tried to take the hat away and Mitchell resisted. It took two adults to get the hat away from him, and Mitchell was placed on in-school suspension for his behavior. During his in-school suspension the last time, Mitchell was assigned to write a paper about why he was in there to begin with. The teacher claimed that this paper contained a veiled threat against her, and she brought it to the principal. It is not clear what the school did in response. In addition to these three incidents, Mitchell was also paddled for cursing on the bus. Despite these incidents, Mitchell was not considered a troubled student or a serious disciplinary problem by the school. He was usually very remorseful for his belligerent behavior and would apologize "ten times over" for his actions.

Mitchell was described by many of his peers as a bit of a bully. His classmates said he was a bragger, a kid who liked to flash gang signs and wear red to signal his supposed affiliation with the Westside Bloods, a gang in Jonesboro. There is no evidence that Mitchell was part of any gang, however, and his friends said that he was only a gang "wannabe." It was rumored that a few weeks before the shooting, Mitchell brought a knife to the school and pulled it on a fellow student. Other students and adults suggest that Mitchell was himself bullied, or at least picked on. One teacher said he was sensitive and wanted more than anything to be popular and well liked, especially by girls. When classmates would tease him, his reaction was sometimes described as out of proportion to the supposed affront.

Mitchell's relationship with his father remained stormy after the boy moved to Bono, and he was apprehensive of his trips back to Minnesota over Christmas and the summer. These visits did not go well; his father would threaten to send him home because he "misbehaved" and in the summer of 1996, he did send Mitchell home after only two weeks because he "was real hostile."

Although Mitchell's mother and his stepfather were viewed as caring, concerned parents, Mitchell may have had a lot of free time during which his parents were unaware of his activities. One Bono resident claimed to have seen him frequently hanging around on his bicycle, remarking that he could stay over at friends' houses overnight without having to call home for permission. Although Mitchell was known for being very polite and well mannered, some felt that this unstructured time allowed him to get into trouble. He sometimes cursed at other kids or made threats, causing their parents to get upset. A community resident questioned whether he was ever disciplined at home for these incidents.

In the days and weeks immediately preceding the shooting, his mother did not notice any sudden changes in Mitchell's behavior but admits that there were some troubling signs that something was wrong. The summer before the shooting, Mitchell was caught molesting a 2-year-old girl in Minnesota and was charged for the incident in juvenile court. When he returned to Arkansas, his mother began taking him to see a psychologist, who concluded that it was probably an isolated incident. At Christmas, Mitchell and his younger brother went back up to visit their father in Minnesota. They took the bus and on the way home were left stranded at the Chicago bus terminal for two days due to bad weather before any adult realized where they were. One of his teachers said that she noticed a change in Mitchell soon after that Christmas break. Mitchell was more reticent and spent less time with her, but she assumed that it was a fleeting or at least normal teenage phase so she backed off. Then, approximately one month before the shooting, Mitchell started to call sex-talk lines and racked up hundreds of dollars of debt on his father's credit card. When Scott Johnson found out, he became furious at Mitchell; he called the police and threatened that Mitchell might have to move back up with him to Minnesota.

In retrospect, we know that Mitchell took this threat very seriously and was feeling "hopeless," as if his life were over. From that point, Mitchell continued to spiral downward. About two weeks before the shooting, Mitchell was kicked off the basketball team (or tried out and didn't make it, according to one school official) for self-mutilation; he apparently engraved his initials into his shoulder. It was also about that time that he was suspended for the baseball cap incident. And to top it off, his girlfriend of three days or possibly a week, Candace Porter, dumped him.[2] With hindsight it is easy to see that Mitchell was sad, angry and was feeling desperate, but no one in his life at that time, either at the school or in his home, was aware of just how troubled he was or how low he felt.

Andrew Golden was 11 years old and in the sixth grade at Westside Middle School at the time of the shooting. He had lived in Bono all of his life, and his father had as well. In fact, many of Andrew's teachers graduated from Westside with Andrew's father, Dennis. Dennis and Andrew's mother, Pat, worked as postmasters in a nearby town and were seen by people in the community as hard-working people with good jobs. Andrew was their only child together. Pat had two other children from a previous marriage and had had a tubal ligation before she married Dennis. But since Dennis did not have any children of his own, Pat had the operation reversed in order to have a child. Andrew was their miracle baby, and friends told the media he was "the center of their world." Andrew was very close to his grandfather, Doug Golden, who was well known in the community since he worked for the Arkansas Game and Fish Commission; it was not uncommon to see the two together around town.

The entire Golden family was known for their avid pursuit of hunting. Andrew was taught to hunt at a very young age; he was given a shotgun for Christmas the year he turned 6. Andrew practiced his shooting at the local range and won awards for his marksmanship. Hunting is a very common activity for boys at Westside, but the Golden family's reputation was notable even for the area. Andrew posed for photographs as a toddler dressed in camouflage with a rifle. The Golden home was the office of the Jonesboro Practical Pistol Shooters Association, a local affiliate of a national gun organization, which is indicative of their enthusiasm for the sport.

Most adults described Andrew, just as they did Mitchell, as an outwardly normal or typical child. If anything he was immature and doted on, maybe even a bit spoiled; his family, some said, gave him anything he ever wanted. Because both his parents worked long hours, Andrew was often left home alone, but his parents called home to check up on him frequently. His half-siblings lived with him for a time but, according to people in the community, they were not treated well by Dennis and the Golden grandparents so they left and went to live with other relatives. There is no evidence that Andrew was ever abused in his home.

Some teachers and neighbors felt that the Goldens may have been overindulgent and failed to discipline Andrew. In his parents' eyes, Andrew could do no wrong. A neighbor reported how Andrew would "curse like a sailor" at a very young age—perhaps 4 or 5 years old—and how his father and grandfather would laugh and encourage it, saying it was cute and would make a man of him. His mother would reportedly throw her hands up and say she couldn't do anything with him since the men continued to encourage him.

Andrew had friends but was probably not one of the most popular kids in school. He played trumpet in the school band. He was an average student and often made As and Bs but was for a time in elementary school in remedial reading and math classes. While Mitchell generally strove to do well in school, Andrew was described as more apathetic about his performance.

Andrew was also not considered a disciplinary problem at school. He was never suspended, and school officials can only remember one significant incident, in first grade, where he got into some trouble. He had brought a toy gun to school, which was taken away from him by a teacher who told him not to touch it. Andrew got another boy to retrieve the gun for him and then, during recess, filled it with mud or a sand and gravel mixture and fired it at a classmate, hitting her in the eye. For this transgression he was paddled. Arkansas is one of 26 states where corporal punishment is still legal, but when Andrew's grandmother discovered her grandson had been paddled, she reportedly became very angry, yelled at the teacher in front of other students, and later convinced Andrew's parents to place him on the "no paddle list." Andrew's greatest offense at school after this incident was reportedly talking out of turn. He could be mischievous and act the class clown, but this behavior was sometimes written off because teachers who had gone to school with his father had known him to be a harmless prankster, too. Andrew's father came to parent-teacher conferences and, when Andrew was in elementary school, his grandmother would often go on school field trips. Most of his teachers described Andrew as a boy who always had a grin on his face and, if anything, would have gone unnoticed.

Neighborhood accounts of Andrew paint a much darker portrait, however. In his neighborhood, Andrew was known as something of a menace, a boy who cursed and yelled at other children, saying that if they came over to his yard he would shoot them with his BB gun. He rode around with a sheathed hunting knife strapped to his leg and reportedly killed cats in his backyard, including one that he starved to death in a barrel. Golden's grandfather also had a reputation among some in the community as being irascible and petulant, and some wondered whether Andrew learned his menacing behavior from him. Andrew's behavior around the neighborhood contributed to his reputation as "mean-spirited." Two cousins who later became his shooting victims were ordered not to play with Andrew after they told their parents they saw him shoot and kill a cat. Despite the fact that the Westside community is small and reportedly closely knit, most teachers and administrators at the school were not aware of Andrew's neighborhood reputation before the shooting.

Although they came from very different family backgrounds, Mitchell Johnson and Andrew Golden shared the description of having something of a Jeckyll and Hyde personality. Both could be described as very sweet or polite, particularly by their teachers and in school, but they also had a more hostile side. Those who were privy to the sweet side were stunned by the shooting and found it hard to imagine that Mitchell and Andrew could be involved. Those exposed to the temper or the abuse of animals were also surprised but considered their roles plausible.

By most accounts, Mitchell and Andrew were not friends, just acquaintances who met on the long bus ride to school.[3] The bus ride provided the boys with a fairly lengthy period of unsupervised time, with the bus driver as the only adult present. Many parents complained of rowdy and inappropriate antics by older children on the bus, to which younger children were exposed. While Mitchell and Andrew were not assigned to sit together on the bus, many believed that their plans were likely hatched there nonetheless. Neither boy visited the other's house, they were in separate grades at school, and they had no common recess or activities together. In fact, adults at the school said they never would have paired the two together, their personalities were so different. But Mitchell and Andrew did call one another at least a few times, and friends said they saw each other outside school.

THE VICTIMS

Four students and a teacher died in the shooting. The students who were killed were Stephanie Johnson, age 12; Paige Herring, age 12; Natalie Brooks, age 11; and Britthney Varner, age 11. Shannon Wright, who had been both Mitchell and Andrew's English teacher, also died; she was married and had a son of her own, 2-year-old Zane. She was hailed as a hero by many in the community and the press for shielding her students from the gunfire, giving her life for theirs.

The wounded were Candice Porter, age 11, who "dated" Mitchell for a few days before she broke up with him; Crystal Barnes, age 13; Whitney Irving, age 11; teacher Lynette Thetford, who had Andrew in her social studies class; Brittany Lambie, age 13; Jennifer Jacobs, age 12, who reportedly dated Andrew and broke up with him before the shooting; Ashley Betts, age 12; Tristan McGowan, age 13, who was Andrew's cousin; Christina Amer, age 12; and Jenna Brooks, age 12, cousin of Natalie Brooks, who was shot and killed.

THE SHOOTING

The shootings took place on March 24, 1998, a Tuesday. Students and teachers had recently returned from spring break and were talking about their vacations and getting back into the swing of things when Andrew Golden, clad in camouflage clothing, entered the middle school at approximately 12:35 p.m., just a few minutes after fifth period had started, and pulled the fire alarm. At least two students saw Andrew pull the alarm and leave the building. The students told their teachers what Andrew had done, but everyone filed out of the building through their assigned exit routes, as required. The 87 students and 9 teachers who exited the west entrance were met with a hail of gunfire.

Many of the details of the shooting were probably planned during the spring break the week before the shooting, but the boys may have been talking about the idea together for months in advance, possibly as early as December 1997. One person close to both Mitchell and Andrew suggested that their initial discussions about the shooting were casual or just "talking big," and then one day, less than a week before the shooting, one of them suggested they really do it and the other agreed. Yet precisely who was the leader and who was the follower in this partnership is in dispute.

On the morning of the shooting, Mitchell and Andrew were absent from school. Mitchell missed the bus and told his mother that his stepfather, who had actually already left for work, would give him a ride. When Mitchell's mother, who was caring for her 2-year-old daughter and babysitting for a neighborhood boy, looked out the window and saw that the van was gone, she assumed that her husband still had it. Instead, Mitchell took the van and drove it to Andrew Golden's home. Andrew's parents, who had left for work, had left Andrew home alone to catch the bus on his own that morning, as they had recently started doing. Instead of taking the bus, Andrew hid in some bushes near his home and waited for Mitchell to pick him up. The boys attempted to get guns from Andrew's house, but the majority of his father's guns had recently been placed in a safe and the boys were unable to gain access to them, even after trying to break the safe open with a blowtorch. They took a .38 caliber derringer, a .38 caliber snub-nose, and a .357 magnum that were not secured and then drove over to Andrew's grandparents' house. They broke into the house with a crowbar and found an arsenal of weapons—a wall completely lined with rifles—secured only by a cable running across them. From the grandparents' shed, the boys found some cable clippers and used them to break the cable and steal four handguns and three rifles.

Contrary to popular perceptions in the community of a carefully executed plan, there were many glitches along the way on the day of the shooting. First, at 13, Mitchell was not a seasoned driver and he apparently had some trouble with the van. It was also low on gas and when the two boys tried to get some, they were unable to operate the pump and were refused service because they appeared too young. For reasons that remain unclear, the gas station attendant did not call the police to report the boys. Mitchell and Andrew went to a second and even a third station before they were finally able to get gas. With their supplies in order, they drove toward the school, parked their van on the cul-de-sac of a street approximately a half-mile from the school, and made their way to a wooded area near the Westside Middle School property.

Many adults at the school believe that the boys intended to do the shooting during lunch recess, but the ground was muddy due to rain the night before and the students did not go outside for lunch. To lure the students and teachers out of the building, Andrew Golden went inside the school and pulled the fire alarm. He then exited the building and joined Mitchell already in position, we assume, in the woods atop a hill overlooking the schoolyard. Only after the majority of the students had exited the building did the boys begin firing. Neither the students nor the teachers realized at first that they were being fired at. Many thought the loud pops they heard were firecrackers, while others thought it was all part of a school play. When they saw their fellow students and teachers fall around them, they finally understood it was an armed attack and began screaming at one another to get down. Since the doors to the school had automatically locked behind them as they left the building, most ran around the school toward the gym.

Mitchell and Andrew fired fewer than 30 rounds and struck 15 people. One student, Stephanie Johnson, was pronounced dead on the scene and four others, including a teacher, were pronounced dead at St. Bernard's hospital in nearby Jonesboro. Ten others were treated and eventually released. According to ballistics reports and the police investigation, Mitchell fired five shots from a 30.06 caliber semiautomatic rifle equipped with a scope, killing at least one, but probably two, and wounding at least three. Andrew fired by far the most shots and was responsible for three deaths and wounding at least two others.

It is not clear why the boys stopped shooting when they did. Some law enforcement officials suspect that when a construction worker on the roof of the new fifth grade wing of the middle school saw the boys shooting and screamed at them to stop, they ended the assault. Mitchell fired a few shots at the construction worker and both boys fled the scene, running toward the van. Another speculation is that the boys stopped because there was no one left standing in the line of fire. The whole shooting probably lasted no longer than a few minutes.

The boys were apprehended by law enforcement approximately 200 yards from the scene and 10 minutes after they had begun shooting. Police officers drove along a service road and reached them as they were coming out of the woods. Officers yelled at the boys to drop their weapons; Mitchell dropped his gun, but Andrew hesitated. They repeated the order, and Andrew then complied. In the van, police found supplies (including ammunition, camouflage netting, camping gear, potato chips, and a pillow with cartoons on its case) and a map indicating that the boys were planning to drive out to a piece of land the Golden family owned in the mountains to camp. According to their lawyers, the boys were planning on hiding out for a few days until things quieted down, when they hoped to return home and everything, they presumed, would be fine. Despite the magnitude of the tragedy, law enforcement officials believe that the shooting could have been even worse. When the boys were apprehended they were carrying a total of 11 guns and several hundred live shells.[4]

According to witnesses, many of the students knew immediately who was responsible for the shooting, and they told their teachers when they reached the safety of the gym, naming Andrew Golden, Mitchell Johnson, and a third shooter, reported to be an older student at a nearby high school.[5] Apparently, Mitchell and Andrew gave many hints about what they were planning to do. The day before they had told some of their classmates that "something big was going to happen tomorrow." And students heard Mitchell say: "Tomorrow you will find out if you live or die," and that he "had a lot of killing to do." Earlier that year, at least one student saw Andrew get up on a table in the cafeteria and say "You are all going to die." The day before the shooting, Mitchell told a friend "You won't be seeing me for awhile because I'm going to be running from the cops." The friend did not know what Mitchell was talking about and ignored the comment. These statements and threats were seen as common coming from the boys and were not taken seriously by most. Their classmates saw Mitchell in particular as a real bragger who said lots of things that he didn't really mean, especially about his supposed gang involvement and the things he had to do as part of the gang. However, at least two students claimed that, when they heard the threats, they told an adult at the school. One student claimed that in October 1997 he overheard Andrew say he planned to bring guns to school and kill people. In a signed affidavit, the student claimed that he went home and told his father, and that the boy and his father then contacted the middle school counselor and told her about the threats.

The school's version of this event is different. According to the school, the threat Andrew made was not against others, but against himself. The counselor called Andrew into her office and asked him about his statements. Andrew claimed that he was just kidding, but his parents were notified of the incident. In the school counselor's affidavit, she claimed that she asked Andrew's mother if Andrew had access to guns. His mother said he didn't—that the guns in the house were locked up and that she would discuss the matter with her husband and that they would take care of it. No further actions on the part of the school were apparently taken in response to this incident. In addition, at least one other student claimed that she had reported the threats to an adult at the school before the shooting and that the adult had done nothing. The school has consistently denied that any adult was ever told about a threat by Mitchell or Andrew against other students.

It is not clear whether the boys intentionally targeted specific individuals. In general, based on the ballistics evidence, people involved in law enforcement and the legal proceedings deny that individuals were targeted. Mitchell has claimed that no one was targeted and that he intended only to aim over his classmates and teachers' heads to scare them. A few bullet holes near the top of the gym wall may support this claim, but at some point, Mitchell lowered his rifle and hit at least four individuals. In addition, according to some students, the boys had a hit list. At least one classmate claimed to have seen a written list that included some teachers and students that Mitchell was angry at, as well as all of his ex-girlfriends, although no list was found during the course of the investigation. Most people connected with the shooting, including investigators, believe that at least a verbal list existed.

It has been suggested that since the boys were approximately 93 yards—almost a football field away—from the individuals they were shooting at, they were too far, given the pandemonium, to have seen whom they were shooting. But Mitchell's rifle had a scope, and Andrew was known as a skilled marksman. We also know that Mitchell hit Candace Porter, a girl with whom he had recently gone out and who had rejected him. Also injured was Jennifer Jacobs, a girl Andrew had recently gone out with and who had also rejected him, although ballistics reports could not confirm who shot her. Some teachers and students believe that some of the others shot may have been girls Andrew or Mitchell wanted to date but who weren't interested in them. In addition, Andrew hit three students who lived on his or his grandfather's street, where he was locally known as a bit of a terror. Moreover, a teacher who was not hit believes the shooting was intended for her, and she claimed to have received confirmation from officials close to the investigation.

Some people in the community believe that girls were specifically targeted, since all but one of those hit were female. The shooting took place during fifth period, when most of the boys were in gym class in a separate building and many of the girls were in choir in the sixth grade wing and were supposed to exit the building by the west entrance. But investigators believe that the boys intended to arrive at the school earlier and the mishaps along the way delayed their arrival, therefore they probably did not intentionally plan to target girls. These assurances from the law enforcement community that no individuals were targeted have not been easy for teachers and other individuals who knew the relationships among the children to accept. To them, the patterns and the reasons these particular individuals became targets is all too clear.

It is also not clear whether Andrew or Mitchell was the instigator in the shooting, and community members—including those who knew the boys well—are fairly evenly divided in their assessments. Those who are close to the investigation, and many in the community, generally believe that Andrew was the instigator because his psychological evaluation seemed to indicate he was the more troubled of the two and he fired more shots and hit more victims. Others who also knew the boys well are equally convinced that Mitchell was the instigator. He was more of a leader; he was older, smarter, and was controlling. While he could be very charming, he had an explosive temper and, by some accounts, an abusive personality. Mitchell was described as manipulative and hence was believed to have coaxed Andrew into helping him because Andrew had the means—access to the guns. Mitchell also had the harder life; he came from a broken home and had suffered abuse in the past. Some who knew the boys felt that their association with each other basically revolved around the planning of the shooting. They felt that each needed the other to carry it out. However, the investigation did not uncover any concrete evidence that sheds light on the interpersonal dynamics of the two.

Some residents also had trouble reconciling the fact that an insider like Andrew, whose family had been in the community for generations, could conceive of such a heinous act, so they find it more plausible to blame it on Mitchell, the relative newcomer. In retrospect, we know that Mitchell seemed to have precipitating events immediately preceding the shooting which could explain his anger at that time; we know little about Andrew, however.

THE CAUSES

Virtually no one in the community seemed to be able to explain why this tragedy happened. When the boys were apprehended by police a few minutes after the shooting, one of the officer's asked "Why?" as much to himself, he claimed, as to the other officer. Mitchell reportedly replied, "Anger, I guess." He then nodded his head toward Andrew and said "He asked me if I would help him do it and I said 'yes.'" The other officer remembers Mitchell stating that "Andrew was mad at a teacher. . . . He was tired of their crap." Since that day, when the boys have been asked the same question, they appeared not to be able to give an answer. Some believe that there will never be an answer to this question; others think that the cause is, at the very least, multifaceted.

In the days and weeks after the shooting, the media put forth many theories in search of an explanation. Some of the most frequently cited reasons include mental illness, a Southern gun culture or culture of violence, a problematic family setting, bullying, and exposure to media violence or media coverage of previous shootings.

Mental Illness

The easiest explanation would be to believe that these boys were simply mentally ill, that there was something terribly wrong with them that could explain why they did this. We have no evidence to support such an explanation. Psychological evaluations were performed on both Andrew and Mitchell, but we were not allowed access to them. According to some who have read the evaluations, there was no conclusive evidence of mental illness, although Andrew was generally thought to be the "darker" of the two. During the criminal proceedings that followed the shooting, Andrew's lawyer attempted to argue that Andrew was legally insane and incompetent at the time of the shooting. The courts found that juveniles are not entitled to an insanity defense, however, and therefore the full argument was never presented in or evaluated by the court. In addition, there is no evidence that either boy was on any form of medication.

Family Problems

Problems within the family—including divorce, domestic or sexual abuse, frequent relocations, fragile family relationships, as well as lack of awareness or involvement in children's lives—are another explanation for school shootings that has been put forth in the popular media. Diminishing adult and parental authority over children is another variant on this theme and was frequently mentioned by community residents as a growing problem. Parents are increasingly losing control of their kids, most often because they work long hours and have little time to spend with them. Moreover, many people in the community feel that parents have relinquished authority for disciplining their children to the school, while the school in turn believes that, increasingly, parents are not supportive of the school's disciplinary decisions.

Problems within the family may offer one of the more persuasive interpretations for the causes of the shooting in this case, although here, as elsewhere, the evidence is mixed. Clearly, there were serious problems in Mitchell Johnson's family history. His parents were divorced, he had made frequent moves, and he was a relative newcomer to a community that is still thought of as a place where people spend their whole lives.

His troubled relationship with his father was a source of much anxiety for him, and the thought of possibly having to live with his father made him feel hopeless. He had suffered repeated sexual abuse, and while it is not uncommon for children to fear telling anyone about such abuse, Mitchell's fears were aggravated by his worries about his father's temper.

Mitchell also seemed to have inherited that temper and was known to overreact and to displace his anger. For example, his teachers said that he felt slighted or put upon by even light teasing or jabs. They also said that he took dating quite seriously, and his friends believed that he would overreact when girls dumped him. Mitchell also seemed to have problems with particular teachers and was known to hold a personal grudge against a teacher who disciplined him. His response to anger, frustration, and abuse was to lash out, frequently not at those responsible for the abuse but at innocent bystanders. When Mitchell was being picked on or bullied, he bullied others, and his fondling of a 2-year-old may have been a response to the sexual abuse he suffered himself. Various people close to Mitchell have suggested that the shooting was a displacement of the anger and frustration he felt toward some family members and his abuse.

Still, Mitchell's life with his mother and stepfather in Bono was more stable than he had ever known before. His mother and stepfather were seen as caring and concerned parents and, by all accounts, Mitchell got along with both of them. His mother was at home and had made efforts to get Mitchell help for some of his problems, at least in connection to the fondling incident. Although Mitchell may have had a lot of unstructured time when his parents were not aware of his activities, it was not necessarily more so than other children. It is not clear whether Mitchell received enough attention and supervision from his family, but family members were certainly available for him and involved at least in his disciplinary problems at school.

In contrast to Mitchell's more clearly troubled family history, Andrew Golden's family background seemed to be remarkably stable. His family was close, and they were well-respected long-time residents of the community. In general, his father and his grandmother were involved in his life at school and wanted him to keep out of trouble; when teachers raised minor issues—like Andrew's making other children feel jealous in kindergarten by selecting who could and could not play with a bag of toy guns he had brought to school—his parents were willing to oblige the teachers and have him keep the toy guns at home. Yet his family was overindulgent and generally gave him his way. In the middle school, Andrew's parents pulled him out of a class when he complained to them that a teacher had spoken harshly to him about not acting up in class and distracting other students. Teachers and neighbors suspected that Andrew was not disciplined for misbehaving. His family may have even encouraged some of his behavior. Andrew was also a latchkey kid and therefore had unsupervised and unstructured time alone at home. It was only in his own neighborhood that he was known as a menace.

We must emphasize that our understanding of Andrew's character is based entirely on interviews with individuals outside his family. The Goldens were not accessible to us and hence we lack the kind of insight that might have been developed had we been able to interview his parents or grandparents.

Bullying

According to a number of people close to the boys—including classmates, teachers, and law enforcement officials—bullying, or at the very least teasing, may have been a factor in the shootings. One of Mitchell's teachers knew before the shooting that Mitchell was being picked on but felt that he was overreacting to his classmates' teasing—that he was being too sensitive. Mitchell liked to brag and when classmates tried to cut him down to size, he would become angry. Concerned about being bullied, Mitchell may have chosen some friends who he felt would protect him, and the "tough" image he tried to project, including his frequent claims to gang affiliation, may have been a protective strategy. At the same time, Mitchell was seen as likely to bully others. He was by no means an outcast, and the anger he expressed at students in the school did not seem targeted at people who might have bullied him.

Andrew was a slight boy, leading some people to think it is possible that he was bullied. Students said that Andrew had friends, and while he wasn't in any clique, he was not considered a loner. While several students did not have the perception that Andrew or Mitchell were picked on a lot, some classmates did tell adults in the community that they were bullied or teased. However, it does not appear that they were singled out for excessive abuse.

Bullying is considered by many to be a significant problem at Westside, especially on the bus and in the hallways, but it is not clear whether it is any worse than at any other middle school. Both students and teachers agree though that much of the bullying that occurs happens out of the sight of adults at the school. While bullying or teasing may have been part of the problem, no one, not even the boys, according to those who have spoken with them, is willing to say "bullying was *the* reason." The boys have said that they felt put upon, but bullying alone does not explain their rampage.

Gun Culture/Culture of Violence

Most of the people in the community understandably resent any representation of Southerners as gun-toting, violent people. Most boys learn to hunt, they admit, but people are generally responsible and furthermore "it is not guns, but people, that kill." Some also argue that guns are prevalent in all rural communities, North and South. The distinction therefore should be made between rural and urban, not between the North and the South.

We found no conclusive evidence to support or discount either claim, but it is clear that Mitchell and Andrew were able to access an arsenal of weapons for this crime. While most of the Goldens' weapons were secured—either in a safe or by a cable—Mitchell and Andrew were able to access more than enough firepower for the shooting. While it is impossible to know the impact that trigger locks might have had, none of the weapons was equipped with them. After this tragedy, civil suits were mounted by the victims' families against the gun manufacturers, and some of them lobbied for trigger locks. Most of the youth we spoke with thought it would be pretty easy to get a gun if they wanted one.

Andrew's proficiency with handling weapons has already been noted; according to media accounts, he was given a rifle for his sixth birthday and had other weapons of his own as well. Andrew was fascinated, possibly obsessed with guns; some teachers say it was his toy of choice as a young child, and the year of the shooting, he drew a picture of two rifles when asked to draw something that symbolized his family for his social studies class; he even did a skit about guns for his English class.

Mitchell was not as skilled with guns, but when officers apprehended him after the shooting, Mitchell was carrying his hunter education card in his coat pocket. Mitchell had access to a gun for hunting purposes, but because his stepfather had been convicted of a felony and was not allowed to have weapons in his home, the gun was stored with a neighbor and was not used for the shooting.

Exposure to Media Violence or Media Coverage

Many residents of the community where the shooting occurred believe that the media's glamorization of violence in general and of school shootings in particular fuel these types of incidents, if not this specific tragedy. Children are influenced by what they see and hear and become desensitized to excessive media violence. In their view, it is no longer a question of simply monitoring what your own children watch; they now fear what the neighbor's child is exposed to because they know all too well that other people's problems can quickly become their own.

What role media violence played in this particular shooting is unclear. According to friends, Mitchell played violent video games, including a shooting game, at friends' houses but did not have access to them at home; his family was too poor. His mother said that he owned only two rap CDs, TuPac Shakur's All Eyez On Me, and Bone Thugz n Harmony's E 1999. A Westside Middle School teacher testified before the Senate Commerce Committee, in a hearing about parental warning labels on albums, that "Mitchell brought this music to school with him, listened to it on the bus, tried listening to it in class, sang lyrics over and over at school, and played a cassette in the bathroom about 'coming to school and killing all the kids.'" According to his mother, Mitchell was not obsessed with violent music or television. One friend mentioned that he often talked about blood and gore and that he liked "gory" movies.[6]

There is no direct evidence to suggest that Mitchell's obsession with gangs came from the media, but such an explanation is possible given his relatively isolated life. He may also have learned about gangs from other children in the area where he lived, as gangs are a problem in the Jonesboro area, or from an older family member who attended an area school, which had a gang problem. We can only speculate as to the reasons Mitchell displayed such interest in gangs; it may have simply been a matter of looking cool, or it may have been a protective measure to stave off any would-be bullies. Whatever the reason, he was compelled to pretend he was part of one and talked about murder and violence as necessary to obtain entry to the gang community. It appears then that in his mind, violence could be rewarded and respected—a view that many attribute to the messages available in the mass media.

We know considerably less about Andrew's viewing habits, except that he was a latchkey kid, whose frequent time alone at home may have left open the possibility for exposure to violent music or television. Andrew's grandfather has told the media that Andrew did play video games with guns. According to adults who knew Andrew, he was enamored with Beavis and Butthead and South Park (as are thousands of other children), and he used to play the clown and mimic the characters in those television shows.

At the time of the shooting, two other school shootings had occurred in the South in the previous six months: one in Pearl, Mississippi, and one in West Paducah, Kentucky. It is unclear whether the boys were aware of either shooting, but people close to the boys do not believe it was a primary motivation. When the police searched their homes and the van, they found nothing to indicate it was a copycat crime and nothing the boys have reportedly said indicates they were directly influenced by the two recent shootings. Whether or not the shooting was a copycat crime, several community members, including one who knew both boys personally, believed that they were influenced by the fame they knew would result from the shooting and felt that the boys relished the attention, even though it was so negative.

Child-Adult Relationships

In addition to the explanations that have been put forth in the national media for the rash of school shootings, members of the Jonesboro and Westside communities have some of their own.

One common theme they expressed is the lack of communication between adults and children. People perceived a widespread disconnect between children and adults, so that children have few meaningful adult figures to connect to. Especially at the middle school age, kids at Westside were more likely to confide in their peers than in any adults. Part of this process is natural; at that age—11, 12, or 13—children normally start to become more independent and develop stronger relationships with their peers. Yet some of the disconnect may be the result of more structural factors. Some teachers feel themselves less able to connect with kids because of the growing number of other demands on them—including disciplinary functions, dealing with "squeaky wheel" students who demand much of their attention, paperwork, and the growing pressure to get their students to perform well on standardized tests. Students may have felt that meaningful adult contact diminished as they entered the middle school. Elementary school students who were used to forming a close relationship with a single teacher found themselves largely on their own in the middle school, where homeroom met only a few times a year to perform administrative functions and was not geared toward building the same kinds of close relationships.

Many teachers argued that they had received little or inadequate training in identifying and coping with troubled children. While there was a full-time school counselor and a part-time licensed therapist assigned to the middle school, children generally need to be referred to them for assistance by teachers or the principal, unless the children or their parents seek the help themselves. With classes of 25 or 30 students that teachers see for only one period a day, only the more extreme cases—children who disrupt the class or show obvious signs—tend to get services. Some students are involved with churches and may have meaningful contact with adults through the church; only a few local churches had full-time youth pastors or ministers, but several had part-time youth leaders. Even so, some church leaders felt that even church services can lead children to be segregated from adults and may not contribute to the connection between parents and children.

There is no strong evidence to suggest that either Mitchell Johnson or Andrew Golden had particularly close relationships with adults outside their families. There were a couple of teachers that Mitchell liked, but he had disciplinary problems even with them when his temper got him into trouble. Even one of his favorite teachers had to assign him to in-school suspension. He did talk to a teacher about being picked on and would sometimes ask her for help with a bullying incident. Mitchell also was part of a youth group at his church, but he had only been attending for approximately six weeks prior to the shooting. Teachers generally described Andrew as a middle-of-the-road kid who didn't cause a lot of problems, and none seemed particularly close with him. Some other students in the same age group as the two boys also said it was rare for middle school students to communicate with teachers or other adults in the school about any problems they were having; they had not done so themselves.

No single reason explains what led Mitchell Johnson and Andrew Golden to open fire on their teachers and classmates. Many factors influenced them, and while none of these alone led to the shooting, the cumulative impact of these forces contributed to the events of March 24, 1998. The boys' family histories and experiences at home seem to have contributed to a state of mind in which other influences, such as teasing or bullying at school and disciplinary or social problems with girls, served as an outlet for their anger, and the media and their access to guns may have contributed to the idea and the means. The lack of meaningful adult contact and authority may have failed to prevent their anger from exploding, and feeling ignored or unheard by adults may lead a child to grasp for the fame that a negative event like this can easily produce. It is likely that neither boy on his own would have carried out this plan; the two together were an unfortunate and deadly combination.

CRIMINAL ADJUDICATION

Within minutes of the shooting, Mitchell Johnson and Andrew Golden were apprehended and taken to the Craighead County Jail's juvenile detention facility. Each boy was charged with 5 counts of capital murder and 10 counts of first degree battery. However, under Arkansas law, defendants under the age of 14 had to be tried as juveniles, and the most serious charge that could be lodged against them was delinquency, which carries a maximum punishment of detention in a Department of Youth Services juvenile center until the age of 18. When they turn 18, Mitchell and Andrew must be moved to a special facility for 18- to 21-year-olds. Since the purpose of juvenile detention is rehabilitative rather than punitive, the boys could only be sentenced to an indeterminate period, up to age 21, in the center, until such a time as they are deemed rehabilitated.

At the adjudication hearing, Johnson entered a guilty plea and read a statement apologizing for his actions: "I really thought that no one would actually be hurt. I thought we would just shoot over everyone's head. When the shooting started, we were not shooting at anybody or any group of people in particular." Golden pleaded not guilty and did not issue any statement. Andrew's lawyer tried to raise issues of insanity and incompetency, but the judge ruled that these arguments could not be raised in a juvenile case because juveniles cannot form the intent that makes an insanity plea possible, and that competency did not have to be determined because the special proceedings of juvenile court are set up to be rehabilitative rather than punitive. A quick trial ensued in which the prosecution presented its evidence. Andrew's defense attorney conceded that he committed the act, so the facts were never in dispute and the trial lasted only three hours. His lawyer then appealed the judge's ruling on his arguments to the Arkansas State Supreme Court, which agreed with the circuit court on the insanity finding but remanded the ruling on competency,[7] arguing that competency does have to be determined, even for juveniles. However, after this decision, Andrew and his lawyer decided not to pursue the competency question further.

The limitations binding the courts, which made it impossible to try Johnson and Golden as adults, were a source of deep frustration for many members of the community, who felt the boys had received a mere slap on the wrist for a heinous and premeditated crime. Family members of the victims were moved to become activists on behalf of legislation that would allow juvenile offenders to be tried as adults for serious crimes. The one law which did emerge from the shooting, the Extended Juvenile Jurisdiction Act, was considered a disappointment. Although it does provide some potential adult sanctions for those under age 14, it actually reduced the number of crimes for which the prosecutor has discretion to try the offender as an adult. Although capital murder had not been among them, there had previously been about 30 or 40 crimes over which the prosecutor had such discretion, but the act reduced this number to about 9. Furthermore, the act is very complex and difficult to implement; multiple criteria have to be met before it can be invoked. Given the availability of appeals, it could take a very long time before any of these cases would be settled. This outcome provided little satisfaction for the families of Westside victims and for many produced a sense of disillusionment with the entire justice system.

Local anger is high over the fact that the two shooters will be able to return to a normal life without so much as a criminal record, since juvenile offenses are not counted as criminal. Community members are universally incredulous that they would be allowed to own firearms once they are released because they do not have felony records. The general frustration with the legal system combined with a widespread view that justice was not served in this case has probably been responsible for the death threats made against Johnson and Golden, which continued even after they were transported to the Department of Youth Services facility at

Alexander, Arkansas, a remote area far from Jonesboro. Because of these threats, special tarps had to be placed around the exercise yard of the high security unit where the boys are currently incarcerated and part of a nearby woods has been cleared to improve visibility and prevent snipers from accessing the area.

Given the abbreviated nature of the confinement, many in the Jonesboro and Westside communities are concerned about what will happen when Johnson and Golden are released, whether they will return to the area, and whether they will be rehabilitated by their stay in the Department of Youth Services facility. The view was commonly voiced in matter-of-fact tones that their lives will be in danger once they are released, generally followed by the opinion that they should not try to take up residence in the vicinity of Jonesboro.

COMMUNITY IMPACT

If community anger was high as a result of the shooters' adjudication sentences, the tension was aggravated further by the media siege that followed the shooting. A total of 70 U.S. and foreign news organizations sent more than 200 reporters, photographers, and support personnel to cover the shootings, completely overwhelming this small city.[8] Out-of-town reporters arrived at the school within an hour and a half of the first report of the shootings. By the following morning, 50 satellite trucks and camera crews occupied the school grounds.

The sheer size of the media combined with the aggressive conduct of some of its members aggravated the trauma of a community already reeling. Although residents generally praised the *Jonesboro Sun* newspaper and KAIT, the local television station, for their coverage and sensitivity, national and international media were criticized for excessive badgering. The evening of the shooting, school officials invited counselors from throughout the community into the school gymnasium to meet with students, parents, and others who needed counseling. Visitors had to be escorted by police and other volunteers past the dozens of crews who thrust cameras in the faces of grieving children and families. Reporters in search of a story were even known to push some of these escorts out of the way to get at a victim. Some media representatives were worse than others—including one crew that put its cameras up to a kitchen window to film a victim's family around the breakfast table and a news producer who posed as a student's relative to gain access to one of the wounded teachers in her hospital room. The families of the shooters, who were bewildered and in shock, were the subject of relentless media attention and were unable to shield their younger children from invasive reporters and cameras. Residents were particularly upset by the license reporters took to interview children without their parents present. The media presence lasted for several months, making daily life and the healing process extremely difficult for those affected by the shooting.

Consequences for the School

Security and Services

Westside Middle School has changed considerably in response to the shooting. Most distressing for its staff and students, they have had to come to terms with the notoriety and negative connotations of the school's name. Immediately after the shooting, the school became a target for violence and threats from strangers coming out of nowhere. A bomb threat three days after the shooting forced the evacuation of the gymnasium, and there have been additional threats against the school since. News of the shooting has attracted various oddball characters to the school—for example, a clown angered because the school denied him permission to perform for the students showed up anyway and was found performing magic tricks in the school cafeteria. A man with a car full of newspaper clippings about different school shootings around the country turned up at Westside as well. The presence of these strangers wandering onto campus caused additional security concerns for the administration. Even several years after the shooting, some parents and students continue to feel that the school is a target of violence because of the shooting.

In the days immediately after the shooting, security was the administration's main concern. Sheriff's deputies were brought onto the school grounds to make the children feel safer. Afterward, a wooden-slat fence was built around the school grounds. Although this was supposed to make students feel more secure, law enforcement personnel and teachers noted that the fence actually provides a much better hiding place for a shooter and would make it easy for someone to push a gun between the slats without being seen. The school adopted a much harsher stance on disciplinary problems and threats. While Westside had always had a "zero tolerance" policy that applied to kids bringing weapons to school or other gross violations of the rules, after the shooting the policy was extended to cover threats that would previously have been ignored, such as those said in the heat of an argument. Some at the school described the policy as "common sense zero tolerance," which still allowed some flexibility for administrators to judge the severity of incidents. But students and staff alike agreed that the climate of Westside is much changed and that transgressions are dealt with more harshly than was the case before the shooting.

After the shooting, Westside, as well as many of the other schools in the Jonesboro area, brought in school resource officers (SROs). These are trained law enforcement officials who spend most, if not all, of their time at the school. Their main objectives are safety and security, through education and open lines of communication with students. Students were thought to be more comfortable confiding in an adult who does not have the same authoritative relationship with them as their teachers. One function of an SRO is therefore to get to know students and establish trust so that students will come to them if they hear of threats or know of students who are in trouble. Students at Westside have confided in the SRO about other students who have brought drugs, knives, and other small weapons into the school. SROs are also on hand for particularly difficult disciplinary meetings with parents, when they refuse to accept that their child has done anything wrong and become angered at the school. Although there was initially resistance to having an officer with a weapon on the campus, most teachers and administrators with whom we spoke, as well as many parents, thought that having an SRO at the school was an important part of maintaining school security.

Westside has also added adult monitors to its school buses since the shooting, in an effort to keep order over the chaotic environment which was said to be the site of much of Johnson and Golden's planning. Other reports attest to the frequent bullying and teasing that occurs on the school buses, and students say the monitors have helped, although the teasing certainly has not disappeared.

The community response to the school's increased attention to security and monitoring has generally been positive. However, one unintended consequence of this increased surveillance and harsh punishment for infractions has been to drive some of the violence out of the school. Knowing that they will be dealt with severely for fighting on school grounds, students have informally arranged meeting places for fights off location, where school officials have no jurisdiction. A nearby gas station and a local field serve as the meeting grounds for these old-style showdowns when someone is challenged to a fight.

A minority has also voiced concern that the presence of a law enforcement official on the school grounds results in more students ending up in the criminal justice system. The SRO can report to the sheriff any rule violation that involves criminal activity, including weapon or drug possession and even fighting in the hallway. Juvenile court officials say that they process more students from schools with an SRO than without one. This criticism of the SROs is relatively rare, but there do seem to be mixed feelings toward the more extensive crackdown on student behavior that comes with these improved security initiatives, particularly zero tolerance. Many people recognize that such a policy is necessary, but there is also a large majority who think that it can be carried too far and greater disciplinary leeway needs to be allowed.

Westside has also increased the attention paid to troubled or isolated children in the school. Along with several other schools in the area, Westside has been the beneficiary of a Federal Safe Schools/Healthy Students Grant, which has provided additional social workers and full-time counselors for the school. One counselor and one social worker cover the elementary and high schools, while another counselor and social worker cover the middle school. Counselors are available to meet with students who are referred to them for behavioral problems or therapy by teachers or administrators, and they meet with some students on an ongoing basis. The social workers make home visits to parents, help provide them with services or information if they are unable to come into the school, and teach students character education, like building social skills, anger management, and coping with bullying.

Dealing with the Trauma

Needless to say, the shooting was the most traumatizing experience ever suffered by the Westside community. From the very beginning, they have had to contend with shock, grief, frustration, lack of privacy, and for the families of Mitchell Johnson and Andrew Golden, shame, remorse, and the sense that they had been judged in the court of public opinion as parental failures. There is no normalcy to be found in the aftermath of such a difficult experience, and yet somehow people have to keep working and children have to go to school.

For the teachers and administrators at Westside, the highest priority initially was to tend to the children's needs. The evening of the shooting, as counselors met and talked with parents, children, and teachers in the gymnasium, an open meeting of about 75–100 people was held in the school cafeteria to discuss a plan for what type of counseling should be offered next. The decision was made to invite members of the National Organization of Victim Assistance (NOVA) into the community to help with the counseling and provide guidance in what people can expect in dealing with the emotional fallout of an event like this. The following day, classes were canceled for students, but teachers were asked to come in. Teachers were divided up into small groups to talk about their experiences and be debriefed on the kinds of behaviors they could expect from their students. The following day, when students returned, counselors and other volunteers were brought into all the classrooms to explain what happened and why they were there, and to open up a conversation about how the students were feeling. Specialists in counseling were also available in the school libraries, and students having serious problems could go and speak with them. These specialists were available throughout the rest of the week, and counselors were on call to go into the classrooms as needed. Counseling was also offered to teachers, police, and staff who felt they needed it.

In a tragedy such as this, community members repeatedly said, everyone is a victim. It therefore became quite difficult to recognize everyone who was in need of assistance, and to provide for those needs given the demands of the situation. The emergency medical personnel that responded to the scene, the cafeteria workers who brought towels for the injured, and the students and teachers who were not in the line of fire but lost a friend or a teacher were all victims in this tragedy as well and needed support.

Like their students, many teachers were also deeply affected by the shooting, but some felt ill prepared to deal with their own suffering because so much emphasis was put on preparing them to deal with their students' reactions. Most teachers we spoke with felt that it was too soon to come back to the school the day after the shooting and be expected to cope with a debriefing. Although they did discuss some of their experiences that day, and limited counseling was offered to them through victims' assistance, they felt in retrospect that this had not been enough to meet their emotional needs after this experience. Many teachers spent so much effort making sure that their students were coping that they pushed their own emotions aside. Teachers, staff, and professional counselors all said that many teachers and administrators were only recently beginning to experience problems as they started to deal with their own grief now, some three years after the shooting. Some teachers are still in therapy, while others claim they would be if only they were able to afford it since not all insurance plans cover it. Teachers and administrators have also suffered from a wide range of emotional and physical health problems since the shooting, including post-traumatic stress disorder, problems sleeping, facial ticks, significant weight loss or gain, and stress-induced epilepsy. Some staff also attribute divorces or problems in their marriages to their emotional states as a result of the shooting.

While there has been no formal monitoring or evaluation of the emotional states of the cohort of students who were in the middle school at the time of the shooting, some counselors and school staff believed that these students are suffering problems to a greater extent than other cohorts. Teachers feared that some of these students' may feel guilty or may still have nightmares, and they have found behavioral and academic problems in their students that they believed are related to the shooting. Some students were still scared and were reminded of the shooting whenever they heard a fire alarm or other loud noise, like a locker door slamming. Some students were still in counseling three years later, while others did not receive any therapy and were never given a chance to talk about their feelings because their parents wanted them to put the shooting behind them.

Others feared that these students might be experiencing behavioral problems as a result of the inordinate attention they received after the shooting. Students who witnessed the shooting were asked to be on national television, were sent gifts from around the world, were taken on trips, and were provided with other activities. Some counselors and parents feared that these activities may have not only kept them from really dealing with their feelings, but that the experience of being treated as a celebrity because of their terrible experience may have given them a sense of entitlement. Some of these students, one parent claimed, use the shooting as a crutch to get what they want.

For its part, the school is striving to put the shooting behind them. As a result, there is a widely held perception that the school does not want to talk about the shooting and is not supportive of those who are still having trouble as a result of it. Teachers who have difficulties performing their jobs due to depression or post-traumatic stress disorder felt they were not supported by the school, but rather were being told that they should be over it. Students who were still affected by the shooting after several years were also told—by teachers as well as other students—that it was time for them to move on. Yet some of those who were most affected felt that they have not had the open discussion that would allow them to move on. From the school's perspective, it was difficult to encourage any formal discussion because people were at such different stages of the healing process. Some were done healing and didn't want to relive the memories, while others still needed to. The school may not be able to cultivate an atmosphere of open discussion, but for those still learning to cope with their trauma, the school's support was extremely important. They sought an acknowledgment that it was alright for them still to be traumatized. When those who have moved on acted as if there was something wrong with those who have not, it deepened the divisions among these groups.

CONSEQUENCES FOR THE COMMUNITY

When the subject of the Westside shooting is raised in the Jonesboro area three years after the fact, it is clear that it still brings up hard feelings and emotions. The community is angry that there was never a sense of closure because Johnson and Golden were given indeterminate sentences, which are seen as too short even at their maximum length. There was also no sense of closure when the Extended Juvenile Jurisdiction Act was passed, because it did not provide them with the solutions they had sought—to enable prosecutors to try more juveniles as adults. The community is still troubled that there seems to be no answer to why the shooting happened, but it has come to a level of stasis in recognizing that they may never know for sure.

But by and large, the Westside shooting is no longer widely discussed. People who are less removed from the events say that it comes up less and less in conversation, and most people want to move on and stop talking about it. People are embarrassed that it is the shooting that put Westside on the national map and would prefer to put it behind them. There is not even much discussion about Mitchell and Andrew getting out. Although there were formal memorials every year, March 24, 2001, was the first year when there was none. Most people just do not want to talk about it any more. Even the civil suits, which have kept the event alive in some other communities, are infrequently discussed in Jonesboro, and have not brought a "second life" to the shooting. Civil suits were brought by the victims' families against the shooters for wrongful death, against their parents for not controlling their children or being aware that a situation like this could happen, against Andrew Golden's grandfather for failing to adequately secure his guns from the known risk of burglary and use by a burglar, and against the gun manufacturers for failing to take available precautions to ensure gun safety—namely for failing to install trigger locks.[9]

Reactions to the civil suits are mixed. In this region where guns and hunting are widespread and popular, there are many who oppose the suits against the gun manufacturers and claim they were not responsible for what the boys did. Some people support the suits out of sympathy and respect for the victims, but this view is generally in the minority. Attitudes toward the suits against the parents are also mixed, as are personal views about the responsibility of parents when their children do wrong. Many feel the parents should be held responsible, but others are sympathetic and can empathize as parents themselves that it is not always possible to control one's children. By and large, however, the civil suits are ignored and community members avoid discussing them, much like the shooting in general.

Yet the attitude that the shooting should be put behind them makes life more difficult for some people in the community. A family life minister and counselor in Jonesboro compares the shooting to an earthquake, with many concentric circles of impact. Community members who were closest to the epicenter of the event—the victims' families, those close to them, and people present during the incident—are those who are most deeply hurt and whose healing process will take the longest. Those who are farther removed from the epicenter heal more quickly and want to put the incident behind them faster. Not recognizing the differential rates of healing that people in the community go through, each side thinks the other is wrong. Those closer to the center feel that others are repressing their feelings and will never get through their trauma if they do not talk about it. Those on the periphery think those in the center are dwelling on the past and need to stop.

Some feel that the community has become closer and more cohesive in the aftermath of the shooting. There were many acts of generosity and caring as neighbors banded together to help one another. One victim's family did not have furniture in their house trailer, and a local business furnished their home for free. A local restaurant sent food to the families and counselors who were meeting at the school on the night of the shooting. The local branch of the United Way established a fund for the victims and collected half a million dollars from around the world. Some rifts within the community were also resolved. There had been a ministerial alliance—a consortium of different churches' leaders who would meet regularly to discuss issues of common concern. In the previous 10 or 15 years, it had divided over whether or not the churches should get involved in the political issue of the county's being wet or dry. After the shooting, the two ministerial alliances decided to put their differences aside and work together, uniting to form the Jonesboro Ministerial Alliance.

Some community leaders do say that they feel this cohesion may only be skin deep. The issues that formed the ministerial alliance have not yet been resolved, and the generosity that arises in the immediate aftermath of a tragedy may fall apart and turn sour later on. Yet many in Jonesboro try to focus on more positive interpretations. Several Westside students noted that their class is now much closer and that bullying and cliques are less of a concern because of their shared experiences of the shooting.

CONCLUSION

No single cause accounts for the behavior of Mitchell Johnson and Andrew Golden. The cumulative impact of the forces that span the individual, the family, the school and society are all involved. Both boys were troubled. Mitchell had a very difficult and stormy relationship with his father and a history of sexual abuse, both as a victim and in turn as an abuser. He had trouble controlling his temper and flew off the handle over slights—being teased by peers or being dumped by a girlfriend. These incidents became a channel for a deeper anger. Indeed, it is possible that the abuse Mitchell suffered as a child intensified normal adolescent male concern over masculinity that was subsequently reinforced by a real failure with a girl. As we have not seen the psychiatric reports, we do not know whether this possibility has been examined in any therapeutic setting.

Andrew had a reputation in his neighborhood as a menace and was reported to have been cruel to animals; he appeared to many in the community to be spoiled by his family, and some speculated that a lack of discipline at home may have led him to think he could get away with anything. Andrew may also have been teased at school. However, nothing in the disciplinary histories of either child suggested to school officials that these boys were particularly troubled. There is also no evidence that either boy had formed close or long-lasting relationships with adults outside their families who might have helped them deal with their problems, but they were clearly well loved by members of their own households.

As a team, Andrew and Mitchell were able to access a veritable arsenal of weapons. Some of those weapons were left unsecured in Andrew's home, and the others used in the attack were secured only by a cable in Andrew's grandfather's locked home. None of the weapons used were equipped with trigger locks. Not only did the boys have access to firearms, but Andrew had the experience and facility with rifles and hunting that allowed them to plan and execute the attack with a degree of proficiency that surprised even the local police.

Finally, while it is not clear what role the media played in this tragedy, many felt that it was a factor. Both Andrew and Mitchell were known to play violent shooting video games with friends, although Mitchell's family did not have such games at home. The extent and the explicit content of their media viewing habits are not known, but a friend said that Mitchell watched and talked about gory movies, and teachers and other adults who knew Andrew said that he watched and mimicked television programs such as South Park and Beavis and Butthead. While there is no evidence that either Andrew or Mitchell got the idea of a school shooting from previous media coverage of the Pearl or Paducah incidents, a few people close to the boys say that they may have been influenced by the fame and attention, though negative, that such an attack would bring. Yet it must be said that millions of American teenagers watch the same films and TV programs, so it is problematic to assign their viewing undue weight.

While residents in the Westside and Jonesboro communities are still unable to explain why this tragedy happened, they are generally quick to tell us, "If it could happen at Westside, it could happen anywhere." Many urged that school districts, communities, and indeed the nation simultaneously work to prevent and prepare for school shootings.

Like a powerful earthquake, the impact of the Westside school shooting was felt far and wide. Few were unaffected by the images of the young boys who were responsible for such a heinous crime. While most people have moved on, those closest to the epicenter are still affected in powerful ways, and their lives will never be the same.

ACKNOWLEDGMENTS

Katherine Newman is the principal investigator on this case study. We wish to thank Martin West, Ph.D. candidate in government and social policy at Harvard University, for his help in analyzing the theoretical perspectives on lethal school violence we relied on and Margot Minardi, Ph.D. candidate in history, for her assistance as well.

NOTES

[1]Bono is the largest of the three with a population just over 1,000 residents. Cash and Egypt count 280 and 112 residents, respectively.

[2]We make occasional reference in this report to "girlfriends" or "going out," but it should be noted that these relationships among very young teens are not equivalent to their later, high school variant. Particularly in a rural area like the Westside school district, prior to legal driving age, boys and girls tend not to spend unsupervised time together. Hence to be someone's girlfriend is perhaps to spend time on the telephone or a few idle moments together in a school corridor. Birthday parties, generally supervised by adults, might be another time for this kind of socializing. But there is no evidence of anything more sustained than this and to "be dumped" under these circumstances is merely to see the end of a fairly sketchy relationship. Nonetheless, the social and psychological consequences of rocky interpersonal relations loom large in the minds of preteens and should not be minimized as a result. There was clearly an element of failure in this for Mitchell and possibly for Andrew Golden as well.

[3]As is characteristic of rural communities with school districts that are large, the bus ride to school is extremely long, often upward of 1.5 hours for the children who lived farthest from the school.

[4]Andrew was apprehended with the following weapons on his person: a universal .30 caliber carbine, a .38 special 2-shot, FIE 380, a .357 caliber revolver, a pocket knife, and over 100 shells. Mitchell was apprehended with the following weapons on his person: a 30.06 caliber rifle, two .38 caliber pistols, a 2-shot derringer, a .38 caliber semiautomatic, two pocket knives, and over 100 shells. At the scene where the shots were fired were also recovered a .44 magnum rifle and 22 spent casings.

[5]The possibility of a third shooter was apparently investigated by the police and found to have no merit, but some of those who were in the line of fire still believe that a third shooter, situated on a hill near the elementary school nearby, was involved and eluded capture. Some community members also speculated that a third person, perhaps someone older, had helped the boys plan the event or had discussed it with them. This issue apparently was not part of the investigation.

[6]So, too, of course, do millions of other American teenagers.

[7]Andrew's lawyer appealed the state supreme court's ruling on the insanity argument to the U.S. Supreme Court, but the Court declined to hear the case.

[8]*Jonesboro: Were the Media Fair?* The Freedom Forum.

[9]The civil suits against the gun manufacturer and against Golden's grandfather have been dismissed. The court agreed that the parties could not have predicted the intervening proximate cause of the burglary and use of the guns by the boys. The suit against the Goldens was settled out of court and a settlement was paid by their homeowners' insurance. The suit against Johnson's parents is still outstanding, as is the suit against the two shooters.

5

No Exit: Mental Illness, Marginality, and School Violence in West Paducah, Kentucky

David Harding, Jal Mehta, and Katherine Newman

At 7:42 a.m. on December 1, 1997, Michael Carneal opened fire with a .22 caliber pistol on a prayer group gathered in the lobby of Heath High School, just outside of Paducah, Kentucky. The 14-year-old freshman killed three students and wounded five others, two seriously. After firing eight shots, Carneal put his pistol on the floor and surrendered to school principal, Bill Bond. The son of a respected attorney and a homemaker and brother of one of the school's valedictorians, Michael Carneal shattered the peace and security of the tightly knit rural community of Heath, Kentucky, and shocked the nation with a brutal instance of school violence.

This chapter is a case study of the shooting at Heath High School, its antecedents, and its aftermath. It is based on more than 75 interviews with more than 100 individuals and participant observation in the school and the community conducted by the authors in May and June 2001. Information from this fieldwork is supplemented by local and national media coverage, police investigative materials, Carneal's own writings, depositions from civil lawsuits, psychiatric and psychological evaluations of the shooter, and an interview with Carneal's most recent treating psychologist, as well as materials from Heath High School and the McCracken County School District. We interviewed legal professionals from both the criminal and civil proceedings that followed the shooting, police officials, victims' families, teachers, high school and middle school administrators, political and religious leaders in the community, parents, and students, both those present at the time of the shooting and those currently in the ninth grade at Heath.

We interviewed Kelly Carneal, Michael Carneal's older sister, but we were not able to interview either Carneal himself or his parents. However, we did review three lengthy interrogations of Carneal by the police, reports and interview transcripts written by Carneal's numerous psychiatrists and psychologists, and an exhaustive 500-page deposition of Carneal taken in preparation for the civil suits. We were unable to interview other Heath students suspected by some in the community to be coconspirators in the crime, although we did read the police interviews and civil depositions they provided.

The events discussed below have been variously described and interpreted by the people involved. While we have done our best to present what we understand to be the facts, school shootings, like other emotionally charged events, produce contradictory accounts that elude complete resolution. In this instance, civil litigation naming many of the people we interviewed was still pending on appeal at the time of our fieldwork, which discouraged the participation of a number of key figures who may someday be able to contribute their perspectives. We have deliberately refrained from using the names of those who were suspected of, but never charged with, participation in the shooting, even though their names were widely reported in the media.

THE SETTING

The Heath community is located a few miles west of Paducah, Kentucky, a city of approximately 25,000 that sits at the confluence of the Ohio and Tennessee rivers. Paducah and Heath are in McCracken County (population approximately 65,000), one of the northernmost counties of the Bible Belt. Long a river town and transportation hub, Paducah is at the center of an economic area that stretches into several counties to the south, east, and west in Kentucky and north into Illinois. Its economic history has seen its share of ups and downs since World War II. Barge and tugboat industries and farming were once the backbone of the county, but the economy has diversified over the past half century. Today there are only a handful of farms in McCracken County, and the main industries include medical services, river shipping industries, railroad manufacturing, chemical manufacturing, paper mills, and the nation's only uranium enrichment plant.

The unemployment rate for the regional labor market area was 6.5 percent in 1997.[1] Today, many of Paducah's downtown storefronts sit empty, rendered obsolete by the strip malls full of chain stores and restaurants on U.S. Highway 60 near the interstate. The county's Information Age Park, an industrial park wired for high-tech firms built in the mid-1990s, is only about 10 percent full. The opening of the United States

Enrichment Corporation plant in the 1950s has brought a steady stream of engineers and professionals to the area, although many are sojourners who move on. It also employs some of the rural working class, the original "Heatherans." This mix of locals and outsiders has been augmented in recent years by families moving out of the city and into newly constructed subdivisions.

The Heath community was described as small and tightly knit by almost everyone we interviewed. "Everyone knows everyone else's business," and gossip travels quickly. Status is often measured by the number of generations one's family has lived in Heath, and people know each other and their families by name. Strangers with northern accents and probing questions stick out, although since the shooting the community has become more accustomed to inquisitive outsiders.

The high school is in many ways the center of life in the Heath community. Community members who graduated decades before still congregate at high school sporting events, choir concerts, and band performances. Parents are heavily involved in students' extracurricular activities, from sports' booster clubs to selling refreshments and building sets for the school play. A year or two before the shooting, when officials proposed combining the three county high schools into two to provide more varied classes and more extracurricular activities, Heath residents were the most fervent dissenters. Old school loyalties run deep.

With between 500 and 600 students each year, Heath is the smallest of the three county high schools, each of which is fed by a middle school and two elementary schools. The predominantly white county school system is separate from the Paducah city schools, which have a considerable black population. Heath's curriculum, like its student body, is a mix of traditional and new. The school boasts three large computer labs, a computer in every classroom, and a classroom with videoconferencing equipment, as well as a greenhouse and an active agricultural education program. Students come from a wide range of economic backgrounds, from trailer parks to million-dollar mansions. Racially the school is almost entirely white, with a handful of blacks, Asian Indians, and Hispanics. Students score above state averages on the Kentucky Core Content Tests, and the dropout rate for the 1999–2000 school year was 2.9 percent.[2]

About 60 percent of graduating seniors go on to college, yet most remain nearby at Paducah Community College or Murray State University.[3] Hence friends made in high school are friends people keep for life, especially among those who begin work right after high school. Like most Kentucky schools, primary responsibility for the school's operations and curriculum lie with a site-based committee, which in the case of Heath High School is composed of the principal, two teachers, and three parents. The County School Board has a minimal oversight and funding capacity.

Faculty at the school reported that the biggest discipline problems are tardiness, unexcused absences from school, and classroom disturbances. The current principal and assistant principal see about eight students per day for disciplinary reasons, although they estimate that about 5 percent of students create 90 percent of the discipline problems. Fights in school are rare, although according to students, there are several fights each year off school grounds to avoid stiff penalties for fighting at school. School discipline, including punishment, is detailed in a countywide Student Code of Conduct distributed to each student every year. Punishments range from a warning to "flex-time" detention to in-school detention to on-site alternative school to a central county alternative school for the most dangerous offenders. There were no violence prevention education programs for either students or staff prior to the shooting. As the principal explained, safety was a "nonissue" before the school shootings. While the school had emergency plans for firearms in the school, the plans were designed with an outside intruder in mind and were therefore not effective at preventing a school shooting by a student like Michael Carneal.

MICHAEL CARNEAL

The figure at the center of this story was a 14-year-old freshman who had been at Heath for less than a semester when the shooting occurred. However, his older sister, Kelly, was well known in the school community since she was an outstanding student (a valedictorian), an active member of the marching band, a regular contributor to the school newspaper, and a member of the choir. Kelly was a senior during Michael's freshman year, and their parents, John and Ann, were heavily involved with the school through support of Kelly's activities. They accompanied the band on field trips, helped at the concession stand, and in other ways demonstrated their support for the school. In this the Carneals were not unusual; relative to the high schools with which we are familiar, the level of parental engagement in extracurricular activities at Heath is exceptionally high. Parents know many kids other than their own, and parents know one another as well. Kelly joked that her parents were at the school more than she was. Participation in the church was equally important to the Carneal family, as it is to most of the families in the area, where religiosity is highly valued and the church is a center of social activity.

John Carneal is a long-time unemployment compensation and injury lawyer, and Ann is a homemaker with some postgraduate education. Paducah is a mixed-class community in which professionals like lawyers are at the top of the social pecking order. However, they were not known as snobs. On the contrary, John Carneal's practice was described to us as "solid" but not overly "flashy"; he mostly represents the hard-working individuals who are considered the bedrock of the town. The Carneals were described by most people as sincere, generous, and actively involved in all aspects of community life, particularly the domains of greatest interest to their children. They opened their home to their children's friends, who were frequent guests at the Carneal family dinner table. Their home was something of a hub for kids, some of whom were experiencing typical teenage conflict with their own parents. Indeed, one of the shooting victims spent time at the Carneal's in part because she was at odds with her own parents.

Hence the atmosphere surrounding Michael Carneal was that of a well-educated household with high expectations for academic performance and an older sister who had excelled. It was a sociable family that appreciated the importance of participation. The Carneals were, in this respect, proper and conventional, in keeping with normative cultural practices of the Paducah community. At home and with his family, Carneal appeared to be a fairly normal 14-year-old boy, but in other social contexts, especially the harsh social world of high school, he was uncomfortable and self-conscious, constantly looking for approval and respect from both youth and adults.

Michael Carneal's friends described him as a jokester and a prankster, always looking for attention and trying to win friends. His small size made him a frequent target of teasing and occasional bullying, although he was also known for teasing others himself. He stole CDs and other items and gave them to students at school because he thought it was cool to steal. He gave students his own possessions and told them they were stolen. Carneal downloaded pornography from the web along with pages from the *Anarchist Cookbook* and sold them or gave them away. He also stole hundred dollar bills from his father's wallet and gave them to students at school.

His teachers described him as intelligent, obedient to authority, forgetful, restless, and less socially skilled than average but someone who had friends. Although he tested at an IQ of 120, his grades slipped in the eighth grade, a slump that did not go unnoticed at home. In the fall of his freshman year in high school, his grades improved again to three Bs and an A. While a respectable performance, it may have been perceived as below his capabilities. Michael Carneal was in the marching band, a focal point of life for many students at Heath in part because it was a competitive activity that involved traveling to other schools. However, his career in the band was not altogether successful, since he and one other student were chosen to sit out of early competitions because the band did not have enough uniforms.

Carneal was also interested in video games and computers and appeared to have spent much time in the middle of the night using the

Internet. He was also an avid visitor to chat rooms and devotee of email. The contents of his hard drive, which were seized by the police, suggest that he made a habit of visiting web sites that were unsavory by local community standards, including some that were pornographic. It is not clear to us how far his Internet habits deviate from that of ordinary teenage boys in this respect. However, his taste for violence—including sites that included how-to instructions for making weapons or rehearsals of violent attacks—fell outside mainstream norms. Certainly his own writings—composed for himself and for classroom assignments—began to reflect a fixation with violence.

Yet Carneal departs from the stereotype of the loner obsessed with computers in many respects. He had several friends from middle school and from band, including shooting victim Nicole Hadley, as well as friends from his neighborhood. These friends accompanied his family on occasional trips, and he slept over at the houses of the boys on occasion. At least in outward respects, and as far as his parents knew, he had friends. What he lacked was a crowd of his own. He was a fringe figure in a number of groups (e.g., band) but was central to none. He did not have very close friends, but it is not clear that this is atypical for 14-year-old boys. Nonetheless, Carneal was clearly searching for a crowd that would define him as more central and undertook various ventures (stealing, giving away pornography) in order to impress one group that seemed particularly attractive to him, the "Goths" or "freaks" as they are described by their detractors.

A few of these youth wore long black jackets and other trappings of Goth clothing and makeup. In this and other ways, the group attempted to stand out from the conventional crowd at Heath. Kelly Carneal described these students as purposefully antisocial: they realized how silly the social pecking order was in high school and refused to participate in it. They were known for rejecting what they regarded as the pious attitude of the prayer group, but they never did more than grumble about it. While this group was certainly noticed for its "statements," school administrators did not regard them as threatening or seriously deviant.

Another way in which Carneal defies the loner stereotype is in his relations with girls. He had had a girlfriend and was friends with several female classmates. Prior to the shooting, he broke up with his girlfriend because he was interested in Nicole Hadley, one of the shooting victims. Freshman boys often find it difficult to locate themselves in the social landscape because they are the youngest in the school and cannot easily compete with older boys. Carneal's slight stature did not improve his chances, but it is notable that he did indeed have something of a social life even though he was a freshman. Hadley took a particular interest in Carneal because she thought she could influence him to "come to God," a direction that he did not embrace. Hence it may be that what at least some girls wanted from Carneal did not match exactly what he wanted from them, leading perhaps to some frustration on both ends.

In the months and year prior to the shooting Michael Carneal began to show signs of mental illness, but it was not diagnosed as such until after the event. However, he did accumulate five discipline infractions during his 71 days at Heath High School. First, he was accused of using a library computer to look at the *Playboy* web site with a friend. Thereafter, at his mother's request, both he and his sister were forbidden from using school computers. Second, he was caught chipping at the paint on the wall with a leather punch tool and received two days of flex-time detention. Third, he was disciplined for scratching another boy on the neck while they were marking each other with pens. Fourth, he stole a can of food from the pantry in life skills class. Finally, Michael Carneal reported to the psychiatrists and psychologists evaluating him that he was caught in school with a pair of plastic "numchucks" he had purchased from a vending machine.

These infractions represent a pattern of fairly minor behavior problems that may have first surfaced when Carneal was in middle school. Heath Middle School administrators reported having no problems with Carneal, although students in classes with him in the eighth grade reported that he had set off a stink bomb at school and that he had taken fish out of a fish tank and stomped on them. The current principal of the middle school confirmed that a stink bomb had been set off at the school but they did not discover who was responsible. Whatever difficulties he might have been experiencing in eighth grade (when his grades slipped) were mild enough to stay "below the radar screen." As such he was never identified as a problem student. On the contrary, the school faculty who knew Carneal said they were completely shocked when they learned he was the shooter.

THE VICTIMS

Carneal appeared to have no cause for ill will toward his victims. They were not the students he described as having bullied or teased him. The school principal described all of them as humble and quiet students who did nothing to draw attention to themselves. Three were killed:

—Jessica James, a 17-year-old senior, played flute in the school band, was a member of the Agape Club, a Christian fellowship group, and attended Kevil Baptist Church. She was described by the principal as a strong student.

—Kayce Steger, a 15-year-old sophomore, played clarinet in the school band, played on the softball team, and was a member of the Agape Club. She was an honor student, worked at Subway, and attended 12th Street Baptist Church. She was a member of Law Enforcement Explorers Post 111 and hoped to be a police officer. Her parents reported that Michael Carneal had asked her out on a date a little over a month before the shooting.

—Nicole Hadley, a 14-year-old freshman, played in the school band and on the freshman basketball team. She was a member of the Heartland Baptist Worship Center and the Heartland Baptist Youth Group. Her family had moved to Paducah from Nebraska the year before the shooting. Nicole was a good friend of Carneal. They had "walked together" at middle school graduation, a custom of some significance in the community whereby graduates choose someone important to them to accompany them during graduation ceremonies. Nicole had been to the Carneal home and felt friendly enough toward Ann Carneal to confide her hopes that Michael would become more religious. Students at the school reported that Carneal probably had a crush on Nicole. Her parents reported that Michael Carneal called her almost every evening, supposedly to discuss science homework, but they believed he just wanted to talk to her. Although she was clearly annoyed with his constant calling, she always talked to him, believing that she could help him find God.

Five other students were injured, none of whom had relationships of any significance to Michael Carneal:

—Shelley Schaberg, age 17, was described by the principal as the school's best female athlete. Voted Miss Heath High School by the senior class, Shelley was homecoming queen. Though her injuries from the shooting prevented her from playing basketball, her college honored her basketball scholarship and she went on to play college soccer.

—Melissa "Missy" Jenkins, age 15, was president of the Future Homemakers of America. She was paralyzed from the waist down in the shooting. She and her twin sister, Amanda, were featured in *Christian Woman* magazine several months after the shooting.

—Kelly Hard, age 16, was a member of the softball team and the Future Homemakers of America. She transferred to the local Catholic school the year after the shooting.

—Craig Keene, a 15-year-old freshman at the time of the shooting, was a member of the Agape Club, the band, and the basketball team.

—Hollan Holm, a 14-year-old freshman at the time of the shooting, was a member of the Academic Team, the Spanish Club, and the Science Olympiad. In his valedictory speech at the class of 2001 graduation, he reminded his class that they had lost not one but two members on December 1, 1997,: Nicole Hadley and Michael Carneal.

THE SHOOTING

According to accounts that Carneal gave to police and to psychiatrists, he stole his father's .38 special pistol from a locked box in his parents' bedroom closet several weeks before the shooting. He later told mental health professionals who evaluated him after the shooting that he considered using it to kill himself, but he did not want to hurt his family. He brought the gun to school, showed it to several classmates, and told them he wanted to sell it. An older boy heard that he had the gun and told Carneal that if he did not sell it to him, he would tell the police. Carneal gave him the gun with the promise that the boy would pay him later, but he never did.

The shooting took place on the Monday after Thanksgiving. According to Carneal, a few days before, he snuck into a friend's father's garage and stole a .22 pistol and ammunition, the gun he ultimately used in the shooting.[4] He had previously fired guns with this young man and his father. Carneal brought this gun to school the day before Thanksgiving break, again seeking to impress his classmates, but the ploy did not work because they said the gun was "small." None of the students reported to school authorities that Carneal had a gun. They would later say that they did not think he had any bullets and did not think he would do anything with it. Carneal often had strange things in his possession and commonly showed them to people for the purpose of getting attention.

The week before Thanksgiving, Carneal warned students that "something big is going to happen on Monday" and even warned some specifically to stay away from the school lobby. No one took him seriously. They thought at worst he would set off another stink bomb or a cherry bomb. Students reported that Carneal had made empty threats and issued warnings many times before.

After school on the Wednesday before Thanksgiving, Carneal went to a friend's house, and they used the pistol for target practice on a rubber ball. On Thanksgiving Day, Carneal and his family, including both sets of grandparents, shared dinner in the early afternoon. Carneal then went outside to rake leaves and later rode his bicycle to the house of the friend from whose father he had stolen the pistol. After a brief conversation with Carneal, the friend went next door to a relative's house to eat Thanksgiving dinner with his family. Carneal took this opportunity to climb into the garage through an open window, found the hidden key to the gun case, and stole a 30-30 rifle and four .22 rifles. He also stole earplugs and many boxes of shotgun shells and .22 ammunition. He carried the weapons home in a duffle bag on his bicycle, which he left outside under his bedroom window and came into the house. He went into his bedroom, locked the door, opened the window, retrieved the duffle bag, wrapped the guns in a blanket and put them under his bed. He would later tell police that Thursday, Thanksgiving Day, was the day he decided to bring the guns to school.[5]

On Friday evening, Carneal, his father, and a family friend went to a basketball tournament at Murray State University. On Saturday afternoon, Carneal put the 30-30 rifle and two of the four .22 rifles in his duffle bag and rode his bicycle to the house of a friend. He showed the guns to this young man and his older brother, a senior at Heath High School, who cautioned the two boys not to get into trouble. In the evening, John Carneal came to the house to pick up his son and his friend, who spent the night at the Carneal home. Michael Carneal left the three guns at his friend's house because he was afraid that his father would discover them if he put them in the car. The two played video games and watched television. Sunday afternoon Carneal did his homework and that evening stole two old shotguns from his father's closet and hid them under his bed.

Carneal's family described the morning before the shooting as a typical Monday morning. Carneal told his parents and sister that the large bundle that he brought to school contained props for a skit he was going to do in English class that day. In reality, it contained the two shotguns and two .22 rifles wrapped in a blanket. He had the pistol, the earplugs, and hundreds of rounds of ammunition in his backpack. He drove to school with Kelly, entering through the back door. As he walked through the band room, the band teacher asked him what was in the bundle, and again he said it contained props for an English project. He walked to the school lobby where the Goths hung out before school.

Each morning before school, a group of about 25 to 30 students gathers in the Heath lobby to say a short prayer. When the leader says, "Time to pray," the students join hands, have a one to two minute prayer, and then go to class. The prayer group draws a wide cross-section of students. Some are athletes, many are band members, and many are freshmen. The small group of students that Carneal hung out with stayed to the side of the lobby and did not participate in the prayer group.

When Carneal arrived in the lobby, he walked through the Goth group, and put his bundle down by the wall. One of the students asked what was in the bundled up blanket and Carneal again said it was his English project. Students later reported to police that one student remarked that the sound of the bundle hitting the floor suggested that it contained guns rather than an English project. After this remark, though, someone changed the subject, and no one talked to Carneal. Carneal later said that he was thinking, "You've got to do this for yourself." He put his backpack on the floor, put in his earplugs, and put a clip in the pistol and pulled it out of his backpack. Just as the group was finishing its morning prayer, Carneal slowly fired three shots and then five in rapid succession, making an arc around the lobby. He would later say that he was not aiming the weapon but simply firing into the crowd. Every bullet struck a fellow student. Kelly Carneal was not in the prayer group but witnessed the shooting. She said later that she would not even have recognized her brother if she had not seen his clothes and his face. His posture was different and he seemed larger than his normal self when he was holding the gun. Carneal told one psychiatrist that he felt like he was in a dream.

Carneal stopped firing when he saw bullet holes in the wall and Nicole Hadley laying on the floor covered in blood, with another friend of Carneal's calling her name. He put the gun on the ground. When the student leader of the prayer group came over to Carneal, yelling at him, Carneal reportedly asked the youth to kill him. The principal, Bill Bond, came running out of his office, pushed the prayer leader away and led Carneal to the office, putting him in the conference room with Carneal's homeroom teacher, who reported that Carneal did not seem to recognize what he had done. He asked many questions, but Carneal would not answer them. The only thing he would talk about was the guns: where he got them, what kind they were, and where the ammunition was. He could not say what he had planned to do. When the school bells rang, students were told to go to class, and they did, including at least two of the victims who were wounded.

One important question that still remains unresolved is whether other students were involved in planning the attack. Two days after the shooting, the county sheriff stated to the media that he believed other students conspired with Carneal prior to the shooting because, he reasoned, Carneal could not have used so many guns himself but must have anticipated others helping him. Many of the police interviews with Carneal and with witnesses were directed at determining whether students from the Goth or freak group were supposed to have picked up the other guns and joined Carneal. Carneal had four other firearms, multiple sets of earplugs, and hundreds of rounds of ammunition with him the day of the shooting. While he denied any sort of conspiracy in formal interviews with the police and his lawyer, he did reportedly talk to a police officer transporting him between one jail and another about how the group had planned the shooting and the other students were supposed to help him take over the school. As hearsay, these statements were not admissible in court and later Carneal retracted the statements, insisting that he had acted alone and that he had just been telling the police what he thought they wanted to hear. In addition, one of the suspected co-conspirators was seen staring at the crime scene and smiling immediately after the shooting. Some of the suspected co-conspirators were evasive in interviews with the police. According to the prosecutor, there was never enough information to charge anyone else with a crime.

Police interviews suggest that Carneal and his friends had fantasized about taking over the school or taking over the local mall. They discussed what they would do and what kind of weapons each would use and where they might get them. However, the group did not discuss any details, and Carneal stated that no one knew what was going to happen on Monday except him. He had warned people that "something big is going to happen," but would not tell anyone, including the Goths, what it was. Carneal may have held out hope that his friends would join him once he started shooting, a viewpoint that may have been influenced by his mental illness (to be discussed below). When pressed on the details of the plan, he admitted there were none and there was no way that his friends would know what to do if they wanted to join him. The McCracken County Sheriff's Office and the Kentucky State Police never uncovered any evidence of any specific information being exchanged about a plan to take over the school. However, there is enough ambiguity in the situation to fuel community speculation that Carneal was not the only person involved or responsible for the outcome.

CAUSES

Why did Michael Carneal shoot eight of his fellow students, none of whom he particularly disliked? In this section, we evaluate the evidence for and against seven interrelated theories: family factors, gun culture and availability, bullying and teasing, peer and social relations, mental illness, exposure to media violence, and adult-child relations. Any complete account would draw on all of them. Ultimately, none of these theories alone is completely satisfactory, since each points to necessary but not sufficient conditions. There are thousands of communities similar to Heath that have not suffered from lethal school violence. And of course it is also likely that entirely different factors may have produced similar tragedies in other communities. Nonetheless, because each of these ingredients was part of what became an explosive stew, they bear further discussion.

Family Factors

On the outside, the Carneal family is the opposite of the stereotypical dysfunctional family. John Carneal held a steady job that provided more than ample income and allowed his wife, Ann, to stay at home to care for their children. Kelly Carneal was very successful, both socially and academically. They regularly ate family meals together and went on family vacations. Michael Carneal's parents were heavily involved in the school and its extracurricular activities. Almost all of the people we talked with had nothing but positive things to say about Carneal's parents and sister. [6] There is no indication that the Carneals were abusive, either physically or emotionally.

Because we were not able to interview either Michael or his parents, our understanding of their internal family life relies exclusively on information gleaned from civil depositions, psychiatric and psychological reports, and our own interview with Kelly Carneal. Our portrait therefore cannot be considered definitive but is suggestive of some of the problems Carneal may have been experiencing as he moved down the pathway toward the shooting. In many respects, John and Ann Carneal were a model of concerned and involved parents who were highly supportive and involved in their children's lives. However, there is also evidence to suggest that there were tensions caused by differences between the siblings, parental expectations for academic performance that Carneal was not fulfilling (particularly relative to his very successful sister), and a general sense (from his perspective) that all the attention went to Kelly. Carneal's minor infractions and slumping grades had indeed generated concern from his parents, since this pattern of behavior was not in keeping with their expectations or family reputation. Perhaps for this reason, Carneal felt that he could not go to his parents for help with the social problems he was having at school or about his fears of imminent harm (described below).

The Carneals were sometimes defensive on their son's behalf or perhaps disbelieving when their son began to misbehave in school. When he was disciplined for looking at inappropriate material on the Internet using a computer in the school library, school officials reported that his mother did not believe that he was guilty. Carneal was also engaging in many behaviors in the home of which his parents were unaware. Evidence from the family computer seized by police showed that he had been up in the middle of the night using the Internet. The computer hard drive contained pornographic and violent materials, and Carneal had printed out this material on the family computer in the living room and sold it or gave it away at school. He had stolen his father's guns and stored them and other stolen guns in his bedroom. Carneal had some of the family's kitchen knives under his mattress and reported stealing $100 bills from his father's wallet.

Critics might argue that John and Ann Carneal were losing touch with their son and that their lack of awareness of his activities was indicative of too much distance between them. It is important to note, however, that part of the adolescent experience involves growing autonomy and privacy. Parents who know absolutely everything about their teenagers' lives would probably be regarded as micromanaging to the point of infantalization. It seems clear that the Carneals were trying to give their son some space to grow up. At the same time, Carneal himself was deliberately evasive and concealing in ways that were obviously effective. He did manage to hide from his family his own growing mental illness and his plans for violence. Despite his parents' involvement in activities at school, in church, and with friends, he made sure that they knew as little as possible about the things he did that had the potential to get him into serious trouble.

Gun Culture and Availability

Some have suggested that the gun culture of the South has played a role in some of the school shootings. This theory hypothesizes that guns and violence are culturally acceptable means of resolving disputes and solving problems in this part of the country and that norms about family honor and masculinity demand retribution for insults. As episodes of lethal school violence have spread to the Western and Northern states, this theory has waned somewhat in popularity. However, it should be examined in the context of this example.

Although Carneal saw bringing a gun to school as a way of getting attention from his peers, we did not find evidence of a prominent role for guns in the local culture of the adult community or in the mainstream adolescent culture of the school. Hunting has declined considerably in Paducah and McCracken County as the area has urbanized, and many students we talked to had never held a gun. The older generation of adults, in contrast, spoke of storing their guns in their trucks during the school day in their youth so they could go hunting after school. One man described showing off his gun to other students in shop class when he was in high school 20 years ago. In contrast, our respondents said that, even before the shooting, weapons could not be publicly displayed in school, would be confiscated if found, and might even result in criminal penalties. Carneal's family did not hunt and his sister did not even know that the family owned any firearms, although Carneal did first learn to shoot a gun at a 4-H summer camp.

However, the availability of guns should be considered a contributing factor in this case. First, a young person in this community who wants a gun can get one. Almost all of the students we interviewed said it would be easy for them to get a gun if they wanted one because either a family member or a friend owned a gun to which they could gain access. Most parents concurred. Over a period of less than two months, Michael Carneal amassed an arsenal of nine weapons and thousands of rounds of ammunition by stealing them from his father and a neighbor. Second, the availability and ubiquity of guns appears to have normalized guns in this community. This may help to explain why none of the students who saw Carneal with a gun in school the month before the shooting said anything to adults about it, even though Carneal told them that "something big is going to happen on Monday."

Bullying and Teasing

When asked by police why he shot his fellow classmates, Carneal's first response was that he was tired of being picked on. Then, and in later conversations with psychiatrists and psychologists, he detailed patterns of harassment going back to elementary school. Some were relatively minor, like having water flicked on him in the bathroom or being called "four eyes" (for wearing glasses). Others were more serious, including one incident in which he was allegedly hit on the back of the head, and another in which he was "noogied" until his head bled. There was also mild hazing that came with joining the band. On one band trip, Carneal was rolled up in a blanket and hit with sock balls by older band members before a chaperone intervened. Carneal reported that he was picked on by older band members nearly every day.

He also reported being called "gay" and a "faggot" multiple times daily after the publication, in eighth grade, of a student newsletter that noted in the "Rumor Has It" gossip column that he liked another boy. Numerous past and present students at the school said that they did not know anyone their age who was openly gay, and that it was a source of shame to have acquired that label. While such a stigma would be hard for an adolescent boy to manage in any part of the country, it is possible that the conservative social mores of the South make the accusation of homosexuality particularly difficult. Carneal told his therapist that this bullying had increased significantly after the publication of the column. This was one of the factors that precipitated Carneal's slumping grades in eighth grade.[7]

There is a considerable discrepancy between what Carneal reported about harassment or teasing and the views expressed in our interviews with his fellow students. From his classmates' perspective, Carneal was not picked on any more or less than other students, and he quite consistently picked on other students himself. Because he was loud, a prankster and a jokester, many of the other students thought he was better able to defend himself than other, quieter students. How do we reconcile these two images of Carneal? We would largely agree with one of the psychologists, who, after several lengthy interviews with Carneal and conversations with a number of his friends at school, concluded: "Michael's experience of being teased and harassed at school stems largely from real events and mistreatments by his peers. However, it was also the case that, because of his personality and mental condition, Michael was sensitive to feeling mistreated, and may have reacted strongly to incidents that other students were able to tolerate." [8] When asked by psychiatrists whether he thought about going to jail, he said he thought jail would be better than school because at least the teasing would stop. Because of his paranoia and general inability to read social cues (see the section on mental illness, below), Carneal probably magnified the impact of very real ill treatment such that it provided a motivation for the shooting.

However, it is important to keep in mind that there is no known history of Carneal being teased by any of the students he shot in the prayer group and that the students he identified as his nemeses were not in the prayer group. We would suggest that, as Carneal himself has said, the shooting was not retribution for past wrongs done to him. Instead it was an attention-getting act that he thought would bring him the power and respect that he deserved.

Peer and Social Relations

Some have painted school shooters as children who are particularly socially isolated, have few or no friends, and have trouble making friends of the opposite sex. Michael Carneal fits this stereotype in some respects but not in others. Although many people said they considered him a friend, he thought of himself as someone without any friends, someone whom no one at school liked. He had an ex-girlfriend who said she loved him, even after the shooting. However, he had difficulty forming close relationships with others and was socially insecure about the relationships that he did have. He was not close to his sister or any relatives outside his immediate family. In middle school he considered suicide on a number of occasions, and in August 1997 he intentionally cut himself on the forearm but told adults the self-inflicted would was from a bicycle accident. Undoubtedly his perceived alienation from those around him should be considered one cause of the shooting.

He also expressed considerable antagonism toward the school's "preps," a loosely defined group of popular students. He had made buttons that said, "preps suck" and had written a story called "Halloween Surprise" in which preps are attacked with grenades and a shotgun by the brother of a boy named Mike. Carneal said in depositions that he objected to the preps because he thought they acted as if they were better than other students. However, while there are some preps in the prayer circle, on the whole it seems that the prayer circle is heavily composed of students from the band, at least some of whom were also Carneal's friends. As one respondent plainly put it, "If he wanted to shoot the preps, he would have been upstairs where the preps [hang out]."

As is true in most high schools, there is social antagonism at Heath between groups whose values or styles clash. The Goth crowd into which Carneal hoped to become better integrated expressed derision toward the "good" kids in the prayer circle. Some students said that the group was hypocritical, because they professed beliefs that they did not live up to. (One respondent cited an example of members of the group signing a pledge to wait for sex until after marriage, even though a number of the signatories were already known to have had sex.) The Goth group signaled its disdain by intentionally talking loudly during prayer time. One friend reported to the police that Carneal had said the previous Wednesday "that the hypocrites in prayer group were going to go down, 'cause he was going to bring 'em down." At the same time, it is important to remember that Carneal was friends with at least some of the students in the prayer circle through band, including Nicole Hadley, whom he later described as his closest friend.

Statements by Carneal and others in the psychiatric and psychological reports indicate that he was envious of other students who were more successful socially and academically, and this broad category included both preps and "good kids." Some of these feelings seem to have been exacerbated by Carneal's perceived inadequacy in comparison to his sister, who was a school valedictorian and popular among her peers. She was also a senior member of the band, whereas Carneal was one of only two of the 62 members who initially did not march in the band competition because there were not enough uniforms. This was a source of embarrassment to him, but later two students dropped out, and Carneal was allowed to march.

Shortly after the shooting, Kelly Carneal told one of the psychiatrists involved in the case that Michael "tries to be as good as me, and he can never size up," although she emphasized that she never felt she really knew her brother.[9] Carneal was occasionally asked by teachers and others why he could not perform as well as his sister had. To escape these unfavorable comparisons, Carneal apparently had begun to adopt an alternative identity that would protect him from these unflattering comparisons. He told one of the psychologists, "Everybody talked how I was not like my sister, so I figured if I was the exact opposite, people would pay attention to me more."[10]

Carneal found a home for this opposite identity in the group of mostly older Goth students, who dressed unconventionally and flouted adult authority in dress if not in action. Some espoused anti-Christian attitudes. The leader of this group dressed in black with a long trench coat, painted his nails black, and was known for having a spork (a spoonlike fork) attached to a chain that he wore. In attempts to win approval from this group, Carneal stole $100 bills from his father and a fax machine from the home of a friend to give to them. At the time of the shooting, the police believed that at least some of these boys were part of a larger plot to participate in the shooting. Carneal and some of these boys did fantasize together about taking over the mall or the school, but there is no evidence that these boys thought that was more than fantasy or that the fantasy involved actually shooting anyone.

One of the psychologists believes that, at the very least, the older boys manipulated and encouraged Carneal to engage in a variety of illegal activities. When he read the psychologists evaluation, Carneal said that while he agreed that he attempted to gain favor among this group, they did not encourage him in his actions. While it is always possible that Carneal is still protecting his former friends, the kid from the "good" family was not simply a victim of manipulation. Instead, given the consistency with which we heard Carneal described as a "wild" youth, one who stole regularly even in middle school, it seems more likely that Carneal was simply trying to win attention and approval from a new peer group. We have no way of knowing with certainty the internal dynamics of this group; particularly how, if at all, they contributed to Carneal's decision to proceed with the shooting.

Overall, it seems that Carneal was at a particularly difficult and fragile moment in the development of his identity—academically, socially, and sexually. Coming from a high-achieving family, he seems to have been ambivalent about his middling academic performance, simultaneously striving to improve his grades from middle school and aligning himself with a group that rejected such standards. While he was becoming friends with the unconventional Goth group, he also retained video game-playing friends from middle school and made friends with band students, many of whom fit the "good kid" stereotype that he derided with his Goth friends. He had a crush on Nicole Hadley, who was in the band, a devout Christian who was trying to persuade Carneal to become more religious.

Carneal believed that he did not have the status that he wanted in any of the social groups with which he was connected. He was not the student his sister was; he was the youngest, newest, and least-well known member of the Goth group; he was one of two students who did not get to march in the band; and his romantic interest in Hadley went unreturned. In short, despite the fact that he had a number of roles in different groups, he had not found a successful niche of his own. Given this context, it seems reasonable to assume that Carneal was neither targeting the Christian students nor the preps who tormented him. Rather, the shooting gave him a very public way of asserting power and winning respect from all of the groups in which he felt only marginally included.

Mental Illness

Michael Carneal had never been diagnosed with a mental illness before the shooting. After the shooting, a history of mental illness on his father's side of the family was uncovered. When he was evaluated after the shooting by forensic psychiatrists and psychologists, two separate defense experts found him to be able to understand the consequences of his actions but mentally ill at the time of the shooting, while the prosecution's team determined that he was not mentally ill. He was diagnosed with depression ("dysthymia") and schizotypal personality disorder by Dr. Dewey Cornell of the University of Virginia and with dysthymia and "traits of schizotypal personality disorder with borderline and paranoid features" by Dr. Diane Schetky of Maine. Drs. Elissa Benedek, William Weitzel, and Charles Clark, hired by the prosecution, concluded that "Michael Carneal was not mentally ill nor mentally retarded at the time of the shootings." [11]

All of the reports, however, detailed similar odd behaviors, paranoia, and trouble interpreting social interactions correctly. Michael Carneal reported unreasonable fears. He thought people were looking at him though the air ducts in the bathroom, and worried that if he touched the floor in his bedroom, he could be harmed by assailants lurking under the floor. He often announced when entering his bedroom, "I know you are in here." He often thought he heard voices calling his name or calling him stupid, but recognized that he might be imagining them. Before the shooting, he told one psychiatrist that he thought he heard people in the prayer group talking about him. He feared going to restaurants because he thought his family would be robbed. As he, his father, and a family friend walked across a quiet college campus the Friday before the shooting en route to a basketball tournament, he remarked, "Boy, you could really get mugged out here." According to the experts' reports, he never told his family about his fears because he knew them to be unreasonable.

Some of Michael Carneal's fears translated into strange behaviors. He covered himself with at least six towels whenever he took a shower and covered the air vents with towels as well. He often slept in the family living room. Knives from the kitchen were discovered under his mattress after the shooting. He reportedly hopped on top of the furniture to avoid touching the floor in his bedroom.

While we cannot determine whether or not Michael Carneal met the diagnostic criteria for mental illness at the time of the shooting, we believe that his paranoia, fears, and misreading of social cues contributed to the shooting. They magnified the extent and meaning of the teasing and bullying that occurred at Heath High School (and probably occurs at all middle and high schools). He misinterpreted group fantasizing about taking over the school and the mall as a genuine desire to do so. Carneal also may have incorrectly assessed the willingness of his friends to participate once the shooting began. His social insecurities led him to believe that he was not respected by his peers and that bringing guns to school would earn him respect. His depression magnified these fears and may have affected his judgment.

However, he was also able to conceal the extent of his fears so that even those living with him had no idea what was going on in his head. Evidently, this is not unusual among adolescents who later develop full-blown cases of psychiatric disorders, which are notoriously difficult to diagnose in 14-year-olds. He believed that anyone he turned to for help would be attacked by the same "demons" who were, by this time, threatening him. This goes some distance toward explaining why he did not turn to his family for help.[12]

Whatever its extent at the time of the shooting, Carneal's mental illness has intensified in the time since, developing into a full-blown paranoid schizophrenia, according to treating psychologist Dr. Kathleen O'Connor. After the Columbine shootings he fell into deep psychosis, blaming himself for those shootings, and twice attempted suicide. He has been extensively treated with medications for his psychosis, because, according to the psychologist, who worked in tandem with a psychiatrist in treating him, verbal therapy was not effective with someone whose problems were so severe: "When someone is as paranoid as he became, talking just wasn't going to get to it. We had to get the psychosis into remission . . . with medication."[13] O'Connor believes that mental illness was a primary factor in the shootings, that Carneal committed the homicides in part because he was unable to continue functioning in normal society, and that the shooting relieved him of the need to do so.

Exposure to Media Violence

Carneal was undoubtedly exposed to a variety of violent video games, movies, and imagery. He had been playing violent video games since he was a child, according to his reports and those of friends. These games included Mortal Kombat and MechWarrior, favorites of millions of American teenagers. Carneal told a psychologist that he preferred "thinking games" and frequently played chess with his father.[14] When asked in a police interview the morning of the shooting whether he had read or seen anything like his shooting spree, he mentioned the movie Basketball Diaries.[15] In the movie, the lead figure takes revenge on a Catholic school priest who had abused him by shooting the priest and a number of his classmates to the cheers of some of his friends. While this was widely reported in the media at the time, Carneal later denied that the movie had played any part in the shooting. Carneal then said that he only brought up the movie because they asked him whether he had seen or read anything like what he had done. He volunteered that although he had seen it several years before, it had not made much of an impression on him. He added that, "it makes me mad" that people are trying to explain his actions in terms of a movie.[16] Carneal had not heard of the shooting earlier that fall in Pearl, Mississippi.

While there is clearly no one-to-one correspondence between exposure to violent video games and behavior,[17] we agree with one of his psychiatrists that "Michael's exposure to media violence can be regarded as a factor which contributed to the attitudes, perceptions, and judgment which led to his violent behavior."[18] One of his teachers recalled that Carneal's solutions to hypothetical problems often involved "shooting someone with a bazooka." Carneal and his friends discussed a number of violent fantasies that were in part based on things that they had seen. His email handle was "Loco," which he said was based on a character in a movie. Thus while it would be far too simplistic to say that Carneal's actions were caused by the movies that he saw, it does seem likely that Carneal's thinking was more generally shaped by these influences and thus could be considered a contributing factor to the shooting.

Adult-Child Relations

Finally, while this is difficult to establish with certainty, as in most communities, it seems there is a profound disconnect between the experiences of adults and a small minority of disaffected teens in the Heath community. While separation from and conflict with adults is and always has been an important part of adolescence, living in a sports-oriented, tightly knit religious community could have seemed particularly constraining to a boy like Michael Carneal. For instance, he wrote in an email to an Internet friend from California the year of the shooting, "Our town really SUCK [sic]. Every year we have this big QUILT FESTIVAL where about an estimated 50,000 old bags in snitty cars that drive about 10 to an amazing 20 miles per hour come to town for a week and we all go [to] town and freak out the old lady quilters. . . . Okay, my point is that there is nothing here." A point of pride and celebration among adults, the quilting festival seemed like an anachronism to a disaffected teen. Carneal was sent to 4-H camp, like many other youngsters in the community, but rejected the discipline and authority that came with it. He went to church and was confirmed, but he had on his computer a downloaded document called "Bible Inconsistencies" which discussed how different passages of the Bible contradicted one another.

Paducah and Heath probably have as much or more social capital as any town in America: there are dozens of clubs and activities that bring children and adults together. This is an enormous source of pride and enjoyment for many in the town, youth as well as adults. At the same time, these strong forms of community may be quite confining for youth who have other interests, by limiting the array of social options available. The broader community's interest in the fortune of the sports teams at the school may have also reinforced the perceived permanence of a social hierarchy in which Carneal, as a marginal band member, was near the bottom. No questions were ever posed to Carneal by local police and lawyers about his perceptions of the community, so it is difficult to evaluate with confidence how important this was to his thinking at the time of the shooting.

CRIMINAL ADJUDICATION

Carneal was charged with three counts of murder, five counts of attempted murder, and one count of burglary. He was 14 at the time of the shooting and was charged as an adult, as is customary in Kentucky when a youth of his age commits murder. Prosecution and defense prepared for a trial. Because the facts of the shooting were not in dispute, the major issue of contention was Carneal's mental health. Mental health experts from both sides examined Carneal, and reached slightly different conclusions (stated above), but neither found that he was legally insane at the time of the shooting. Since an insanity defense was not possible and a trial would be an ordeal for the many witnesses, the defense agreed to a plea bargain the morning before the trial was to commence.

Carneal pleaded guilty but mentally ill to all charges and received life without parole for 25 years, the maximum sentence permissible under law given his age. The mentally ill plea does not reduce Carneal's culpability under the law, but it does flag him as someone who would need psychiatric treatment during his incarceration. Carneal's plea was also pursuant to *North Carolina* v. *Alford* (known as the Alford plea), which allows the defendant to avoid admitting guilt while acknowledging sufficient evidence for conviction. The Alford plea is frequently utilized in sexual harassment cases to protect the defendant's reputation. According to one of the Carneal family attorneys, employing the Alford plea in this case might be advantageous in the civil suits that were looming.[19]

Tim Kaltenbach, the prosecutor in the case, said that he accepted the deal because Carneal had agreed to the maximum sentence. Most of the people in the broader community with whom we spoke felt that the sentence was fair or had no opinion about it and were glad that it had been settled without a trial. However, some of the families of the victims felt that they had not been sufficiently consulted by the prosecutor. They were also unhappy that the Alford plea did not require Carneal to admit guilt, and they argued that the resolution of the matter should have included Carneal taking full responsibility for his actions. There was also some sentiment among the families of the victims and others in the community that Carneal had received favorable treatment by the legal system because his father is a lawyer. In the end, he received the maximum sentence, but the perception of favorable or insider treatment was one motivation for the civil suits that were yet to come.

COMMUNITY IMPACT

Everyone that we talked to in Paducah emphasized that the community "came together" after the shooting. In a devoutly Christian community, much of the public emphasis was on healing and forgiveness. The day after the shooting, students at Heath High School taped a huge banner along the front entrance to the school saying, "We Forgive You Mike." As one would expect in a community rarely touched by criminal violence, the Heath community saw the event as the result of a deeply troubled youth gone wild rather than any larger social problem. Almost everyone we talked to was unable to offer an explanation for the shooting and viewed it as something that could never be explained.

The immediate and public emphasis on forgiveness put tremendous pressure on the families of the three slain girls, still reeling from the loss of their daughters. For them, a long period of introspection, guilt, and acceptance of personal responsibility on Carneal's part is required before forgiveness can be dispensed. Short-circuiting this process made them feel they were being pushed to put the shooting behind them before this moral account was cleared.

The families of the victims were showered with monetary gifts and sympathy, and friends and neighbors brought food. The day after the shooting, more than 200 students attended the prayer circle, and three days later 2,000 mourners filled the largest local church for the memorial service honoring the three slain girls. One student put together a web page honoring the trio, including photographs, tributes written by other students, and an address where donations for a memorial fund could be sent. An older gentleman told us that the spontaneous unity that emerged in the town after the shooting could compare only to the feeling of patriotism that he remembered as a child during World War II. A local business donated t-shirts with the slogan, "We Believe in Heath."

School officials remarked that the best decision they made after the shooting was to open the school for students, faculty, and parents the day after the shooting. This decision indicated confidence in the safety of the school and allowed the Heath community to come together following the tragedy. Attendance was 90 percent. Counselors were brought in for the students, parents, and teachers. After a few days, school officials decided to ask counselors from outside the school district to leave. Students responded best to adults they knew well, and teachers became informal counselors. Over the next week, the school day gradually returned to its regular schedule and instruction resumed. Yet teachers reported that it was not until the next school year that the school returned to anything approaching normal.

Many in the community united against the presence of the media, particularly the national media, who bore the double onus of being outsiders to the community and reporting on the worst incident in the town's history. The national media outlets, in fierce competition with one another, staked out the school, the local barbeque spot, and the courthouse, generally disrupting already fraught daily routines. Even more egregious, in the view of many we talked to, was the badgering and harassment of students to give interviews, often without parental consent and sometimes without even the students' own consent. One student described how a reporter refused to accede to his request not to be interviewed and chased him across the school parking lot. In one response to the media onslaught, neighbors who lived on the street of the Carneal home built a blockade to keep the media away from the beleaguered family.

The Carneals were also a subject of much sympathy in the community. Many of those we talked to spoke movingly about what they imagined it would be like to be the parent of a child who had done something so horrendous, saying that in some ways it would be worse to be the parent of the shooter than of one of the victims. The parents' previous reputation in the community and Kelly's all-around success shielded the family from many of the negative judgments that otherwise surely would have emerged. Numerous people reported that on the morning of the shooting, before she learned the identity of the shooter, Carneal's mother raced to school with blankets to see if she could help. The Carneals had roots in the community that went back several generations, and the family attended the church that Michael's grandparents attended. For all of these reasons, many in the community, even those who did not know the Carneals personally, extended their web of empathy to include the Carneal family.

Civil Suits and the Second Life of the Shooting

The parents of the three slain girls had little sympathy for the Carneal family, however. They commented that the Carneal parents were given a "free pass" in the community because of their social status and that the police and the prosecution had not investigated the case aggressively enough. According to the victims' families, they hired their own lawyer in order to learn who was involved, why it happened, and to force Michael Carneal and his parents to take responsibility for his actions. They also hoped that the suits would help prevent future shootings by illuminating the causes of the Heath shooting and by putting a wide variety of people and institutions on notice that it was their responsibility to prevent them. Among those against whom they brought suit were Michael Carneal, his parents, and the neighbor from whom Carneal stole the guns; students who had seen Carneal with a gun at school before the shooting; students who had heard that something was going to happen on Monday; students who may have been involved in a conspiracy; teachers and principals at Heath High School and Heath Middle School; and the producers of the Basketball Diaries, the makers of the point-and-shoot video games that Carneal played, and the Internet pornography sites that he visited.

The families felt that students allegedly involved in a conspiracy had not been fully cooperative with the police and that the suit would force these people to answer questions that they had not previously addressed. They also thought that Carneal's parents had missed warning signs, such as towels over the vents, a history of vandalism, and the disappearance of family guns and knives that should have indicated that Carneal needed to be closely monitored or given psychological help. The complaints alleged that the schools had not noticed or addressed Carneal's scholastic decline in the later years of middle school and in the months before the shooting and had not raised concern over violent stories that he had written. The complaint also faults the school for not formulating any plan to prevent school shootings, despite several past instances of school shootings in Kentucky.

Not surprisingly, given the number of people sued, there was a significant community backlash against the families who brought the suit. The families reported receiving some hate mail, being stared at in public, and being avoided by some of their old acquaintances. One of the teachers sued was still in his teacher training program at a local university at the time of the shooting and successfully countersued. This story was brought up by many as an example of the excess and carelessness of the handling of the suits. Some thought that the families were not actually interested in discovering the truth and were simply trying to win a large monetary judgment. Others felt betrayed because they felt they had reached out to the victims in their time of need, only to have them turn around and bring suit. A large majority felt that the suits were inhibiting the already very difficult healing process, making it impossible for the community to move forward. Although a fair number supported the entertainment industry suits, they thought that pointing fingers at others in the community was inappropriate.

Michael Breen, the lawyer for the families, countered that it was exactly this unwillingness to pay attention to problems that had caused the tragedy in the first place. In Breen's view "accountability is always painful," but by bringing attention to those at fault, schools, parents, and the entertainment industry will become aware of their responsibilities, which may help prevent future shootings.

Thus far the courts have found overwhelmingly against the victims' families. With the exception of a $42 million dollar judgment against Michael Carneal himself, all of the other cases were dismissed by the judge before trial and are on appeal.[20] The quick adjudication of these cases has reinforced the sense of many in the community that the suits were groundless, but the victims' families say that there have been positive results from the publicity that their suits have generated. As an example, they point to the fact that some large national retailers no longer sell point-and-shoot video games, and education professionals are paying close attention to prevention of school violence.

Preventative Changes at Heath High School

There were a number of changes at Heath High School and Heath Middle School in response to the shooting. While the principal at the time of the shooting said that before 1997 school shootings were "not even on the radar screen" of issues that principals needed to worry about, elaborate mechanisms are now in place to prevent and react to potential shootings in McCracken County. Because school shootings are such rare events, it is very difficult to tell if these changes are "working" or even what sort of indicators might be used to judge their success. We limit our analysis here to what students, parents, faculty, and administrators think of their effectiveness.

Beginning the day after the shooting and continuing to the present day, Heath has posted teachers at the entrances to the school in the morning to search students' bags for weapons. Some of the teachers expressed considerable ambivalence about this role. One said that on the first day she apologized to each and every student for the invasion of privacy and lack of trust that the searches embodied. Students are also required to store their bags in their lockers once in the building. Over time, these procedures have become routine, and most students and teachers said that they make them *feel* safer, even if they do not think the searches are thorough enough to stop a determined criminal.

Other changes, which originated from the school board and not the site-based committee, have met with a much less positive reception, including building fences around the school and requiring every student to wear an identification tag. The combination of the fences and the identification tags led a number of students to independently voice the complaint that the institution felt more like a jail than a school; one student dubbed it "Heathcatraz." Given the inefficacy of the fences at actually keeping people out (as a number of students routinely demonstrated by going over or under the fences), students worried that in the unlikely event of a future shooting, the fences would do less to keep the shooter *out* than to keep innocent students *in*. School officials in turn noted that Carneal brought the guns in through the band room, and that the fences give school personnel needed control over access to the school.

The identification tags were an object of particular bewilderment and outrage among the students. They accurately pointed out that ID tags seemed premised on the idea that outsiders were the likely source of problems, when, based on tragic past experience, it was students at the school who should be the primary concern. Some students and parents criticized both the identification tags and the fences as knee-jerk responses by the school board to give the appearance of action without actually addressing the issue.

One change that was almost universally praised is the addition of school resource officers (SROs) to McCracken County high schools and middle schools. Before the shooting, there was little to no police presence at Heath, and officers were called to the school no more than once or twice a year. The school resource officer at Heath, who was hired in the August following the shooting, is a former Paducah city police officer who works for the school full-time and is in charge of maintaining security.[21] Like traditional police officers, SROs carry a radio, gun, handcuffs, and a club, but they try to blend into the school by wearing a "soft" uniform of slacks and a collared shirt. In our observations, the SRO at Heath has successfully integrated himself into school life and has befriended a number of the students. Tips he has received from students have led to several arrests for drug and other contraband violations. While he related that initially some in the community were suspicious of an armed presence in the school, over time he has become an accepted, liked, and valued resource at the school. Both he and the former principal emphasized that students were more apt to trust someone who does not report directly to the school hierarchy, and as such the school resource officer is sometimes able to get a better sense of what goes on among students, especially less academic students, than teachers and administrators. Teachers were also very supportive of the resource officer, because it relieved them of some of the more serious discipline duties that they did not really want. One teacher said the SROs might be useful in preventing school violence: "An ID badge and a fence won't stop a potential shooter, but a security officer might."

The school also made some less immediately noticeable changes that administrators hoped would help address the type of social and psychological issues that prompted the shooting. The principal at the time and almost all of the 18 teachers that we spoke with emphasized that, even before the shooting, there was a focus among the teachers on the social well-being of their students, but that since the shooting the level of attention to social issues has increased, and more specific strategies were put into place to help teachers with this task. For instance, since the shooting more professional days have been devoted to training teachers to identify students who have serious emotional or psychological problems. Heath has also extended the freshman orientation period, and the academic advising time of the school day has been used to teach all students the *Seven Habits of Highly Effective Teenagers.*[22] In addition, a half-time guidance counselor has been added to the staff who is responsible for counseling students and helping freshmen transition to high school.[23] (The primary guidance counselor mainly does scheduling and college counseling.) Finally, an outside therapist visits the school at least one day a week to meet with students.

Still Reverberating

It is remarkable the degree to which an event that happened more than three years ago continues to affect the lives of many in the community, including those who were not close to the epicenter. For example, the head of a large business umbrella organization told us that major employers continue to have trouble persuading highly skilled workers to move to Paducah because of its association with the shooting. The youngest of the high school classes that attended Heath at the time of the shooting graduated in June 2001, but some of the siblings of those students who were in middle or even elementary school at the time report that they are afraid of going to the high school. There are enduring reminders of the shooting, such as a plaque in front of the school in memory of the slain girls and the ever-present identification tags, which ensure that each new group of entering ninth graders are made aware of the awful events of December 1, 1997.

For the students who were freshmen at the time of the shooting, friends and classmates of both Nicole Hadley and Michael Carneal, the shooting was the defining event of their high school careers. On the eve of their own graduation three years later, they looked back at the shooting as an event that brought the class together in a bond of innocence lost, sparking a corresponding commitment to reject the petty meanness and exclusion common to adolescence. They addressed some of these issues head on, forming a thriving peer mediation group. But the tragedy also left psychic scars on those closest to the shooting. Many students went to individual and group therapy for years after the shooting, including some who are still in treatment, and at least one went abroad for a semester in part to escape the memories of the horrible event.

When asked what they regretted about their handling of the aftermath of the shooting, the school administrators noted that they did not give enough support to the teachers and other school staff. Teachers were expected to be there for their students but also needed time to grieve and heal. Many teachers have children of their own attending the school and live in the community, so they were some of the most affected by the tragedy.

Of course the greatest loss is to the lives of those who were shot and to their families. The students who survived have been remarkably resilient. Of those who have had media accounts published about them, Melissa Jenkins went to college, Hollan Holm was a valedictorian, and Shelley Schaberg, unable to fulfill her lifelong dream of playing college basketball because of nerve damage that affected her hand-eye coordination became a varsity college soccer player. The families of the victims who died, by contrast, have been devastated by the loss of their daughters. Whatever the result of the civil suits, the families said, there will always be a huge void in their lives that nothing can replace. And the personal rejection they have encountered in the community as a consequence of the civil suits has left them isolated from some former friends.

CONCLUSION

In the course of our interviews with adolescents, we are reminded once again of how "adolescent society," as James S. Coleman famously dubbed it 40 years ago,[24] continues to be insulated from the adults who surround it. While many of the values of adults and children are shared and the hierarchies of the adult world are mirrored in the adolescent world, the social dynamics of adolescence are almost entirely hidden from adult view. The insularity of adolescent society serves to magnify slights and reinforce social hierarchies; correspondingly, it is only through exchange with trusted adults that teens can reach the longer-term view that can come with maturity.[25] No one knows this better than the teachers at Heath; we could not put it better than the words of a beloved long-time teacher: "The only real way of preventing [school violence] is to get into their heads and their hearts. Everyone in the building needs to have one person on their side."

ACKNOWLEDGMENTS

Newman is the principal investigator on this case study. We wish to thank Martin West, Ph.D. candidate in government and social policy at Harvard University for his help in understanding many of the theoretical issues raised in this case study. We also like to acknowledge the assistance provided by Margot Minardi, Department of History, Harvard University.

NOTES

[1]Greater Paducah Economic Development Council.

[2]McCracken County School District.

[3]McCracken County School District.

[4]In a deposition taken in February 2000 for the civil suits, Carneal said that he had actually stolen this gun approximately a month before the shooting. It is not clear which account is correct.

[5]In the same February 2000 deposition, Carneal stated that he had planned the shooting at least a month before it took place, perhaps contradicting his earlier account. We believe that this confusion reflects a series of decisions and uncertainties on Carneal's part. He began planning the shooting before he actually decided to carry out his plan. Indeed, he testified in the deposition that after arriving at school the day of the shooting, he decided to leave the bundle of weapons in his sister's car, but she reminded him that he was forgetting his "English project."

[6]The exceptions were the families of the victims, whose portraits of the Carneals were less flattering and the emotional timbre far more angry. This is not hard to understand given what these families have lost.

[7]Interview with Kathleen O'Connor, October 23, 2001. O'Connor was Carneal's treating psychologist when he was incarcerated in the juvenile detention center.

[8]Dewey Cornell, "Psychological Evaluation of Michael Carneal," September 3, 1998, p. 10.

[9]Elissa P. Benedek, William D. Weitzel, and Charles R. Clark, "Report of Psychiatric and Psychological Evaluation," July 17, 1998, p. 15.

[10]Cornell, September 13, 1998, p. 13.

[11]Benedek et al., p. 26.

[12]Interview with Kathleen O'Connor, October 23, 2001.

[13]Interview with Kathleen O'Connor, October 23, 2001.

[14]Dewey G. Cornell, "Michael Carneal Evaluation," February 1, 1998, p. 17.

[15]Police interview, December 1, 1997.

[16]Benedek et al., p. 12.

[17]A review of the research by an expert panel of the American Psychiatric Association suggests that there is a link between video games and behavior. Cited in James Garbarino, *Lost Boys* (New York: Anchor Books, 2000), p. 115.

[18]Cornell (September 3, 1998), p. 16.

[19]If Carneal had admitted guilt, it would have been easier to pursue a wrongful death suit against him in civil court.

[20]Since Michael Carneal himself has no assets or income, the judgment against him can only be collected if the victims' families win a suit against the Carneal family insurance company, a case that is still pending at the time of this writing.

[21]At some of the schools this also includes monitoring the security cameras, but the principal of Heath decided not to install cameras at the school.

[22]Sean Covey. 1998. Simon and Schuster.

[23]Guidance counselors are assigned by a formula based on the number of students at a school. An increase in enrollment at Heath is a central reason for the addition of the half-time counselor.

[24]James S. Coleman. 1961. *The Adolescent Society.* New York: Free Press.

[25]Of course, adults can also reify social hierarchies and often do. However, by virtue of their distance from adolescent society, they at least have the potential to provide guidance and direction, and thus it is with them that one places responsibility.

6

Shooting at Tilden High: Causes and Consequences

John Hagan, Paul Hirschfield, and Carla Shedd

Joseph White shot and killed Delondyn Lawson at Tilden High School in the last fatal Chicago school shooting nearly a decade ago. This event was portrayed by extensive news coverage as random and senseless and by a jury trial as a first degree homicide that was inexcusable as self-defense. However, this shooting was not random, and while it could not be justified as self-defense, it evolved out of a gang-related dispute with socially structured roots. This case study shows with an analysis of news media reports, census statistics and trial transcripts, and through personal interviews with community and school participants, lawyers, news reporters and public officials, how the Tilden school shooting was connected to major historical changes in the economic and racial context that led to entrenched gang conflict in the South Side community where it occurred.

This case study further shows how the trial of Joseph White became a political object lesson in the need for metal detectors in South Side secondary schools, a policy that evolved into a wide range of "zero tolerance" and disciplinary and exclusionary measures that diffused across the country, with Chicago Mayor Daley as their persuasive advocate. The Tilden High shooting was embedded in social conflicts that reveal it was neither random nor senseless in the way initial news coverage suggested, while the trial had much broader consequences than was indicated by the conviction for first degree homicide itself.

AN AMERICAN TRAGEDY

No student has been fatally shot in a Chicago public school since 1992. The last two fatal shootings occurred in high schools on Chicago's South Side during a 10-day period from November 10 to 20, 1992. A youth accidentally shot himself at school in the first incident. The second incident that is the focus of this study resulted in the death of one 15-year-old boy, the serious wounding of a second youth, and the shooting of a third youth in the foot. The youth who was killed, Delondyn Lawson, was identified in the first front page news stories (November 20, 1992), while the shooter and other two victims were not.

The initial stories came on Friday, the day after the Thursday shooting. They reported the shooting as a response to a skirmish about a gambling debt. The gambling involved a dice game earlier in the week in a school washroom. The shooter was described as having "fired a handgun about four times randomly toward the skirmish." Two unidentified suspects were reportedly arrested, and it was noted that metal detectors at the school were not functioning on the day of the shooting because they were being used on randomly chosen days to minimize cost and disruption.

The next day, Saturday, the story was again on page one and the headline read, "Even Safety of Schools Shattered, Student Slain, 2 Others Wounded in Hallway at Tilden" (November 21, 1992). The story was built around the victim's mother, Linda Lawson, who asked, "What was I supposed to do with a 15-year-old? I drove him there in the morning, and I was there at 2:02 p.m. to pick him up. What else can you do? You have to send them to school." The story noted that Delondyn Lawson had been fatally shot in the back, and two other youths were injured.

Joseph White was identified in this story as the alleged shooter and as a 15-year-old freshman at Tilden now charged in adult court with first degree murder. He had fled the school pursued by other youths and a security guard and hid under a nearby back porch before being found and arrested. The story again mentioned the dice game and now briefly the possibility of gang involvement. Delondyn Lawson was described as a "great dancer and funny," a boy who liked football and video games and who was trying to stay away from gangs.

This second story also marked the entry of Mayor Richard Daley into the matter. The mayor responded to the random use of metal detectors, which was defended by the police and schools. "They have to realize you have to run them every day," the mayor insisted, "because if we run them in the federal building and in the state criminal justice system, you can't get [guns] in there." The mayor asked, "What's more important? Children are more important than anybody else in society. And that shooting, the death of a young child, directly affects everyone." The mother of a sophomore girl at Tilden elaborated the headline of the story by observing that "It's bad enough they're [the gangs] taking violence into the neighborhoods, but when they're taking it into the schools, it's bad." She emphasized that schools are different kinds of places: "They're just here to learn. They're not here to be dying."

The third day, Sunday, brought another page one story focused on the apparent randomness of the deadly victimization, saying "Student Was in the Wrong Place at Wrong Time" (November 22, 1992). Joseph White again was identified as the charged assailant who was in jail and had been denied bond in court on Saturday. The story then focused on Delondyn Lawson and his family. His mother described Delondyn as a child who cried easily and who recently had been attending funerals for boys he knew at a rate of one or two a month. "His friends are constantly dying around him," his mother reported, "They're getting shot on the corners. Every month somebody he knew in his age group was dying." Delondyn's former school principal called it an American tragedy.

The news story reported that while he was walking between classes "a bullet tore through Delondyn's back and through his heart," killing him almost instantly. Delondyn had been staying with his aunt and then his father until his mother had retaken custody of him in the preceding weeks. She was trying to help Delondyn keep out of trouble and get his grades up, in part by picking him up after classes each day and then tutoring him for two hours of schoolwork. Other members of the family said that he wasn't in a gang but his friends were. "The guys who are in the gang grew up with him," a relative explained. "It's not like he don't know them. He's got to go through them. These are kids that he's seen all his life."

The conclusion of the initial news coverage of the Tilden High shooting came in a page seven story the following Tuesday (November 24, 1992). (For a summary of the media coverage regarding this case, see Box 6-1.) The family of Joseph White had retained Chicago attorney Robert Habib to represent their son, and he appeared in Cook County Circuit Court on Monday seeking to bar the news media from further reporting of the boy's name, because he was a juvenile. Yet White had already been charged as an adult, and raising this issue simply resulted in his name appearing again in the same sentence as the report of the judge's refusal to suppress his identity, followed by a repetition of the report of the Saturday refusal of the appeal for his release on bond. Habib noted that the state's attorney asserted in the bond hearing that "White just walked in . . . and started shooting." As a result, "you had that image right off the bat, that Joseph White had made an unprovoked attack in the school, literally just walked in and started firing."

BOX 6-1 MEDIA CHRONOLOGY

October 12, 1989.	"Youth Slain in Front of Classmates." *Chicago Tribune.*
September 26, 1991.	"Once Safe Havens, Schools Now in Line of Fire." *Chicago Tribune.*
March 21, 1992.	"Police Fire Two Police Officers Over Incident with Teens." *Chicago Sun-Times,* p. 4.
June 5, 1992.	"Daley Offers Metal Detectors to Schools." *Chicago Tribune.*
October 22, 1992.	"Daley Prods Schools on Metal Detectors." *Chicago Sun-Times,* p. 4.
October 23, 1992.	"School Metal Detectors Offer no Protection for Kids Outside." *Chicago Tribune.*
November 20, 1992.	"Boy Killed in Shooting at Tilden High School." *Chicago Tribune,* p. 1.
November 21, 1992.	"Even Safety of Schools Shattered, Student Slain, 2 Others Wounded in Hallway at Tilden." *Chicago Tribune,* p. 1.
November 22, 1992.	"Student was in Wrong Place at Wrong Time." *Chicago Tribune,* p. 1.
November 23, 1992.	"Teen Charged as Adult in Killing." *Chicago Defender,* p. 1.
November 24, 1992.	"Body Count Rises in the Public Schools." *Chicago Tribune,* p. 18.
November 24, 1992.	"Tilden Tries to Heal Wounds, Counselors Help Students Cope with Fatal Shooting." *Chicago Tribune,* p. 7.
November 27, 1992.	"Metal Detectors in School Set Off Alarms." *Chicago Tribune,* p. 1.
December 1, 1992.	"It May be too Late to Sound an Alarm." *Chicago Tribune,* Bob Green Editorial.
December 5, 1992.	"Bullets Rip School's Illusion of Safety." *Chicago Tribune,* p. 1.
June 14, 1993.	"Death of an Innocent Bystander: The Peacekeeper in his Family, Delondyn Lawson is Cut Down at Random by a Classmate Firing Wildly." *People* magazine, pp. 48–49.
January 20, 1994.	"Trial Opens in Killing at Tilden." *Chicago Tribune,* p. 5.
January 22, 1994.	"Teen Guilty of Slaying in Tilden Hallway." *Chicago Tribune,* p. 5.
March 15, 1994.	"Prison Term in School Shooting." *Chicago Tribune,* p. 3.
September, 1998.	"Chicago Public Schools: Expulsions Rise, But Safety Issues Persist," by Natalie Pardo, *Chicago Reporter.*
March 11, 1999.	"Off-Campus Arrest Could Boot Kids: Expulsion Policy to Stiffen." *Chicago Sun-Times,* p. 3.
February 15, 2000.	"Boy Charged in Shooting of Classmate." *Chicago Tribune.*
February 16, 2000.	"Vallas, Daley Encourage Broad Gun-Detector Use." *Chicago Tribune.*
April 7, 2000.	"Vallas Wants Poverty Cash for Security." *Chicago Tribune.*

Habib later encountered the judge who had denied the suppression of Joseph White's name. The judge confided that, "You know, quite frankly, had there been no publicity on the case, I probably would have granted your motion. But at this point, we'd look like total fools given the fact that everybody in Chicago knows his name." Habib felt he had probably already lost the case at this point. "The dominant issue was the fact that he was convicted in the media, before we had a chance." He lamented taking what he called the approach of "an attorney of the old school": "The last thing you did was try your case in the media. That was a huge mistake that I made when I look back at it. When I watched the attorneys in the O.J. Simpson case jump in front of the TV cameras every chance to try to portray him as being innocent or whatever, I congratulated them." Habib believes he didn't do enough of this, but he did encourage White's mother to assume a public role.

Following the Monday court appearance, Joseph White's mother, Karen, appeared before television cameras to answer reporters' questions. "Any parent with kids understands there are no model kids," she began. Picking up on the American tragedy theme introduced by Lawson's former school principal, she then observed that, "It's a tragedy for both and all parties concerned—for the families and the kids. He is a victim of a tragic situation that cannot be altered." Karen White then echoed Mayor Daley's pleas for the regular use of metal detectors. An editorial in the *Chicago Tribune* also took up Daley's theme, observing that "had metal detectors been used routinely at Tilden, Delondyn Lawson's killer might have been deprived of his weapon—or at least forced to use it elsewhere."

Karen White's public comments began to suggest the way in which she was caught up in this story along with her son. She was off work on the day of the shooting and, because her son was identified so quickly by other youths as the shooter, the police were at her doorstep within the hour. "I was washing my hair when somebody was beating on the door. When I went downstairs and opened the door, there was just like police everywhere." She was told of the shooting and that Joseph was at large and armed with a gun. "The only thought that I was thinking was that my son was going to be killed. They said he had a gun and that if he came back, I needed to let them know because he was in danger of being killed if they saw him with that gun." Karen could not reach her husband at work but was able to get her sister-in-law to drive her to the school.

She was terrified. "Over in that area there have been a lot of police shootings and there were so many boys that had gotten killed, you know, so I was thinking that Joey was going to get killed." When Karen White arrived at the school the police had apprehended Joseph across the street. "They were laying him up against a car and they were handcuffing him. . . . My first thought was just being thankful to God that he was alive." They took him to the 51st Street police station.

Karen followed and when she arrived "everything was just pandemonium." Joseph was already locked in a cell, and she was allowed to join him. "When I saw him he was scared, he just looked wild-eyed because he didn't know, you know, he did that [the shooting] but he did not know the full ramifications of his actions. Joey, when I saw him, I saw that my baby was scared out of his brain and confused looking, you know, he was just hysterical." There was little she could do. "We really didn't have a chance to talk, they were trying to ask him questions, they read him his rights and then told him that he had a right to an attorney and that if he didn't want to speak about anything that he didn't have too." Karen contacted attorney Habib through a friend and then stayed with Joseph. "We stayed there until maybe 11 or 12 o'clock at night and finally the district attorney came down and they told me that they were indicting him for murder." Karen returned home after "the worst day of my life."

Less frequent but continuing news stories appeared over the following days and weeks. Habib did not push for the case to go quickly to trial, hoping the intervening time would allow the effect of the pretrial publicity to subside. However, the appearance of an article in a June 1993 issue of *People* magazine about Delondyn Lawson and his mother, Linda, focusing on the loss of her son, brought new attention to the shooting to a national audience (June 14, 1993). Linda Lawson revealed that Delondyn had not wanted to go to school on that Thursday and that she had worried that he was becoming involved with gang members. She wondered now, "Was he having problems he didn't want to tell me about?"

Still, Delondyn was described as a "helpmate with a tender heart" and Joseph White as the youth who "had pulled a small semiautomatic pistol from his waistband and blasted away until he ran out of bullets." Linda described how she learned from a neighbor that her son was a victim in the shooting and rushed first to the school and then to the hospital. "When I came through the door, I knew, . . . no one had to tell me." She continued, "Seeing my baby laying up there, that's a feeling I can't describe. . . . I think the thing that hurt me most was that I wasn't there with him [when he died]." Linda's sister, Cathy, looked for answers: "Do you blame the boy or the boy's parents or the school?"

White's lawyer, Habib, knew that in this atmosphere his client was going to be convicted and receive a prison sentence, and he therefore initiated plea negotiations in an attempt to reduce the charge and the sentence. He was stonewalled with the response that there was too much adverse publicity about the shooting to plea bargain. "I tried several times in conferences with the state's attorney, and they just came back and said, 'No, we can't do it. The supervisor says no way. Because of the pressure on us, we cannot give you a plea on this case.'" When the alternatives were exhausted, the case finally went to trial more than a year after the shooting, on January 19, 1994.

TILDEN IN CONTEXT

Tilden High School is located at the intersection of 47th Street and Union Street in the south side area of Chicago known historically for the Union Stock Yards. The gates to the Yards were opened in 1865 and stayed open for more than 100 years. The area surrounding is still called New City, the name given to it by sociologists in the 1920s. New City contains two historic neighborhoods that emerged to the east and west of the Yards. The area to the east that includes Tilden High School is known as Canaryville. Its history through the early part of the 20th century was dominated by the packinghouse district and its adjoining neighborhood to the west.

The area to the west and south is known as Back of the Yards. This is the area that gained notoriety as a desolate industrial slum in Upton Sinclair's 1905 novel, *The Jungle*, which was published just after the major packinghouse strikes of 1904. Back of the Yards became more notorious after a 1919 race riot and the major packinghouse strike of 1922. Southern blacks were brought to Chicago as strikebreakers in this part of the century, anticipating some of the change and conflict to follow. The population of this part of Chicago peaked in about 1920. The older, white ethnic and more established Mexican American families and institutions in the Back of the Yards area have been represented since 1938 by the Back of the Yards Neighborhood Council (BYNC), which was founded by the famous community organizer Saul Alinsky and other local leaders.

The end of World War II brought major changes to the New City area. Meat packers began to buy animals directly from farmers, and interstate trucking replaced the use of the railroads in the transportation and delivery of live animals and frozen meat. The packinghouses of the area steadily declined as a focal point of the meat industry. By the early 1970s, the Union Stock Yards had closed its gates and the economic underpinnings of this area were permanently altered. There have been various efforts to revive the economic life of the area, including the designation of the Stock Yards District as one of two Chicago urban enterprise zones in 1984 and the building of a new shopping mall with several large retail anchor stores in 1990. Although these efforts have been cited as a national model of urban redevelopment, the results are still more promise than reality, and the New City area remains economically troubled.

Social change accompanied the economic alteration of life in New City through much of the last century. When Joseph White's family moved in the spring of 1971 to their home at 324 West 51st Street, about a mile from Tilden High, the surrounding area was still overwhelmingly white and only 4 percent black. However, as indicated in Figure 6-1, by the time Joseph was approaching kindergarten in 1980, the area was less

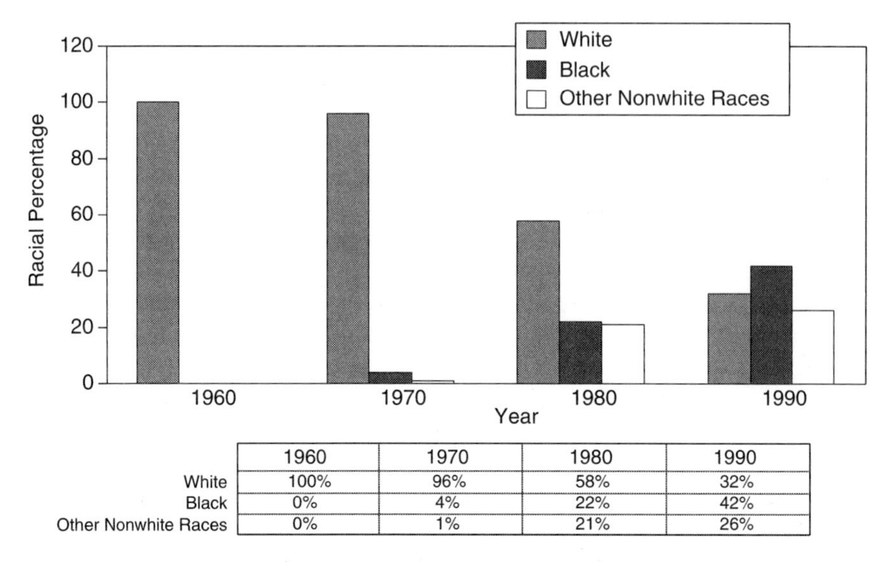

	1960	1970	1980	1990
White	100%	96%	58%	32%
Black	0%	4%	22%	42%
Other Nonwhite Races	0%	1%	21%	26%

FIGURE 6-1 New City population by race.
SOURCE: Chicago Area Geographic Information Study (2001).

than two-thirds white, nearly one-quarter black, and about one-fifth other, mostly Mexican American. As Joseph approached high school in 1990, New City was less than one-third white, nearly half black, and about one-quarter Mexican American.

The White family was experiencing this era of rapid social change firsthand, and it involved more than changing numbers on a graph can tell. As blacks arrived in increasing numbers from the southern states, they settled in the South Side of Chicago, and low-cost public housing projects began to rise. Real estate operators steered blacks who could afford to buy homes into South Side neighborhoods and simultaneously scared whites into leaving, profiting on the rapid turnover. The resulting plunge in real estate values contributed to the impoverishment of the new residents. Meanwhile, banks "redlined" areas for which they would deny loans. Businesses began to leave these neighborhoods, and city services declined. A cycle of disinvestments contributed to the ongoing process of decline. Chicago was emerging in the latter half of the 20th century as one of the most racially segregated cities in America.

As Figure 6-2 shows, 1990 census figures indicate that the numbers actually do not reveal the extent to which New City is racially segregated. There are census tracts within New City that are almost entirely white as well as tracts that are almost entirely black. The tract around Tilden High retained a large white population into the late 1980s, even though by then the majority of youth attending the school were black.

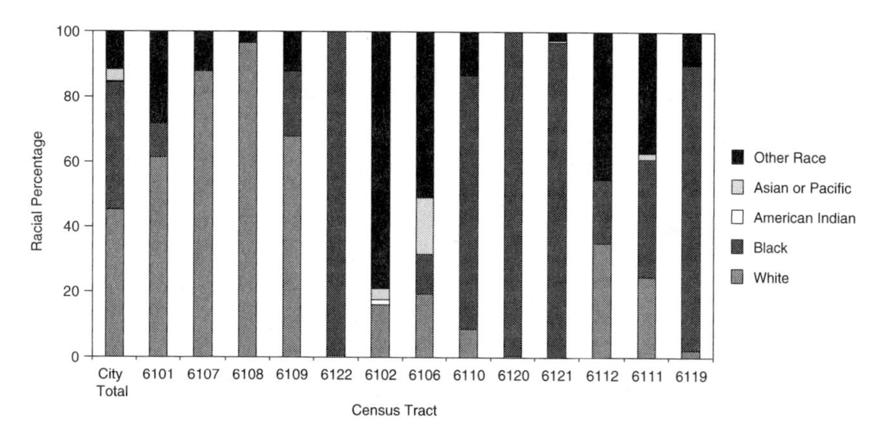

FIGURE 6-2 1990 New City census data by race.
SOURCE: Chicago Area Geographic Information Study (2001).

The White family moved into a low-income housing development of new homes in 1971, when Joseph's mother was only 21 years old. Her grandparents had joined in the northern migration that led from Tennessee to Chicago, where Karen White was born. Almost all of her immediate family members have lived in this city. All the families moving into the adjoining new houses with the Whites were also black. Karen had recently married; she and her new husband felt the area offered the advantages of a new community, with a nearby church and elementary school, where they would be accepted and could successfully raise a family. They have now been married for more than 30 years, raising several children in this home. Karen's husband is disabled from several strokes; they now live mostly in Texas to be closer to their daughter and her children.

The changes we have described brought inevitable racial tension to the New City area and to Tilden High School. Tilden opened in 1889 and once ranked as an advantaged institution in the educational hierarchy of south Chicago. At the turn of the century, "The students were special, out of the ordinary, and they knew it. Their chances to succeed were better than most of the friends they grew up with, and they would soon move on to a different rank in society" (Slayton, 1986). However, Tilden changed as the community around it experienced an economic and social transformation over the century. The Canaryville neighborhood made news when in the winter of 1982 the mayor of Chicago, Jane Byrne, ordered the dismantling of four heavy iron gates erected during the 1960s to separate white from black areas.

Karen White was a supervisor in the accounts department of a large Chicago newspaper when Joseph entered high school. She had achieved a measure of success and felt hopeful for her family. Nevertheless, soon after moving into her new home, Karen White learned about gang problems in the community. "It was the Mickey Cobras that were in the area. I didn't know a lot about gangs; I was never affiliated, but moving into that area and through my kids and their friends, we found out the area we lived in was the Mickey Cobra area."

The transformation of New City and Tilden High is well summarized, in terms of its racial implications for young people, by the classic work of Gerald Suttles in another south side Chicago neighborhood. Suttles (1968) observes that racial and ethnic groups become identified with sociospatial locations in urban areas, and these identifications have strong implications for interactions between and within groups. In earlier times there was intense racial conflict, including the race riot of 1919 in the Stock Yards over the hiring of strikebreakers (Chicago Commission on Human Relations, 1922). While this produced enduring racial divisions that would continue to plague Chicago (Halpern, 1997), this situation ultimately settled into a pattern of community isolation and ethnic insularity reflected most prominently in persistent residential segregation (Cohen, 1990). In addition, there was historical significance to a contradictory kind of integrated employment that ultimately emerged in the packinghouses of the New City area: "Indeed, in many urban areas the packinghouse was the largest interracial institution, one of the few places where blacks and whites interacted on a daily basis" (Halpern and Horowitz, 1996:2).

Suttles (1968:31) describes the kind of isolation and insulation that emerged in the surrounding neighborhoods: "First, each group is a socio-spatial unit. Second, inclusion in these groupings is mutually exclusive. Third, opposition is between 'equivalent' units. Fourth, the order by which groups combine preserves the equivalence of oppositional units." The reflection in New City of the social order that Suttles describes consists of racially homogeneous gangs that more often enter into conflict with one another than across racial lines.

Each of Tilden's low-income neighborhoods was dominated by a different gang. At the time of the shooting, at least six gangs had a significant presence in the black neighborhoods around Tilden: the Black Gangster Disciples, the Mickey Cobras, the Blackstones, the Vicelords, the Latin Kings, and the Satan Disciples. White students also belonged to neighborhood gangs. Prior to the wave of disciplinary reforms in Chicago, Tilden High School was a place where gang members could freely assemble and carry out recruitment and money-making activities. More generally, schools provided a setting where rival gangs could interact and compete for shares of various illegal markets. In the early 1990s, Tilden was an extremely volatile environment, and fights were a nearly daily occurrence.

Most of the clashes at Tilden occurred between gangs of the same race. An exception, according to the school police officer, was a clash that occurred shortly before the shooting. One of the members of a black gang reportedly slapped the girlfriend of the leader of Tilden's Latin Kings. Outnumbered by the rival black gang, the leader solicited the help of adult Latin Kings from the nearby Latino neighborhood. Soon after, these adult gang members surrounded the school with shotguns and prepared to seek revenge against the various black gangs. We interviewed the school police officer about his recollection of this incident:

> Q: What did you do?
>
> A: I pleaded with the gang leader to call them off.
>
> Q: Did you call other cops?
>
> A: If I put out a request for assistance, I could expect one car at the most. If I put out a 10-1, which is a serious officer distress call, I might get two cars. So I had to handle it myself.

The Latin Kings' leader heeded the officer's pleas and called off the rival gang. This experience, the school police officer surmised, nonetheless indicated to the black gangs that in the face of a serious deadly threat they could not rely on authorities but must instead rely on each other. This experience of a common enemy briefly fostered an atmosphere of relative harmony among gangs from the same Gang Nation, which for the purpose of this analysis most notably included the Mickey Cobras and the Blackstones. In this way, race may have been an important factor in explaining why two rival gangs were exhibiting a sort of camaraderie by gambling with each other in the school bathroom (Yablonsky, 1997). In the overarching organizational structure of the Gang Nation, the Blackstones and the Mickey Cobras were temporarily friendly affiliates who fell under the branch of the "Peoples." In times of war with the rival branch, the "Folks," these gangs would come together. A gang officer at the 51st Street police station confirmed that such alliances still occur today. The school police officer who worked at Tilden High in the early 1990s concluded that the gambling episode that precipitated the shooting was borne out of this tentative truce between black gangs. He nonetheless recalled his own surprise as he witnessed members of the different black gangs, the rival Mickey Cobras and the Blackstones, all gambling together in the bathroom.

This was the complicated and gang-dominated world of young people who attended Tilden High School. Students had to negotiate the boundaries of gangs and neighborhoods in their everyday movements to and from school as well as in it. The "ordered segmentation" of different groups, which was simultaneously spatial and social in nature, was confronted by these youth on a daily basis. George Knox (1992:134) makes this point in noting that the "school environment brings rival gang members in close proximity to one another and blurs haphazard turf lines which leads to confrontations and challenges within school, on school property and on the streets surrounding the schools." This is a complicated world in which today's enemies may be tomorrow's allies. It was difficult if not impossible for most youth at Tilden to completely avoid these gang entanglements. As he moved into high school, Joseph White was clearly becoming a part of this fast-changing world of gangs.

Karen White was aware that although much of the open racial conflict that had earlier characterized New City and its ongoing transition had subsided by the time Joseph was ready to attend high school, the gang problems within the black community had become an equal or greater threat. She recalled: "I knew that Tilden was a racially mixed school but, . . . by the time Joseph was coming along the racial tension was not as great. . . . We had learned to coexist together." Instead, the White family encountered the community and the school's more persistent problem with gangs. Karen White observed, "The gang activities were the problem, that's what my concern was because in our area there were the Mickey Cobras on that end, but in Tilden there was a whole other faction." The other faction was the Blackstone Rangers, the largest and oldest gang in Chicago, called by the father figure of American gang research, Frederick Thrasher (1927:278–9), "a moral lesion on the life of the city" (cited in Barrett, 1987:219). Karen White worried that "Tilden was riddled with gang conflict, so that was more of the parents' concern than the racial part of it."

It is nonetheless also important to identify the suppressed nature of racial conflict in this situation. One salient reason why racial violence probably remained a within-group phenomenon in this Chicago community, as elsewhere in America, was because of the severity of sanctions for interracial violence, and sometimes even interracial contact. For years it was known that sometimes when Chicago police picked up black youth whom they suspected but could not charge with crimes in white neighborhoods like Canaryville, they would drop them off at a location in the neighborhood where they were likely to be beaten by local residents (see Suttles, 1968). In the year of the Tilden shooting, the Chicago Police Board fired several police officers who had left two black youth in the Canaryville neighborhood, where they subsequently were assaulted by a gang of white youths (March 21, 1992).

A related incident was reported in the Canaryville neighborhood on a summer evening in 1985, when two young black adults detoured around road construction onto South Union Avenue (July 12, 1985). They came up behind an apparently stalled vehicle in front of Tilden High School and were about to ask the driver if he needed help when a group of white youths surrounded their car. "They were screaming, 'Nigger, what are you doing here?'" and "Then they came at us with the bats. All the parents were just watching on their porches." Every window of the car the black youths were driving was smashed, and the front end of the car as well. One of the youths was struck in the face with a bat. "Lots of stuff like that happens all the time," a Canaryville resident told a newspaper reporter on the scene, "If you get a bunch of guys sitting on the corner and a colored guy rides by, they will chase him."

Finally, it is important to bear in mind that the early 1990s was a period of economic recession and historically high rates of violent crime in America, which especially affected young black males, especially in areas like New City and schools like Tilden High. It also bears noting that it took a unique social history to produce the high rates of violent crime that America experienced by the early 1990s: it began with a violent revolution, institutionalized the right to bear arms, celebrated a lawless frontier, incorporated slavery, included a civil war, allowed a proliferation of guns, facilitated a large gun industry, witnessed the emergence and growth of large and impoverished urban ghettoes, and produced a nation in which there are more guns than people (Hagan and Foster, 2000).

Meanwhile, in June 1992 seasonally adjusted unemployment rates reached their contemporary peak at 7.8 percent. The recession had technically ended, but the recovery left unemployment unabated, especially in places like south Chicago. A teacher described the situation for many youth at Tilden starkly: "a lot of these kids from blue-collar families, it didn't look like they had much of a future. You know, they had seen their parents laid off left and right." It shouldn't have come as a complete surprise, then, that the latter half of the 20th century brought a significant increase in American gun homicides and violence that peaked in the early 1990s along with unemployment.

As Blumstein and Rosenfeld (1998) point out, the aggregate U.S. homicide rate moved to its near-term peak from 1985 to 1993 and then began to decline to mid-1960s levels. The peaking levels of homicide for young black males were devastating, rising among 18-year-olds from about 20 to 60 per 100,000 population between 1985 and 1993. Handgun homicides alone accounted for all the growth in homicides in the United States after 1985 (Blumstein and Rosenfeld, 1998). The youthful wave of black male homicides in the late 1980s and early 1990s prompted warnings from some of a coming "bloodbath" (Fox, 1995) in which "superpredators" would be terrorizing American cities for years to come (Bennett, Dilulio and Walters, 1996). Subsequent declines proved these predictions alarmingly mistaken, but the source of the apprehension is nonetheless clear.

Americans became extremely apprehensive about the movement of guns and gang violence into public schools in the late 1980s and early 1990s. Chicago was no exception, although the peak in gang violence may have occurred a year or two later than in large East Coast cities. This may have reflected the rate of movement of the crack cocaine epidemic across the country and into the Midwest. Figure 6-3 shows that Chicago street gang-related homicides with black male victims rose steadily from 1987 to 1994 (Block and Martin, 1997). Figure 6-4 shows there was also an increase in arrests in or near Tilden High School from the mid-1980s to the early to mid-1990s.

So by almost any measure, the time of the Tilden shooting was a high point in criminal violence among black youth. The special concern was that this violence was coming into the schools. The Illinois Criminal Justice Information Authority published results from two surveys conducted in 31 public high schools in 1990 (Stephens, 1992). The results indicated that 1 in 12 public high school students in Illinois reported being the victim in the past year of a physical attack while in or going to and from school. About the same proportion—1 in 12—reported sometimes staying home from school for fear that someone would hurt or bother them. These proportions would certainly have been higher among Tilden High School students when the shooting occurred in November 1992.

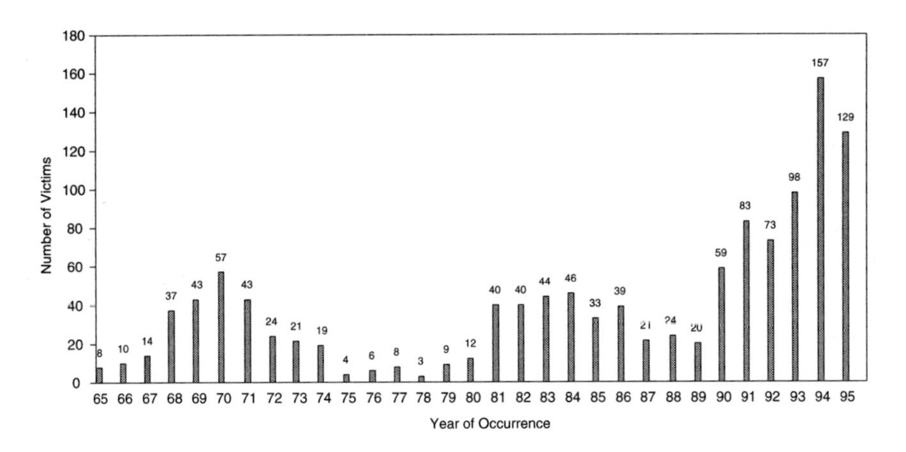

FIGURE 6-3 Chicago street gang related homicides, black male victims, 1965–1995.
SOURCE: Chicago Homicide Dataset, a collaborative project of the Chicago Police Department, the Illinois Criminal Justice Information Authority, and Loyola University, Chicago.

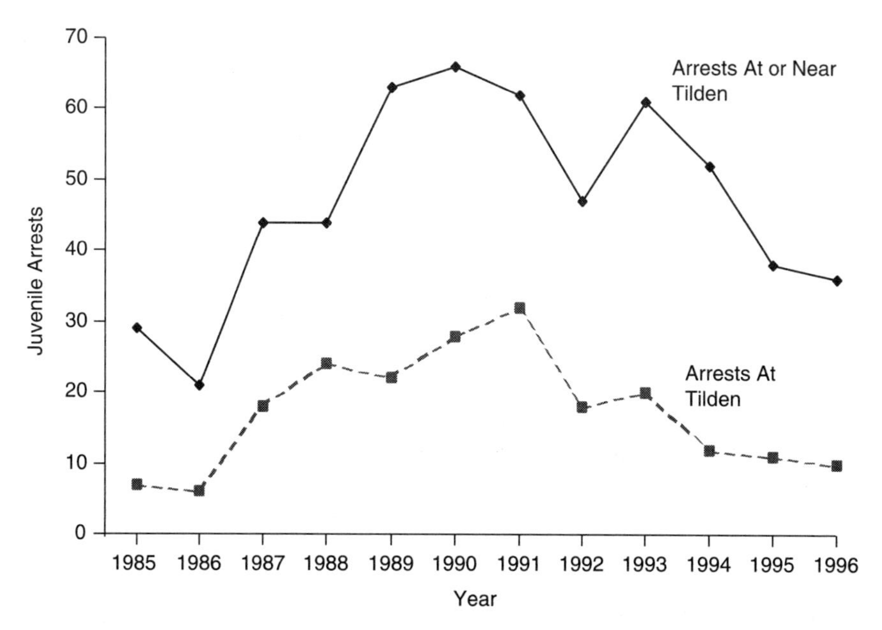

FIGURE 6-4 Arrests in or near Tilden High School, 1985–1996.
SOURCE: Chicago Police Department, Youth Arrest Records, 2001.

This is the world in which Joseph White and the three victims of his shooting moved. The gangs and guns left few untouched, and the impact was often swift and severe. The year before the shooting, Joseph was receiving good marks in school and had won a trophy as the most valuable player on his basketball team, but by spring 1992 he was already at least a passing part of the gang scene, and Karen White felt increasingly powerless to keep him away from it. He was taken into custody and charged with 22 other youth by the Chicago police on an evening in May at a South Side park where approximately 50 youth associated with the Mickey Cobras street gang had gathered. The officer reported that "they were throwing bricks and stones and bottles at other people that weren't dressed like they were, and they were trying to keep them out of the park." This case was not taken to court, but the month before the November shooting Joseph was spotted with another youth removing cartons containing stereos from a railroad boxcar. The stereos were later recovered from Joseph's home, and he pleaded guilty in juvenile court. The two surviving victims of the shooting also came into conflict with the law. One was convicted and received a prison sentence for armed robbery, and the other came into contact with the police for drug activities.

Meanwhile, Tilden High School was besieged with gang activity and hallway violence. A teacher who was at Tilden prior to the shooting recalled that gang-connected fights were common in the hallways, "the kind of fights that would empty classrooms out, you know, you've got your kids in the classroom and all of a sudden something goes on in the hall and then the whole class just runs out." The principal of Tilden, Hazel Steward, reported that gang members were conspicuous by their presence: "They roamed the halls; there were gang fights everyday, teachers constantly going to lock the doors." A school policy was adopted of locking students into the classrooms during class time, "the teachers went in and were supposed to lock their doors, to close out the chaos in the halls."

A teacher at Tilden emphasized the intragroup nature of this rising tide of violence, which matches the more abstract description provided earlier by Suttles (1968). He observed that "from what I could tell, it was two separate black gangs that were fighting each other, and then two separate Hispanic gangs that were fighting each other. It was more amongst, kind of fighting for dominance, you know, within the racial groups." The violence that was engulfing South Side neighborhoods in Chicago was not random in the way implied by the initial news reports of the November shooting at Tilden High.

THE TRIAL

The most telling evidence in the trial came from Joseph White himself, and the following discussion is based on his testimony, with some elaboration as indicated from interviews. Joseph had acknowledged that he was a member of the Mickey Cobras gang and began by recounting the gambling dispute that led to the shooting. The dispute was over a Monday morning dice game in the upstairs locker room or bathroom at Tilden. The first witness for the prosecution, Dewaun Glover, had joined the game and was losing $10 to $20 at a time. Glover indicated that he was losing money that came from the head of his gang, the Blackstone Rangers. Other members of the gang had joined the group when Dewaun tried to reclaim the money he had lost. Before a fight broke out, a police officer arrived and took Joseph and Dewaun to the school disciplinarian's office. Both youths were suspended for three days.

While he was away from school on suspension, two youths reported to Joseph that "if I didn't bring the money that I was going to get whipped or banged." Joseph recalled in court that "I knew that they were going to beat me up, but I didn't know whether they would have a gun." His mother recalled that Joseph had refused to let her become involved: "Joey didn't want to appear cowardly or scared, that's the reason he didn't want me to go up, he told me later he didn't want me to be involved because he felt that we were powerless against the gangs." Another youth in the neighborhood who had heard about the dispute approached Joseph with a small, semiautomatic pistol, which he loaded and test fired.

Joseph bought the weapon and carried it to school, thinking he was most likely to be jumped by members of the Blackstone Rangers as he walked through their territory on the way to or from school. He wasn't likely to go to the police for help. His mother later explained, "young boys in that area, the police and them were at odds. So it wasn't like he felt he could go to the police for help." Joseph reported being chased by members of the gang on the way to school, but the shooting occurred later as he arrived in a hallway at the top of a stairwell inside the school.

Joseph was walking with a girl who testified on his behalf. When they arrived at the top of the stairs, Delondyn Lawson was standing with Duwaun Glover in an area of the hallway that the Blackstones commonly occupied. Duwaun crossed in front of Joseph, and he remembered him saying something like, "Hey, man, what's up with my money?" Another unidentified youth reportedly said, "Man, we didn't come here to talk. Let's do what we gone do so I can put this (guy's) head up in his locker." Joseph continued "and then I began to get hit, and I fell. And when I fell, I didn't fall flat; I kind of braced myself with my hands." The cluster around him included Delondyn, Duwaun and other Blackstone gang members. Attorney Habib asked, "Did you think you could run away at this point?" and Joseph answered that "I had no way of getting out. I tried, but I . . . couldn't."

The ensuing shooting by Joseph White can possibly be explained by the male posturing that is prevalent in both youth and gang culture. In their research from an earlier era in Chicago, Short and Strodtbeck (1965) noted that "gang rivalries were often focused on turf, men defending 'theirs' against the encroachment of another man. Such incursions are a sign of 'disrepect' which leads to men needing to display toughness. Fights were often a matter of chance combined with a tough guy image" (p. 87). Duwaun Glover testified that when he asked Joseph where the missing Blackstone money was, Joseph replied, "I ain't giving your money back, pussy." Joseph's refusing to give the money back and signaling disrespect for Duwaun by calling him a derogatory name ("pussy") probably played a role in provoking the skirmish that escalated into Joseph pulling his gun and shooting into the cluster of students around him.

A considerable amount of time was taken in connected testimony to establish whether the shots that Joseph fired came "defensively" from the ground or more "aggressively" from a standing position. In his own words:

> When I fell, I fell on my arm, and I caught myself. I never was really flat on the floor. And then after—after I—a few seconds, that's when I reached for the gun and begin to get up. And before I got all the way up, I shot the gun. And a guy jumped; either he was jumping to get to the floor so he wouldn't get shot, or he was trying to knock the gun out of my hand. And the gun went off two other times, and at that point I turned around, and I ran.

He insisted in response to questions that his intent was not to kill anyone. "They were never aimed to no particular person. And the fact that the guy got killed, it was not on purpose, sir. My intent was to get the crowd up off me." Joseph said he then ran not because he felt guilty but because he was terrified, "I was scared. I never felt like that again in my whole life. I was afraid."

In his closing arguments, Joseph's attorney argued that the shooting was essentially an act of self-defense:

> Is there anyone of you who would have said, no, I would have stayed there and got beaten? He did what any reasonable person would have done at this point. He took out the gun to defend himself.
>
> Joseph White may be guilty of something, he had no right to bring this gun onto school property. That's a separate charge, though, unlawful use of a weapon on school property.

The prosecution responded:

> How can you say he's afraid when somebody else is hitting the ground, or running, because somebody is yelling he has a gun, and that's why he did not truly then believe he had the right to self-defense, that he was mad, and, unfortunately, he had that loaded gun in his pocket.

The prosecution view was that:

> There is only one person in the whole group that has a dead body to his credit, and that's that guy standing right over there, Mickey Cobra, who settles a dispute with a gun, and you cannot let him do that, and that's going to be your decision, and what you will decide is your message to the rest of society about what is, or is not, going to be tolerated.

The bottom line to the defense was that if the jury did not come back with a verdict of not guilty based on self-defense, then in the alternative they could reach a verdict of guilt based on a lesser charge than first degree murder, for example, second degree murder or voluntary manslaughter.

THE VERDICT

The trial that began on a Wednesday was finished on Friday and the jury retired to reach its verdict. Hours went by without the jury's return. This gave some grounds for hope on Joseph's behalf, but the hope was against the odds. "The dominant issue," Habib later reiterated, "was the fact that he was convicted in the media, before we had a chance." He had not pushed the case to trial, so that "hopefully it would be out of the public mind." But this was an idle hope, for "as soon as we started talking to the jury, picking the jury, almost everybody remembered the case."

Furthermore, Habib found that attitudes toward the case were particularly strong. He observed that "a lot of my clients are South Siders. And the reaction from the community was intensely anti-Joseph White on the grounds that if no matter what else, the idea that he had brought a gun to school was just—I had clients telling me, 'How can you take this case?'" This same view quickly emerged when jury selection began. "The reaction that he had brought the gun to school was looked at as, forget it, he's guilty, that's it." The feeling was that at least to that point the gangs had done their shooting outside school, "but nobody had done a shooting in the schools." This clearly influenced Habib's approach to jury selection: "the sentiment in the community on the South Side was so severe that we made sure when we picked the jury we had a suburban, white middle-class jury."

For Karen White, however, Delondyn Lawson's mother, Linda, was also a jury in her own right. Karen encountered Linda Lawson outside the courtroom.

> I apologized to her. I knew there was nothing, absolutely zero, that I could say or do that could minimize her loss or bring her son back. I was trying to talk to her, to let her know that in no way did I condone this action, if I could do anything in the world to change things, I would. I also let her know that Joey would, because Joey told me, "Mama, I didn't want to kill that boy, I killed somebody." Not only did he take a life, he destroyed all our lives in that single miscalculation. She wasn't mean to me or anything. We both have a terrible sense of loss and like I explained to her, you are there because your son was a victim, he died, but I was there also because my son was a victim and he died, because he lost his innocence.

Karen White knew that her son had already lost a year of his life, and he was likely to lose much of the rest.

The jury finally returned at about 10 o'clock Friday evening. They had deliberated at length and come back with the maximum possible verdict: Joseph White was convicted of first degree murder, two counts of aggravated battery, and one count of unlawful use of a weapon. The sentence exposure on the first degree murder conviction was 20 to 60 years in prison.

As hopeful as Robert Habib was of his client's chances based on the self-defense argument, he conceded that this defense would have been stronger if he could have shown that the Blackstone gang members were also armed. Karen White is still today convinced that Joseph's attackers were well armed, "They had bats, they had car antennas, and they had weapons in the school. They were not coming to him for an old-fashioned fistfight, I think that went out years and years ago." Habib remarked in retrospect,

> Joseph, in one way, was really honest, all right. And had he testified that some of these boys coming after him had knives out, along those lines. His whole thing about self-defense, that he had a right to pull out the gun to shoot at this boy, would have been much stronger. But he never would say that. I asked him several times, "Did anybody pull a knife? Did you see any weapons or knives?" He never did say that.

Habib added that it was an indication of how violent times then were around Tilden that when Joseph sprinted away from the school with gun still in hand, his assailants ran right after him. "That's how tough these kids were, you know, they'd just seen one dead. None of them at least had any guns on them themselves. And they just still took off running after him. So you were dealing with kids that could be very violent." This was the context in which Joseph would be sentenced.

THE SENTENCE

Joseph White was sentenced in the early spring 1994, nearly a year and a half after the fall 1992 shooting at Tilden High. Karen White spoke to the court on behalf of her son. She began by expressing her sorrow for the Lawson family and the loss of their son. She expressed the love of her family for their own son. She explained that "he acted under his perceived notion that he was in mortal danger for his life" and that "given the depressed ghetto area in which this situation took place, it is not at all unlikely or unimaginable that a teen would resort to violence with a weapon." Karen White concluded that her son had made a terrible mistake. "And myself, I have made some mistakes, and I know you have to pay consequences for them. But I don't see what his life or most of his life behind prison walls can serve."

The sentencing judge stressed the issues of personal responsibility and protection of the community. He observed that Joseph White had engaged in earlier acts for which he had suffered few consequences. He focused first on the theft of the stereos from the train boxcar, for which he had pleaded guilty and was otherwise essentially unpunished. The absence of punishment for this theft clearly disturbed the judge, who commented that "of course, all the 15-year-olds, they are well-versed in what happens in this county. And 15-year-olds know that when they steal, nothing happens to them. They go back home." He then noted that Joseph was also a member of the Mickey Cobras and that he apparently had decided "within the confines of that gang in his school, that allowed him certain leeway and rights regarding his conduct with other boys who belong to other gangs." Finally, the judge noted that Joseph had fathered a child that he clearly could not financially support. The judge completed the three strikes analogy by concluding "I think society is tired of people who are 15 who make these kinds of judgments."

The judge retold the events leading up to the shooting and wondered why Joseph had not tried to find other ways of solving the gambling dispute. He asked the apparently rhetorical question, "Why didn't Joseph White go back to school and pay back the $40, go to the principal, apologize for what happened?" An unidentified voice from the public gallery echoed this question, "Why didn't Joseph do those things?" The judge continued by dismissing the self-defense claim, explaining "self-defense does not allow you to go to school with a loaded revolver and then claim later that unarmed people are threatening you and that you are defending your own life." He again noted of Joseph that "he did not seek help from authorities."

An unidentified male voice again was heard and the judge asked the speaker to identify himself, saying, "I don't know who you are. You need to tell us."

> MR. WRIGHT: My name is Kenneth Wright, and I am Joseph White's uncle. I don't condone his wrong. I have one thing to say. The teacher did take the money. If she had came forth with that money—
>
> THE COURT: She who?
>
> MR. WRIGHT: His teacher. They were in the hallway shooting dice; the police came; they broke and ran. They left the money on the floor. She took it. They thought he had it.
>
> THE COURT: A teacher from the school?
>
> MR. WRIGHT: Yes, sir, and she was fired.

Joseph's uncle obviously wanted the judge to have a better sense of why this young man felt so trapped in his situation and unable to turn to his parents, the school, or the police for help. Without condoning Joseph's violent solution, his uncle wanted the judge to consider the circumstances that could mitigate his sentence. "I wanted you to get that part," his uncle pleaded and then conceded, "he had no business with a gun, good enough."

For the sentencing judge, the gun and the resulting killing in a school were everything. He explicitly dismissed the youth or other circumstances of Joseph White as mitigating factors. "It is time for everyone to understand," the judge concluded, "that those people who choose to take guns to settle disputes are accountable for what they do, be they 12 years old, or 15 years old, or 50 years old." Joseph was sentenced to 45 years in the Illinois Department of Corrections on the charge of first degree murder, with sentences on the other charges to be served concurrently.

BEFORE AND AFTER TILDEN

The shooting at Tilden High was an event whose impact was felt well beyond this particular school and its neighborhood setting. This incident was only one violent episode among a series of at least partially connected events that shaped Chicago public school policy on student safety. While there is an inevitable interplay between actors and institutions that in complex ways form the substance of social history, this shooting nonetheless stands out as particularly important in several respects, bringing several salient issues in school reform to a head and marking a turning point in school security and disciplinary policy and practice in Chicago.

Two identifiably different pathways of influence emerged out of the Tilden incident. The first involved mobilization on the part of prominent policy makers in the mayor's office and on the Chicago Board of Education. The second pathway involved change from a more grassroots level as school principals, local school councils, and other actors at the school level sought to prevent a recurrence of the violence at Tilden.

We noted earlier that youth violence rose to alarming levels in Chicago, as in the nation, during the late 1980s and early 1990s. This issue became among the most important in Chicago politics during this period and involved a particular focus on youth gangs and their growing effects on and in the city public schools. Many citizens were concerned that the schools had become sites of both gang recruitment and conflicts. Gang fights inside schools were common, as were gang-related shootings near schools.

The attention of the crime-vigilant media were focused intensively on schools in fall 1989 when a student at another South Side school, Harper High, was fatally stabbed in front of his geometry class (October 1989). This killing prompted Reverend Jesse Jackson and at least one Chicago alderman to call for the installation of metal detectors in all city high schools. Jackson made his case in a speech to students at Harper High. "They have metal detectors to protect pilots and passengers from armed terrorists," Jackson observed. "I'm convinced that high school classrooms should be as safe as airports. Teachers and students should be as safe as pilots and passengers."

The apparent connection between problems of rising youth violence in the community and in the schools also attracted the interest of recently elected Mayor Richard Daley. As a former state's attorney, Daley was ready and eager to tackle this issue. School safety represented the convergence of two of Daley's core campaign issues—crime and education. Efforts to increase the safety of students promised to draw the support of both the teachers and parents of public schoolchildren.

The Daley administration sought a policy solution that could be implemented systemwide and, it was hoped, bring rapid and dramatic results. Such a solution would require widespread compliance and therefore the support of targeted schools as well as the Board of Education—which was charged with setting the overall budget and implementing education policies for the school system as a whole. The support of individual schools was necessary because, under the school system's 1988 decentralization mandate (Illinois Public Act 85-1418), each school's local school council (LSC) was given discretion over most matters of policy implementation—including school security and discipline.

The prevailing attitude among the LSCs was reflected in the comments of Sheila Castillo, executive director of the Chicago Association of Local School Councils, who told the *Chicago Reporter*, "Each [school] should decide which measures are appropriate, which could include at-random searches, selective use of detectors, or anti-gang programs." So it would have been politically imprudent for the mayor or the City Council to impose specific measures on the entire school system without first obtaining a measure of support from the LSCs as well as the Board of Education

The support of the LSCs and the board were by no means guaranteed. First, many local school councils were carefully nurturing their emergent autonomy—especially with respect to discretionary spending—and were therefore disinclined to concede authority to the central administration. Similarly, the board, until 1995, consisted of 15 members who were not appointed by the mayor. This fact, along with the presence of public input before and at each meeting, meant that the board felt accountable to the general public as well as the central administration. One board member, who served during the period 1990 to 1995, reported that she conscientiously fielded calls from the public "all the time." She accordingly described the activity of the board as "very political," resulting in long debates on many issues.

Thus, implementing any systemwide education reforms in Chicago required a broader base of support than is necessary in most policy domains. This task was especially challenging given that Mayor Daley was proposing school safety solutions that were largely unheard of in other cities. In the mayor's own words, "Chicago led the nation and addressed head-on the challenge of improving school safety—taking steps that at the time many felt went too far." The first such far-reaching and controversial initiative was a threefold expansion of Chicago police patrols in public high schools in 1990. This initiative received the necessary budgetary commitment from the board and encountered relatively little opposition from the LSCs. According to a board member, a few LSCs actually did oppose police officers in their schools. However, they were outnumbered by people, such as the principal of Tilden, Hazel Steward, who welcomed police involvement in response to serious problems in keeping order. It is also noteworthy that the LSCs did not incur any direct costs for the added patrols—which may also explain why more concerted opposition to the police never materialized.

Mayor Daley's highest profile initiative—adding walk-through metal detectors to all city high schools—was a completely different matter. Metal detectors were at that time a major issue, in spite of the support the idea also had received from Jesse Jackson. A major concern was that implementation of the metal detectors impinged on the schools in ways even more intrusive than the police. Opposition to metal detectors flowed from several other concerns. Some argued that metal detectors projected the image of schools as dangerous places and drove a wedge of distrust between students and staff. A board member noted that "some of the parents didn't like the idea." They objected "that this wasn't a good image for the schools." Some student advocates further argued that detectors represented a violation of student rights. However, the most telling opposition came from school administrators, who questioned whether metal detectors were needed in their schools, whether they merited the costs in terms of manpower (i.e., trained personnel to operate them and teachers and volunteers to secure unmonitored entrances to school buildings), and especially whether the time taken away from instruction was justified to clear students through the detectors each time they entered the schools. The process of moving students through the detectors could take more than an hour each morning.

Mayor Daley and his subordinates adopted several strategies to neutralize the political opposition to metal detectors. First, the mayor, with the assistance of the school patrol units that he helped establish, amassed and disseminated the statistical information necessary to show that weapons and violent crime in schools were a serious problem. For example, in fall 1991 (September 26, 1991), the school patrol unit reported through a news release that during their first year of operation "they made 153 arrests for carrying guns and 380 arrests for carrying knives and the like." The next year they reported confiscating 192 guns (June 5, 1992). The confiscations—many of which resulted from metal detector use—offered evidence of the potential utility of the detectors to prevent serious violence.

Armed with these fresh figures suggesting the need for metal detectors, Mayor Daley pursued a second strategy of minimizing the costs of the metal detectors to the school system. He (June 5, 1992) announced a plan in the early summer of 1992 to purchase from city funds two walk-through metal detectors and four handheld detectors for any Chicago high school that requested them. This initiative did not utilize funds from the Board of Education's limited budget and thereby preempted fiscal objections on the part of board members. The board responded by voting in September to amend their search and seizure policy and by requiring schools to approve safety and security plans that provided for the use of metal detectors. Tilden was among the first high schools that accepted Mayor Daley's offer.

While Daley managed to win the support of the school board, the facts about gun seizures and the city's coverage of costs were not sufficient, even with board support, to persuade many LSCs to install and operate the metal detectors. As of fall 1992—one month before the Tilden shooting—27 of the city's 75 public high schools and special schools had not yet taken up the city's offer to provide metal detectors (October 22, 1992). Mayor Daley responded by publicly naming each of these schools at a news conference, thus implying that the school administrators—and not the police department—would henceforth be considered responsible for weapons-related injuries or deaths occurring inside their schools.

A principal (October 22, 1992) of a school that had not chosen to install the detectors was quoted in the media as saying that walk-through metal detectors, as opposed to locker searches and handheld metal detectors, presented "staffing problems." Another principal (October 23, 1992) observed, "we thought it would send the wrong message, that we won't trust them and that this is a prison," adding, "we made a decision not to do it for reasons we thought were valid, and now we have to back off on it because of a guilt trip from the mayor."

According to Thomas Byrne, the deputy chief of the school patrol units, these 27 schools were not the only ones resisting the metal detectors. He observed that some schools had accepted the metal detectors but were generally not operating them themselves. Rather, these schools operated their detectors selectively on the occasion of random searches that were conducted, with the help of the school patrol units, by the Board Security, who generally utilized their own metal detectors for this purpose.

Tilden High School was among the latter group. Hazel Steward, who was Tilden's principal through this turbulent period, explains that Tilden's own metal detectors were in operation about once a month prior to the shooting, because "it took a lot of extra staff and police." She added, "It wasn't mandatory, so we did it periodically; that was the way it worked." She explained that it would have been very expensive for the school to hire and train extra security staff to run the detectors each day, requiring funds the school believed it could not justify spending in this way.

Schools that were not participating in the metal detector program at all, and schools like Tilden that were participating partially, were forced to reconsider their policies with the shooting death of Delondyn Lawson. Occurring only one month after Mayor Daley had sounded the alarm about metal detectors, the fatal Tilden shooting underscored Daley's arguments for the use of metal detectors and fulfilled his warning that the schools rather than the city would be held accountable for such incidents. The mayor (November 21, 1992) immediately placed responsibility for the fatality on the school by implying that the shooting could have been prevented if on that day and all others the school's metal detectors were fully utilized. The school superintendent, Ted Kimbrough (November 20, 1992), earlier had been less judgmental, suggesting that the random metal detector searches employed at Tilden at this time were "standard procedure."

After the Tilden shooting, however, no Chicago school principals were quoted as saying that metal detectors were still unnecessary for their schools. The Tilden incident magnified the threat of in-school deadly violence in the minds of formerly skeptical principals, or at least took the political wind out of the sails of any opposition to metal detectors. Nor did any locally initiated alternatives to the metal detector policy receive attention in the press. Rather, metal detectors were generally hailed as a readily accessible and reasonably effective—though unpleasant—solution that could immediately allay growing fears in the community. The press portrayed the metal detectors as a necessary, though not a sufficient mechanism, to cope with the real and severe threat of school violence. A lone voice of dissent was *Chicago Tribune* columnist Bob Greene, who believed that metal detectors would promote safer schools, but at a considerable cost. Metal detectors are "our most awful failure," Greene lamented after the Tilden shooting, "both as a symbol of today and a signpost to our children's future."

Deputy Byrne credits the Tilden incident with laying to rest any lingering doubts about the use of metal detectors in the city's high schools. His view was that after the Tilden incident, everybody jumped on board. "From November 1992 to the present they're in full compliance with the metal detector program. Up until that point it was a tough sell. . . . The metal detector program flew off from that day on. All the excuses, all the talk about manning them, and the searches, all the different problems, from November 1992 and on, it was no longer a problem." Tilden's Principal Steward remembers the period somewhat differently, recalling that about five high schools still declined to install metal detectors after the shooting and complied only when they were required to do so. The Board of Education (*Chicago Reporter*, 1998) spent about $200,000 of its own funds on metal detectors from 1993 to 1998.

The support of the Board of Education and the LSCs was buttressed further by the Chicago City Council. Less than a week after the Tilden shooting, the council approved Mayor Daley's resolution that urged schools without metal detectors to take advantage of the offer to provide a detector to any school that wanted one. The *Chicago Tribune* reported some aldermen at the meeting "blasted school officials who have not installed the devices or are using them selectively."

STILL LATER

The implementation of metal detectors was the first of a series of security and disciplinary initiatives endorsed by the school superintendent and the Chicago Board of Education. In 1995, the Board of Education adopted a new Uniform Disciplinary Code that mandated suspension and expulsion for certain offenses. The new policy was in response to the reauthorization of the federal Gun-Free Schools Act, which required states to pass laws requiring schools to expel for at least a year students who brought weapons to school. The Chicago Board of Education, like others across the country, saw this as the time to revise the Uniform Disciplinary Code, creating mandatory minimum penalties, often including extended suspensions for a variety of offenses, and requiring police referrals for serious instances of assault, battery, disorderly conduct, and other minor offenses and all instances of gang activity, drug violations, and more serious offenses. The resulting set of mandated punishments became collectively and popularly known as the zero tolerance policy.

In subsequent years, the board further increased the penalties for certain offenses. Paul Vallas, who was Mayor Daley's 1995 choice for superintendent of Chicago's public schools, took a leadership role in advocating tough new penalties and policies. In spring 1997, with the encouragement of Vallas, the City Council made student involvement in drug and other serious offenses committed off school grounds subject to expulsion (March 11, 1997). In addition, in 1996 the board passed a resolution mandating that each school adopt a dress code and consider school uniforms as an option. A sizable majority of Chicago public schools now have strict dress code policies.

In February 2000, in the wake of the accidental shooting of an 11-year-old student at Duke Ellington Elementary School, Superintendent Vallas urged the daily use of metal detectors in elementary schools and ordered the purchase of 1,000 additional handheld metal detectors. By April 2000, not long after the discharge of a gun inside Parkside Elementary School, the proposed number of metal detectors had reached 4,000, to be combined with 178 additional off-duty police officers (allowing at least one off-duty Chicago officer in each elementary school), with the money to support these commitments to come from federal antipoverty funds.

These school security initiatives reflect and reinforce an administration, under the proactive leadership of Paul Vallas, that centralized control over local school issues, especially from schools that, including Tilden, the administration had now placed on probation. It seems reasonable to ask whether the central administration would have felt empowered to impose or propose these measures on individual schools if the Tilden shooting had not so vividly set the disciplinary agenda in Chicago. There is no way to unequivocally answer this question. Still, the strong response from the mayor and the board to the Tilden shooting at least set a precedent of centralized control that was repeated regularly during the 1990s.

THE IMPACT

To more fully understand the impact of the Tilden incident for Chicago schools, one must examine the response of particular schools. Notwithstanding the significance of the centralized initiatives, in the decentralized context of Chicago public schools, change must also find support at the local level. Under the State Public Act 85-1418, each Chicago public school is permitted considerable discretion in determining how to prevent and discipline misconduct and crime. Furthermore, schools still vary in the manner and extent to which they implement mandates from the city and the central administration (Bryk et al., 1998).

A logical starting point for our analysis is Tilden High School. Other schools responded in their own way to the incident and are discussed later. Meanwhile, the shooting dramatically and immediately changed several aspects of the school climate at Tilden. Most notably for our purposes, the shooting fostered an almost immediate emphasis on school security and violence prevention.

Changes in security were evident nearly instantly. After the shooting, Principal Steward reestablished control of the situation by ordering everyone to report to their next classroom, under supervision of their teachers. She then dealt with the arriving authorities, who included the police and the superintendent, as well as the press. Steward explains that she kept the students in the classrooms all day in order to prevent any retaliation for the shooting. According to Deputy Chief Thomas Byrne of the school patrol units, keeping the students in the classroom was imperative, so all students could be screened by metal detectors on their way out of the school building to find Joseph White's gun. Despite these efforts, the murder weapon was never recovered.

From that November afternoon forward, the metal detectors became fully operational at Tilden. Steward was quoted in the *Chicago Tribune* explaining that it took one and a half hours for all the students to file through the metal detectors on the Monday morning after the shooting, delaying the start of classes by more than an hour (November 27, 1992). Extra officers were assigned to Tilden after the shooting. Steward hired the staff necessary to run the detectors every day.

While increasing school security was also Steward's priority prior to the shooting, after the shooting she recalls immediately recruiting more volunteers from the community, who were also now more forthcoming. By January 1993, she had used the school's discretionary funds to increase the number of paid full-time security "volunteers" (who were paid the minimum wage) from about 10 to about 42. Significantly, Steward reports recruiting these volunteers from the surrounding community, emphasizing that "I had more volunteers than I did staff, but it was because I used brothers, uncles, fathers, aunts, mothers, so they knew the kids" Steward reports complete support from teachers and the LSC for this action, even though the money could otherwise have been spent on improvements in instruction.

The increase in policing and security coincided with a surge in arrest activity after the shooting. During the 49 school days immediately prior to the November 1992 shooting, 5 arrests were made at Tilden High School. In the same number of school days following the shooting, 17 arrests were made.

Notably, some further short-term security assistance came from an unexpected source—the white community surrounding Tilden High School. She explains, "the community surrounded the school. . . . For several weeks the community showed up for security. . . . It was led by a Methodist minister. . . . They showed up in support of the school. . . . They patrolled outside the building, acting as security."

Steward also integrated the larger community into her programmatic efforts to prevent violence. On the Monday after the shooting, grief counselors were deployed to the school. Sixty students volunteered for grief counseling on the Monday after the shooting. Steward recalls that four crisis teams were operating in the school.

Several initiatives evolved into ongoing prevention programs at Tilden. In response to the shooting, Steward instituted a Rights of Passage Program for teachers and students. She maintains that this program, which emphasizes positive role-taking and leadership skills, empowered students and teachers to unite against gangs and seek alternatives to gang activity. Steward contends that problems of violence already had begun to decline prior to the Tilden shooting, but that these efforts precipitated further declines.

Steward insists that other common weapons against school violence, suspensions and expulsions and more stringent screening of potential students, were not part of her arsenal. Tilden actually reported a decline in suspensions during the 1994–1995 school year. Nonetheless, a former substitute teacher who returned to teach at Tilden a couple of years after the shooting suggests that Steward found less formal ways to exclude gang members from the school. "I don't think that they felt that they had to take just any student that came in off the street," he explained, "And so I felt like there was a lot of . . . screening of students."

Steward disagrees and maintains that the gangs actually did not leave the school but rather "went underground" and shifted their activities to outside the school. She insists that the primary disciplinary changes she instituted were "hall sweeps" to remove students lingering in the halls. These students were informed of the school's rules for hallways and classrooms and about the consequences of breaking them. "They could not bring their activities into the building," she observes, "and if they did, then there were consequences including transferring out. . . . They didn't want to go anywhere . . . and they knew I was serious."

Some of Steward's efforts apparently caught the attention of others in the Chicago public school system. She recalls that other principals sought her help in implementing policies and programs in their schools. For instance, she reports she was asked to give workshops for other principals in "how to react when a crisis occurs." Furthermore, other schools adopted some of her programs, including Rights of Passage. In fall 1995, Steward was appointed to the position of regional superintendent—a position that allowed her to mentor other principals on a full-time basis. As regional superintendent she was also authorized to oversee the disciplinary decisions, such as expulsions, that were imposed at schools in her region.

BROADER IMPLICATIONS

The previous sections addressed the question of whether the tragic shooting incident at Tilden high school had a lasting impact on the school and the system as a whole. We conclude based on our interviews with the principal, teachers, and school board members that the incident clearly expanded the prevalence of metal detectors and the frequency of their use. We also conclude that the incident gave the mayor a new position of increased political leverage over school disciplinary policy matters. Installing metal detectors in all the high schools was an important accomplishment for the Daley administration. As explained, the metal detectors were the first of many successful efforts on the part of the mayor's and the superintendent's offices to increase their authority with regard to security and discipline in schools. The publicized school shootings, even some of those that took place outside of Chicago, incrementally widened a window of opportunity for the mayor's and the superintendent's offices to advance systemwide school safety initiatives.

How do we make sense of the response to the Tilden shooting? Two sociologists, Burns and Crawford (1998), have analyzed official responses to well-publicized school shootings as a "moral panic." We do not believe that this fully or accurately captures what happened in Chicago; nonetheless, this perspective raises important issues.

Sociologists argue that a moral panic can occur when a large segment of the public and its leaders show a heightened concern about a problem and identify a particular behavior or group of people as the cause of the problem. The behavior or group thus pinpointed then becomes the target of policy intervention in response to public opinion and pressure, which is often mediated through the press. Burns and Crawford (1998) describe the policy response to school shootings in the 1990s as a moral panic, because the threat of shootings in schools is often exaggerated in the media, and because this exaggerated picture can stimulate a largely punitive policy response.

The events preceding and following from the Tilden shooting do not fully fit the moral panic analysis. It may be true that the Tilden incident led to exaggerated impressions of the problems of actual weapons in schools or their frequent use. While the locker and metal detector searchers and other police activities did uncover a sizable number of knives and guns, and while surveys did indicate that students frequently brought weapons to school (Illinois Criminal justice Authority, 1991)—there was little evidence that students, especially innocent bystanders, were frequently victimized by students with weapons in Chicago schools. Killings almost never occurred in these schools, and they were not the kind of random shootings early accounts suggested. At Joseph White's trial, it became fully apparent that this was a gang-related incident with gang-involved causes and consequences that were specific and targeted. The title of a *Chicago Tribune* editorial two days after the shooting, "Body Count Rises in the Public Schools" and portrayals of the victims Delondyn

Lawson and Ching Wilson as random bystanders, helped to form a distorted picture (November 24, 1992).

A more convincing case could be made that routine fights were a serious problem that were disrupting learning in many schools. Indeed, a serious melee involving police and students caused a wall to collapse in another school on the same day as the Tilden incident. Accordingly, many board members were pushing for a greater emphasis on teaching nonviolent conflict resolution rather than for metal detectors. The net result is that while concerns about shootings, stabbings, and killings may have been somewhat exaggerated, concerns about violence in the high schools were not.

The specific push for metal detectors is also not entirely consistent with the moral panic framework, in that it appeared to be a relatively measured response to concerns about violence in the neighborhood and guns in the school. Recall that the call for metal detectors was first made by Jesse Jackson and only subsequently taken up by Mayor Daley. Furthermore, in choosing this policy response, the mayor also sidestepped what many might have expected from a nonminority politician and former criminal prosecutor: a call to try Joseph White and others in the adult courts. This transfer to adult court was a decision made by prosecutors and with no public comment from the mayor.

Nor does the moral panic framework adequately account for the long-term course of action on the part of the mayor and the board's central office. It is clear from our observations that Mayor Daley and School Superintendents Kimbrough and later Vallas did not push for metal detectors merely to allay the public concern about school violence. Rather, the metal detectors were more clearly a part of the longer-term ambition of the administration to centralize the setting of school security and disciplinary policy and to make schools accountable to the administration. Other organizational and political interests were also served by the metal detectors. Some of our respondents emphasized that the detectors, which provided more arrest opportunities, helped to make the school police feel safer, while also providing more work for off-duty police officers—which helped the mayor to gain support from the Chicago Police Department.

Subsequent actions of the mayor and the superintendent also do not fit neatly into the moral panic framework. It seems that some incidents, such as the accidental shooting in the Duke Ellington Elementary School and the discharge of a weapon in Parkside Elementary School did not generate an outpouring of public concern that could be characterized as a moral panic. Still, these incidents did provide further examples of unresolved problems that the mayor and the superintendent could point to in reinforcing their calls for more security in the Chicago public schools. The school board security initiatives in the later 1990s, especially the zero tolerance policies, occurred after the heightened anxiety over the Tilden shooting had largely subsided.

It might have made more sense in the context of reduced and declining violence to reassure the public—after each of these accidental shootings, or after each incident on the national stage, such as the one at Columbine High School in Littleton, Colorado—that Chicago schools were now largely free of shootings, and indeed never did experience the kinds of random rampages experienced in less urbanized American settings. This possibility merits more attention than it has yet received in drawing lessons from the Tilden High School and subsequent Chicago experiences.

Still, it is noteworthy that nearly 800 weapons were confiscated in Chicago public schools between January 1987 and October 1990. Since the Tilden incident, no student has been killed—and apparently only one shot (accidentally)—inside a Chicago public school (February 15, 2000). Deputy Byrne attributes this fact to the metal detectors as well as to the concerted effort on the part of the school patrol units, which began before the Tilden shootings, to make arrests for crime in or near schools. Byrne summarized his view of the issue by observing that "God only knows how many crimes we prevented through all the guns that we confiscated and arrests that we made." At least one other member of the school board we interviewed also attributes the absence of any school shootings to the metal detectors. Indeed, this board member also credits the metal detectors with declines in problems of gang intimidation, as students harbored less fear of gang members.

The nine-year hiatus in fatal shootings may seem remarkable for a public school system in a city the size of Chicago, especially in light of recurrent national anxieties about school shootings since the mid-1990s. Still, the claim that metal detectors deserve the credit cannot be accepted uncritically. First, one must rule out the possibility that this outcome would have been achieved in the absence of any policy changes. While this possibility may seem remote on its face, it bears note that killings in schools and other in-school shootings actually were rare in the several years prior to the Tilden incident as well. A search of the archives of the *Chicago Tribune* since 1985 revealed only one murder inside a Chicago public school prior to the Tilden incident—and these were years in which the juvenile homicide rate in Chicago was higher than it was during most of the 1990s.

Second, one must rule out competing explanations for the absence of school shootings. As mentioned, Chicago police officers had been assigned to patrol the city's high schools since September 1990. In addition, the Board of Education doubled their security budget, helping schools hire more security guards. Many improvements to the physical conditions of school facilities also occurred during this period. The wave of exclusionary policies, which began with the zero tolerance disciplinary code in 1995 and included other policies, such as increased alternative schooling for youths with disciplinary problems, may also have played a role. However, as Hazel Steward observed, exclusion of students was very common before the zero tolerance initiative began. In the earlier period, exclusion was accomplished informally, as problem students were simply "told not to come back" and were recorded as dropouts.

The observations of Principal Steward at Tilden High School remind us that individual schools adopted a variety of approaches to steer their students away from gangs, drugs, and violence. A full explanation of why Chicago public schools now seem so remarkably safe must take into account not only the policy response from above, in the mayor's and superintendent's offices, but also from below. The leaders in the trenches of Chicago public schools responded to the tragedy of Tilden High School with courage and innovation.

Joseph White is still serving the first quarter of his 45 year sentence for the Tilden shooting, at Menard State Prison in Southern Illinois, 300 miles south of the school where it happened. Even if he is paroled after serving half of his sentence, the child he fathered before being incarcerated will be an adult before Joseph White leaves prison. This part of his life is lost forever. His mother was the first to say the shooting was a tragedy for everyone involved.

REFERENCES

Barrett, J.
1987 *Work and Community in the Jungle: Chicago's Packinghouse Workers, 1894-1922*. Urbana: University of Illinois Press.

Bennett, W., J. Dilulio and J. Walters
1996 *Body Count: Moral Poverty and How to Win America's War Against Crime and Drugs*. New York: Simon & Schuster.

Block, C.R., and C. Martin
1997 *Updated Graphs: Major Trends in Chicago Homicide: 1965-1995*. Chicago: Illinois Criminal Justice Information Authority.

Blumstein, A., and R. Rosenfeld
1998 Explaining recent trends in U.S. homicide rates. *Journal of Criminal Law & Criminology* 86:10–36.

Bryk, A., P. Sebring, D. Kerbow, S. Rollow and J. Easton
1998 *Charting Chicago School Reform: Democratic Localism as a Lever for Change*. Boulder, CO: Westview Press.

Burns, R., and C. Crawford
1998 School shootings, the media and public fear: Ingredients for a moral panic. *Crime, Law & Social Change* 32:147–168.

Chicago Area Geographic Information Study (CAGIS)
1990 New City Data Source. Available at: http://www.cagis.uic.edu/demographics/demographics_intro.html [Accessed 8/13/2002] and *New City Community Area Map*. Available at http://www.cagis.uic.edu/cgi-bin/camap_get [Accessed 8/13/2002].

Chicago Commission on Human Relations
1922 *The Negro in Chicago: A Study of Race Relations and a Race Riot.* Chicago: University of Chicago Press.

Cohen, L.
1990 *Making a New Deal: Industrial Workers in Chicago, 1919–1939.* Cambridge: Cambridge University Press.

Fox, J.A.
1995 Presentation to the meeting of the American Academy for the Advancement of Science. Atlanta, GA.

Hagan, J. and H. Foster
2000 Making corporate and criminal America less violent: Public norms and structural reforms. *Contemporary Sociology*, February, 44–53.

Halpern, R.
1997 *Down on the Killing Floor: Black and White Workers in Packinghouses, 1904–1954.* Urbana: University of Illinois Press.

Halpern, R., and R. Horowitz
1996 *Meatpackers: An Oral History of Black Packinghouse Workers and their Struggle for Racial and Economic Equality.* New York: Monthly Review Press.

Illinois Criminal Justice Authority
1991 *Trends and Issues 91: Education and Criminal Justice in Illinois.* Chicago: Criminal Justice Authority.

Knox, G.
1992 *Schools Under Siege.* Dubuque, IA: Hunt.

Short, J.F., and F. Strodtbeck
1965 *Group Process and Gang Delinquency.* Chicago: University of Chicago Press.

Sinclair, U.
1995 *The Jungle.* Chicago: Grosset and Dunlop.

Slayton, R.
1986 *Back of the Yards: The Making of a Local Democracy.* Chicago: University of Chicago Press.

Stephens, R.
1992 Congressional testimony: Weapons in school. National School Safety Center. October 1, 1992.

Suttles, G.
1968 *The Social Order of the Slum: Ethnicity and Territory in the Inner City.* Chicago: University of Chicago Press.

Thrasher, F.
1927 *The Gang.* Chicago: University of Chicago Press.

Yablonsky, L.
1997 *Gangsters: Fifty Years of Madness, Drugs, and Death on the Streets of America.* New York: New York University Press.

7

What Did Ian Tell God? School Violence in East New York

Mindy Thompson Fullilove, Gina Arias, Moises Nunez, Ericka Phillips, Peter McFarlane, Rodrick Wallace, and Robert E. Fullilove III

Thomas Jefferson High School, in East New York, was the scene of two episodes of school violence during the 1991–1992 school year. On November 25, 1991, Jason Bentley shot a teacher, Robert Anderson, and shot and killed a fellow student, Daryl Sharpe. On February 26, 1992, Khalil Sumpter shot and killed Tyrone Sinkler and Ian Moore. Although the first shooting had been a shock to the school, the neighborhood, and the city, the second shooting was something more than that. Following on the heels of the first episode and occurring on a day when the mayor was scheduled to speak at the school, the incident took on enormous weight. In brief, the intrusion of violence into the school was read as the breaching of one of the last sanctuaries in a city wracked by violence. Whoever was to blame—and many candidates were proposed—it was surely a terrible and intolerable state of affairs that had come to pass.

That the shootings had a logic of their own, albeit one related to the surrounding violence, was lost in the apocalyptic rhetoric that gripped the press and drove the discourse about the events. Lost in the hubbub was the remarkable work of a few calm people who understood the big picture and kept trying to make it better. Equally lost was the pain of the youth—not just the shooters or their victims, but a whole generation of young people—who were coming to maturity at that peculiar moment in the history of East New York.

In the following pages, we tackle three areas of investigation. First, we describe the two episodes of fatal violence and their aftermath. Second, we explore the context in which those events occurred. Third, given that this paper is written 10 years after the events, we take a retrospective look at the meaning of the events.

A NOTE ON METHOD

This project employed situation analysis,[1] a theoretically derived qualitative method, to conceptualize data collection and analysis. For these purposes, a "situation" is defined as a complex interpersonal episode. Situation analysis is used to specify the setting, the actors, their roles, the rules governing their behavior, key scripts they follow, and cultural assumptions governing action in order to arrive at a "definition of the situation." A central theorem governing situation studies is the Thomas theorem, which states, "If men define situations as real, they are real in their consequences."[2] Accordingly, perception is accorded great weight in understanding the processes that drive human behavior.

Triangulation is a key method in situation studies. The mathematical concept of triangulation refers to using trigonometry to locate a third point using bearings from two known points, a fixed distance apart. As applied in situation studies, triangulation refers to the collection of different kinds of data, from different points of view, in order to arrive at the "definition of the situation." Typically, such exercises are guided by theory: in this case, the work is informed by George Engel's biopsychosocial model,[3] which postulates that illnesses and disorders occur within nested systems, hierarchically ordered such that higher-level systems control lower-level systems. The systems we investigated, from highest to lowest, are: the city, the neighborhood, the small group, the family, and the individual.

For purposes of this study, a multicultural team, including three young people literate in "hip hop" culture, was organized. It was assumed—correctly as it turned out—that young people would be essential to the effort of comprehending the culture of the place and the time.

The team collected information from multiple sources representing divergent points of view. Box 7-1 is a chronology of the events described in this case study. We spoke formally and informally with many people who were aware of the incidents and/or life in East New York in the period of interest. Formal interviews were conducted with 55 people, including Khalil Sumpter, Jason Bentley, and Joseph Fernandez, then Chancellor of New York City Schools. We were heartened by the helpfulness of many members of the East New York community, as well as by the high level of cooperation we received from all of those involved with the legal system: judges, attorneys, and members of the police force. A few people who held key positions of responsibility at the time of these events refused to speak with us, including: David Dinkins, then mayor;

BOX 7-1 Chronology of Key Events in the East New York School Shootings

4/90	Khalil Sumpter and Tyrone Sinkler arrested for robbery; Sumpter got probation; Sinkler got one year detention. This led to ongoing "beef."
10/91	Jermaine Bentley and Jesse Thompson had a dispute over alleged "disrespect" shown to Thompson's sister. This led to "beef."
11/25/91 Monday	Jason Bentley, intent on protecting brother Jermaine from Thompson, fired three shots, missing Thompson but killing Daryl Sharpe and wounding Robert Anderson.
11/29/91 Friday	Daryl Sharpe's funeral
2/92	Sumpter perceived that Sinkler was repeatedly threatening him and his family.
2/25/92 Tuesday	Sinkler stepped on Sumpter's shoe—they had an angry confrontation. Later, Sinkler allegedly fired shots at Sumpter.
2/26/92 Wednesday	Sumpter fired two shots in second floor hallway at Thomas Jefferson High School, killing Sinkler and Ian Moore. Mayor David Dinkins was on his way to the school at the time of the shootings. Later that day Marlon Smith, friend of the deceased young men, shot himself.
2/28/92 Friday	Marlon Smith died at Brookdale Hospital, Brooklyn.
3/1/92 Sunday	Mayor Dinkins announced $28 million school safety initiative.
3/2/92 Monday	Funeral for Tyrone Sinkler.
3/3/92 Tuesday	Funeral for Ian Moore. Reverend Johnny Ray Youngblood asked, "What is Ian going to tell God about us?"
3/4/92 Wednesday	Funeral for Marlon Smith.
8/3/92 Monday	Bentley sentenced to 3 to 9 years in prison for manslaughter.
9/7/93 Tuesday	Sumpter sentenced to 6 2/3 to 20 years in prison for manslaughter.

Bruce Irushalmi, then the head of school security; and Carol Beck, then the principal of Thomas Jefferson High School. Lena Medley, principal of Thomas Jefferson High School at the time of this study, also refused to participate. We also did not interview any members of the victims' families as we were unable to obtain contact information for them.

In addition to interviews, a number of outstanding books have been written on the East New York community. Luck was with us: journalist-teacher Greg Donaldson's book, *The Ville: Kids and Cops in Urban America,*[4] based on his experiences in Brownsville and East New York in the 1991–1992 period, included detailed reporting on the neighborhood of East New York, Thomas Jefferson High School, as well as the two episodes of school violence described here. Other books used to develop this case study include: *Upon This Rock,*[5] *Getting Paid,*[6] and *Will My Name Be Shouted Out?*[7] We collected news articles from *The New York Times, New York Newsday, Time, Newsweek,* as well as major newspapers from other U.S. cities. We obtained a copy of the record of Khalil Sumpter's trial; Jason Bentley's case did not go to trial, so no comparable record was available. Statistics on crime, school performance, and housing projects were collected from the appropriate city agencies.

Analysis of the qualitative data was conducted using Atlas.ti,[8] a software package developed for the analysis of qualitative data. Detailed notes of all formal interviews were entered into Atlas.ti and coded for key themes that emerged from the data. Conceptual mapping was used to develop the relationships of the themes to each other. Ecological analysis was conducted using Infoshare, a geographical information system program for New York City that includes census and health data organized by zip code.

A draft of the case study was written and refined based on two processes. First, we assessed the accuracy of the manuscript in light of the raw data, including interview notes, articles, statistical information, and other literature, looking particularly for any data that might refute statements made in the text. Second, we sent copies of the draft to interviewees who had agreed to review and comment on it, including both Jason Bentley and Khalil Sumpter. The final version incorporated new information and comments from readers. The authors are entirely responsible for the content of the final document.

This case study has several limitations. First, the brief period of fieldwork limited our ability to understand the complexity of the East New York community. Second, the absence of some key actors and victims' families from our dataset created a bias toward the "told" story. Third, the passage of time has obscured some parts of the story but made others clearer. To some extent, we are telling the story as it is known now, as opposed to how it happened then. We were aware of all of these issues throughout the data collection and analysis and attempted to balance the effects when possible.

The institutional review boards of Columbia University, the New York State Psychiatric Institute, and the National Research Council approved this study.

THE EVENTS

"Let's Finish It Now"

Jason Bentley grew up in East New York, a bright child protected by loving parents.[9] He did well in school through the first six grades. His life, however, was deeply affected by living in a neighborhood that was engulfed in violent drug wars. Jason saw violence in the neighborhood from a very young age. He was, as well, surrounded by a crime-dominated street world that had a powerful allure for boys. An event in eighth grade stood out in his mind: he had the opportunity to go to a Catholic high school, but he decided not to take the scholarship test. This was, as he saw it, a decision to follow his brother on what he called in 2001 "the wrong path": the path of the streets. At about the same age, he started to carry a gun, thinking that he needed it for protection.

In October of Jason's freshman year at Thomas Jefferson High School, a small dispute began. As he recalled, a friend had tried to "talk to a girl." When she said she wasn't interested, the boy called her a "bitch." She reported this to her brother, incorrectly naming Jason's older brother, Jermaine, as the person who had "disrespected" her. This led to "beef"—a word used in East New York to refer to interpersonal disputes—between Jermaine and the girl's brother, Jesse Thompson. The beef escalated over the ensuing month. Jermaine was not particularly competent by the standards of the streets of East New York. Jason, who was not only loyal but also protective, was brought into the beef on his brother's side.

Jason, who had been cutting class a great deal, remembered going to school on November 25, 1991, with the intention of going to class and attending to school work.[10] His brother Jermaine's midmorning call for help changed that plan. Jermaine needed his brother's aid to confront Jesse Thompson in the hallway on the third floor of the school. At first the two were fighting, and others were interceding to break it up. Just as the combatants were pulled apart, Jason thought he heard someone say, "Let's finish it now." Jason, watching Thompson reach for his book bag, thought that Thompson was planning to get out a gun and shoot his brother and him. Jason pulled out his gun and fired off two shots. "I did what I had to do," he told us.[11] One shot hit a fellow student, Daryl Sharpe, who was nearby in the hallway.[12] Another hit Robert Anderson, a teacher who had stepped out of his classroom to investigate a noise that sounded like firecrackers.[13] Jason Bentley ran from the scene but was quickly apprehended by the police.

Daryl Sharpe was taken to a local hospital, where he was pronounced dead a few hours later.[14]

Robert Anderson was also taken to a hospital and treated for the wound from which he would eventually recover.[15]

At Thomas Jefferson High School, Principal Carol Beck called a special assembly of the students and announced what had happened. Students were shocked and dismayed. School was dismissed early.

In the immediate aftermath of the shooting, offers of help came in from community organizers, religious leaders, and the political and educational leadership of the city. Grief counseling and increased security were immediately instituted. A grief room at the school provided a site for ongoing discussion of the events of November 25. Mourning had a prominent place in the life of the school, antedating this event. A large number of Jefferson students had died due to violence in the community—between 30 and 75 in the preceding five years, the number varying in different reports[16]—and Principal Beck had wanted a place for young people to talk about the traumatic events that were occurring.

In the grief counseling room, Jermaine Henderson, 17, spoke of the moments after the shooting. "[Daryl] kept saying, 'Get me up.' The more he talked, he kept losing his voice. He pointed toward his chest. I unbuttoned his collar. There was a hole in his neck. . . . If he had stepped the other way, that could have been me in the newspaper today."[17] Henderson himself carried a bullet in his head and another in his shoulder from random street violence. One report noted that 50 percent of Jefferson students had wounds from violent injury.[18] In our fieldwork 10 years later, people on the street spontaneously lifted their shirts to show us the marks left by violence in the early 1990s.[19]

Sharpe's funeral was held on Friday, November 29, at Messiah Baptist Church in Brooklyn. It was reported in *The New York Times* that Reverend Elijah Pope offered comfort to the mourners, who included some officials from the Board of Education.[20] He then commented that many city agencies were missing: "Do we have anyone here from the Mayor's office? Do we have anyone from the Governor's office? Anyone from the Police Department? From Lee Brown's office?" None of those major city agencies was represented. The minister interpreted this as a sign of disrespect and lack of concern. Representatives of various agencies offered excuses to the press.

A newspaper report noted that a woman, who said she was Sharpe's English teacher, read from an essay he had written:

> I have many visions of my life. I sometimes wish it could always be a good life. I, too, wish to be rich and successful. But when I look beyond the material things and at the world the way it is, I say to myself, "Why am I so selfish? I love feeling good about myself, but I feel so much better when I do a good deed."[21]

A second assembly, in memory of Daryl Sharpe, was held on December 9, 1991. At that assembly, senior Shawn Cameron was reported to remark, "It could have been anybody on the floor, but it was one of my closest friends. This crime could have been prevented. If the city would have listened to our cry for help, the gun that the person possesses should have never entered the school."[22]

Members of the clergy spoke, as did Chancellor Joseph Fernandez and president of the Board of Education, Carl McCall. The appearance of rap star Doug E. Fresh introduced a note of pandemonium, as girls' squeals filled the auditorium and undermined the solemnity of the event.[23]

Jason and his family received little attention in the press, except for an article in *The New York Times*.[24] That story described Jason's parents, Rudolph and Sally, as hard-working people who lived in an "oasis of neatly tended private homes with red and white aluminum awnings." They were members of St. Michael's Catholic Church. Reverend Brendan P. Buckley was quoted as saying, "If I knew Jason came from a family that wasn't a support structure like his is, then I would say, 'Ah.' But that's not the case. What is the lesson here? I don't know. I would hope that the pain itself would be part of the process of learning. We must pray for an end to the violence."

Buckley's question was echoed by many. The incident, though shocking and curious, quickly slid into obscurity.

The Day the Mayor . . .

Like Jason Bentley, Khalil Sumpter grew up in the violent atmosphere of East New York.[25] Like Jason, Khalil was bright and did well in school, falling off at approximately the same age. Donaldson described Khalil's behavior at Thomas Jefferson High School, noting, "His records show that he is a very smart kid who has decided that academic achievement is counterproductive to his goal of gaining and holding props."[26] Instead, he was described as playing the role of a class clown. In addition to his ambivalence about schoolwork, Khalil had emotional problems, probably beginning with his parents' separation, which occurred when he was 5. More recently, the stabbing of a friend had affected him deeply. Finally, Khalil, like other youths in East New York, lived cheek-by-jowl with violence: as one example, he lived around the corner from the home of Yusef Hawkins, the young black man whose death at the hands of an angry white man had tormented New York City in 1989.[27]

That winter, Khalil was constantly worried by a long simmering feud with Tyrone Sinkler. The two boys were arrested for a robbery in 1990. Khalil got off, but Tyrone had to spend a year in youth detention. Tyrone believed that Khalil had "ratted him out" to the police. His "beef" with Khalil escalated that school year: he threatened Khalil, and, according to Khalil and his family, called his home and threatened his parents, as well. Against this backdrop of chronic tension, two events on February 25th appeared to have been the immediate catalysts for the shooting. The young men had an angry encounter at school during which Tyrone stepped on Khalil's shoe, and, later that day, according to Khalil, Tyrone fired some shots at him.

Tyrone Sinkler was much bigger than Khalil, and ferocious whereas Khalil was not. Khalil feared for his life, and he feared for the safety of his family. As he explained his situation, "You can put it off, but sooner or later you have to play your cards." Supplied with a gun by one of his friends, Khalil slipped into the school through a side door.[28] He heard that Tyrone and his friend, Ian Moore, were "looking for you," an East New York euphemism for "search with intent to harm." Khalil ran into them on the second floor of the school. Fearing that they intended to shoot him, he pulled his gun and shot Tyrone in the head and Ian in the chest. A bystander was quoted in the newspapers as saying, "He was aiming at both, he hit both, he killed both."[29] Like Jason, Khalil fled but was quickly arrested.

An eyewitness, whom we shall call Bill, described the scene to us. As he walked in, he saw Ian and Tyrone with some other guys.[30] He remembered going up the staircase and seeing Khalil with his friend Dupree. He greeted them and turned to talk to a young woman, who instantly pointed out, "Ooh, there go Ian and Tyrone, ooh, there go Khalil. They about to get into it." Bill thought the three were going to scuffle. He thought he heard Ian or Tyrone say, "Yo, whaz up" to Khalil. The next thing he knew, Khalil had pulled a gun out of his jacket. As soon as he saw the gun, Bill grabbed the back of his head, started ducking, and ran to his coach's office for help. The coach was the first adult on the scene. Seconds later, a security guard arrived to secure the crime scene. "I was hysterical because I was that close," Bill remembered. Despite the intensity of the moment, he pointed out that he did not dwell on it. There were so many murders; these were in many ways indistinguishable from the others. "Back then," he explained, "people [murdered each other] over a beef. It just wasn't nothing."

The drama of the day was intensified by the fact that David Dinkins, then mayor of New York City, was on his way to the school. A heightened level of security was in place, and several security guards and police officers were within yards of the site of the killings. In all, 25 security personnel and police officers were on the school grounds.[31] The association of the murders and the mayor was burned into people's minds. In the course of our interviews and conversations, dozens of people, when asked if they remembered the events, began their comments by noting that the mayor was on his way to the school when it happened. That the day had meaning for people was underscored by the comments of one expatriate New Yorker who commented, "Of course I remember that day. That's the day I decided to leave New York. Not because of those murders, but because of another murder near my son's school which went unnoticed because everyone was so preoccupied by what had happened with Mayor Dinkins."[32]

The mayor was briefed about the murders by Inspector Patrick Carroll, who was then the commander of the 75th Precinct. On his arrival at the school, Dinkins joined Principal Beck for a school assembly. He urged students to choose nonviolence as a way of life. *Newsday* reported that he told students, "If you know that somebody's got a weapon, you make sure that you pass that information on. You might think that that's being a snitch. Well, you consider the alternatives. We've got two young people lying dead today and it might not have happened had it not been for that gun."[33]

The horrors of the day did not end with the Sumpter shootings. That evening, Marlon Smith, a friend of the two young men who had been killed, shot himself. He died two days later at Brookdale Hospital.[34] Marlon was eulogized on March 4, one day after Ian, and two days after Tyrone Sinkler.[35] A Thomas Jefferson teacher shared with us that there were actually five funerals that week for Jefferson students. In addition to Ian, Tyrone, and Marlon, he recalled that two others had died in unrelated events.[36]

The funeral of Ian Moore, held at St. Paul Community Baptist Church, seems to have received the most attention of the five.[37] By contrast with the obsequies of Daryl Sharpe, a host of luminaries joined the 1,000 mourners gathered at the church. At the service, a poem written by Ian Moore was read:

> What I fear began when my grandmother died
> Obviously, it was the fear of death
> Death is something I just can't handle
> When she died, it was so unbelievable
> I fear death because I don't know
> What will happen when I go

It is something I can't face
When I die, will I be thought about?
Will my name be shouted out?
Death will come at anytime
No matter how far you're up the ladder.[38]

Mayor Dinkins's comments emphasized nonviolence. The mayor said, "Ian Moore is done with the trouble of the world, gone home to God."[39] Reverend Johnny Ray Youngblood, senior minister at St. Paul, remembered that he bristled at that thought.[40] In his remarks, Reverend Youngblood pointed out that Ian's death was not unique. "The tragedy is that what used to be unusual is now usual."[41] He explored the roots of the current state of affairs, and raised the issue of adult responsibility.

> I heard the mayor say something that shook me. He said that Ian has gone to be with God. That's frightening. That's frightening because if Ian has gone to be with God, what's he going to tell God about us when he gets there? If God has entrusted our children to us and we are his babysitters what are these young children gonna report about us when they stand before God?[42]

In pointing out that the "unusual is now usual," Reverend Youngblood was describing the rigors of life in East New York. Before proceeding to investigate the aftermath of the shootings, it is essential to sketch some of the key features of that troubled neighborhood.

East New York "Back in the Day"

Three features stand out as defining East New York for purposes of this case study. First, it was internally fractured, with clusters of massive housing projects forming an archipelago of mutually antagonistic residences amidst acres of burned and abandoned housing. Second, it was flooded with guns and drugs. Third, it was isolated from the rest of New York.

The fracturing of East New York occurred in the late 1960s, as a result of redlining and disinvestment by public and private organizations.[43] The area quickly burned down. At its nadir, East New York had 250 acres of vacant land. Much of what burned were private houses and apartment buildings. What remained was the public housing. As depicted in the map (Figure 7-1), 10,549 apartments were built in 11 massive blocks, scattered throughout the East New York-Brownsville area (see Table 7-1). These housing units were always separated in space, but the massive destruction of housing meant that they became separated by dead space, most starkly in the area next to Thomas Jefferson High School, which was universally referred to as the "dead zone."[44]

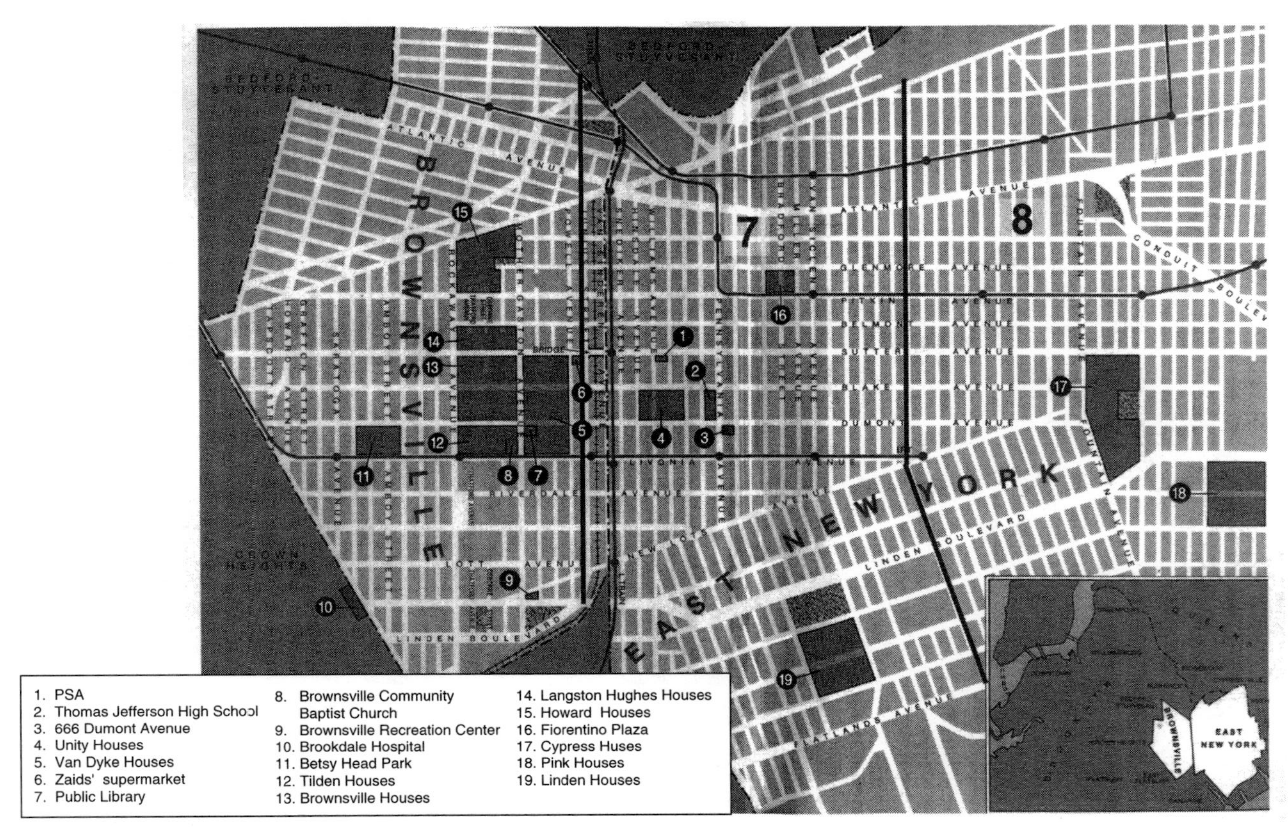

FIGURE 7-1 Map of the East New York section of New York City.

TABLE 7-1 Housing Projects in the East New York-Brownsville Area[a]

Development	Buildings	Apartments	Population	Completed
Unity Plaza[a]	5	482	1,310	9/30/73
Unity Plaza[b]	3	187	476	11/30/73
Van Dyke I	22	1,601	4,330	5/31/55
Van Dyke II	1	112	126	4/30/64
Tilden	8	998	2,850	6/30/61
Brownsville	27	1,319	3,762	4/16/48
Langston Hughes	3	508	1,357	6/30/68
Howard	10	814	1,978	12/31/55
Cypress Hills	15	1,442	3,526	5/31/55
Pink Houses	22	1,500	4,011	9/30/59
Linden	19	1,586	4,023	6/30/58

[a]Sites 4, 5A, 6, 7, 9, 11, 12, 27.
[b]Sites 17, 24, 25A.
SOURCE: New York Housing Authority (2001).

Drugs and guns, a feature of community life since the beginning of the decline in the 1960s, took on new proportions with the entry of crack into the community in the mid-1980s. By 1993, East New York led the city in the number of homicides. Arrests related to drugs and weapons possessions were strikingly high. The battle among drug gangs for sales territory terrorized the residents, inspiring a culture of violence that crippled civic life and distorted child development.

All of these problems were aggravated by the community's isolation. When Reverend Johnny Ray Youngblood was called to serve as minister of St. Paul, one of his friends told him, "You are being sent to one of God's Alcatrazes."[45] This image of a brutal island prison captures a piece of the isolation that characterized East New York. It is a neighborhood away from the major expressways. It lacked cultural, recreational, or commercial facilities that might tempt outsiders to visit. It was far from the center of either New York or Brooklyn. There was no reason for the average New Yorker to ever see East New York. It was equally possible for an East New Yorker to never leave the neighborhood. People who worked with the youth of East New York repeatedly commented on their surprise that the youth never left the area.[46] Conversely, because there was no employment in East New York, parents had to leave for work. This had many consequences for the community, among them that the mothers, who tended to be the organizers in comparable poor communities, were unable to be involved in local issues because of their long commutes.[47] The isolation meant that rescue was delayed or nonexistent. The school shootings, by contrast with many area problems, focused a spotlight on the area and brought people and resources theretofore missing. Some of that attention is described in the next sections.

IN THE AFTERMATH OF THE SHOOTINGS

Assessing Violence in New York City Schools

The shootings at Thomas Jefferson High School were of interest to public health professionals, who had begun to conceptualize violence as a health problem. At the time, homicide was the leading cause of death among New York City youth 15–19 years old, and the second leading cause of death in that age group nationally.[48] Obviously, the increase in violence among youth undermined well-being and diminished life expectancy. In March 1992, the New York City Department of Health, under the leadership of Margaret Hamburg, asked the U.S. Centers for Disease Control (CDC) to assess violence in New York City high schools. Richard Lowery, of the CDC, spent a month in New York City conducting the study. He developed and field-tested a survey instrument that was administered by school officials. He remembered that school officials were clearly shocked by the violence at Thomas Jefferson High School. His study did not focus on that school, however, because the school officials were interested in data representative of the whole school system, rather than an anecdotal report on one incident.[49]

The 1,399 students surveyed were from a representative sample of high schools stratified by presence or absence of a metal detector. Lowery's questionnaire asked them about weapon carrying, beliefs about weapons, and threats in the 1991–1992 school year. Threats of physical harm were reported by 36.1 percent of the students, while 24.5 percent had been involved in a fight. More than one in five (21 percent) reported carrying a gun, knife, or club. Students who had been in fights and those who carried weapons were more likely than their more peaceful counterparts to believe that these weapons were (1) effective ways to manage difficult situations and (2) methods endorsed by their families. The editorial note that accompanied the report concluded, "This survey of NYC public high school students suggests that violent behaviors reflect the personal attitudes of students and the attitudes students attribute to their families."[50]

The study also found that students were less likely to bring weapons to school if they attended a school with a metal detector. This was interpreted as providing support for the use of metal detectors on the basis that they might reduce, although they did not eliminate, weapon carrying in school. The report suggested a variety of other interventions that might be undertaken, including encouraging parents and community groups to teach youth nonviolent conflict resolution. In Lowery's view, the physical security offered by metal detectors was only a small part of what was needed. The CDC worked with the Board of Education and the Department of Health to put together comprehensive antiviolence recommendations that all could support.[51]

Improving School Safety

That New York City schools were unsafe was not news to the chancellor or his staff.[52] They had been asking for increased funding for metal detectors and other programs, without success. In addition, they had worked hard to convince principals to accept these programs. Carol Beck had been particularly opposed to the institution of metal detectors, fearing that they made school seem like a prison. The Sumpter shooting shifted attitudes dramatically: those unwilling to fund the security programs became ready to do so, and those opposed to their use then were now ready to accept them.

At the time of the shooting, having a "metal detector program" meant that a team of security personnel came to the school on a rotating basis and checked students using handheld instruments. On February 27th, the day after the Sumpter shooting, Chancellor Joseph Fernandez told a meeting of high school principals that he would ask the city for money to put a metal detector team at each of the city's 120 high schools.[53] According to newspaper reports, metal detectors were "immediately" placed at Jefferson on a daily basis.[54]

Politicians worked quickly to secure funds. State assembly speaker Saul Weprin, a graduate of Thomas Jefferson High School, joined with other leading politicians in the effort to obtain financial support. By March 1, 1992, Mayor Dinkins had announced a $28 million plan to improve school safety. The plan included the following elements:

- Metal detectors will be used on all school days at Thomas Jefferson, Erasmus Hall, George W. Wingate, and Samuel J. Tilden High School in Brooklyn and James Monroe High School in the Bronx. These high schools have had high incidences of violence and weapons seizures.
- After a Police Department audit of the system's 120 high schools, three schools will be added to the upgraded security program.
- The Police Department will conduct daily patrols inside and outside these schools. Each school will also appoint a security coordinator.
- The Department of Mental Health and the Department of Youth Services will develop supplemental programs at the 40 high schools and in their communities dealing with violence prevention and crime.
- There will be additional after-school programs at these schools and violence prevention training for teachers and students.

- The Board of Education and other agencies involved will submit reports on the school safety program to deputy mayors Barbara Fife and Norman Steisel every two weeks.
- The city comptroller and the School Construction Authority will be asked to speed up the purchase of metal detectors.[55]

The United Federation of Teachers, the New York City teachers' union, argued for other safety measures as well. They urged that "dangerous kids" be placed in "special settings." They demanded the transfer of 25 to 30 students out of Thomas Jefferson High School. This was a controversial proposition, not least because none of the students involved in the shootings, whether shooter or victim, appeared to fit the description of "dangerous kid." The chancellor was not in favor of such a program and resisted wholesale removals without due process.

School transfers were a tool that parents considered as a way of protecting their children. Mass transfers were feared in the immediate aftermath of the second shooting. In fact, *Newsday* erroneously reported, "More than 400 students asked to transfer out . . . in the wake of the shooting deaths Wednesday."[56] Carol Beck was quoted in the same piece as saying, "We got through a very trying day. There were many parents who came, some apprehensive and concerned, but they saw the strength of the students who were here. We aren't acting so quickly, snatching the children and discharging them." Her comments acknowledged the very real fear of mass transfers. That the transfers did not materialize may be attributable in part to the massive outpouring of support for the school provided by celebrities, politicians, local community leaders, and others. One student, Eric Alexander, was quoted as saying, "At first we thought we were on our own. I wanted to see who cared, who is for real, which teachers showed up. It made a difference."[57] In the weeks after the second shooting, only 10 students applied to transfer.

Perhaps the most innovative idea—one that was proposed prior to the shootings as part of his efforts to respond to high levels of violence—was Chancellor Fernandez's idea to create 50 small schools, to replace very large schools like Thomas Jefferson, which had had as many as 4,000 students enrolled. Fernandez received substantial support from New York's business and foundation communities and was able to establish 40 new schools, often organized around innovative concepts.[58]

Building Social Cohesion

A wide array of community efforts was undertaken to build social cohesion. Among these were a series of antiviolence marches and rallies, led by local organizers and attended by celebrities, including Bill Cosby,

Cicely Tyson, and Spike Lee.[59] The marches advocated for multiple improvements in the community, including metal detectors, more programs for youth, and getting guns off the streets.[60] At the first march, on March 2, nearly 800 students, parents, and residents, together with Bill Cosby, Mayor Dinkins, and local clergy, gathered in front of the school. Mayor Dinkins called for an antiviolence crusade: "I say what we need now is an anti-violence movement. We've got to stop killing each other. We've got to shield our babies. We've got to stand in front of them. We must."[61]

A remarkable year-long intervention, led by Councilwoman Priscilla Wooten, Board of Education member Irene Impellizzeri, and Principal Beck, took groups of 150 students on retreats to the Fallsview Hotel, a famous resort in the Catskill Mountains.[62] These weekend encounters posed a simple question, "What do we have in common?" Councilwoman Wooten remembered that it was difficult for the students, who came from different ethnic backgrounds, housing projects, gangs, and lifestyles, to answer that question. By the end of the weekend, students were able to recognize that they did have something in common: they were all members of the Thomas Jefferson High School community. The idea that the behavior of each would reflect on all created a new foundation for cooperation.

But it was not easy for Wooten and her team to achieve this goal. It required, on one hand, enforcing discipline, and on the other hand, ensuring students that the adults cared. Wooten adeptly combined the two in managing incidents large and small. For example, she worked with the Fallsview team to organize a prom for the young people, which included an "ice cream sundae room" with 25 different flavors of ice cream. When no one wanted prom night to end, she took them swimming at 2 a.m. By contrast, she dealt swiftly with infractions of the rules. Young police officers, who accompanied the retreats, were always on hand to take those who crossed the line back to the city. On one occasion, she got so angry with the young people that she sent everyone to their rooms. The next day she received flowers from all the youths by way of apology. "I cried," she told us. "You have to let them see your tears."

Newsday reported that, a year after the shooting, a new positive attitude existed at Jefferson. Principal Beck attributed most of the positive change in the school to the program of retreats. She described, "[Students at the retreat] sit down and talk about issues that relate to the whole rite of passage concept. They talk about the destructiveness of violence. They use films, videos, and role playing."[63]

Another important effort to stem violence was the East New York: United for Safety Project. This program, funded by CDC, brought together a diverse group of organizations, including Victim Services, the New York City Department of Health, the Cypress Hills Development

Corporation, the East New York Urban Corporation, and the United Community Center, as well as an evaluation team from New York University.

Diverse efforts were undertaken as part of the United for Safety Project. One that we learned about in detail was a community education project, under the leadership of the Health Department. In the first phase of the project, the agency's public education coordinator conducted focus groups to understand attitudes toward violence and to develop effective educational materials. The focus groups made it clear that children, teens, and adults were responsive to different messages. While the younger children responded to messages like "Guns will shatter your dreams," accompanied by the image of a cap and gown under broken glass, the teens made it clear they needed something more extreme. The poster they decided to go with carried the message "Buy One, Get One Free" illustrated with the image of a young boy with a gun and a coffin. The poster targeting adults was "Parents Teach Children to Fight." The visual aid included photographs of families wearing picket signs with slogans such as "People Fighting for a Safer Community" and a subheading, "Fight Violence, Not Your Neighbor." Once the team had developed the posters, they placed them strategically throughout the community (see Figure 7-2).

Adjudication of the Bentley Case

Jason Bentley was initially charged with murder in the second degree, a charge that carried a minimum sentence of 5 years to life and a maximum sentence of 9 years to life. Bentley's case came before Judge Michael Juviler because of a question about the admissibility of a statement Bentley had made to the police. Because the police had denied Bentley access to a lawyer, Judge Juviler found the statement tainted and suppressed it. Without that evidence, which was an important part of the case, the district attorney and the defense attorney agreed with Judge Juviler to lower the charge to manslaughter in the first degree. Judge Juviler pointed out that, according to witnesses and the shooter's statement, there was a conceivable, but weak, self-defense justification. Apparently, this was not pursued by the defense team. Bentley pleaded guilty to the lesser charge and was convicted on June 22, 1992. On August 3, 1992, he was sentenced by Judge Juviler to 3 to 9 years in prison. He spent the first part of his sentence in a youth facility and then was transferred to an adult facility, where he finished out his term of imprisonment.

Adjudication of the Sumpter Case

Khalil Sumpter's case was controversial from the outset. The first newspaper accounts reported that he was "a 15-year-old boy bearing a

FIGURE 7-2 Poster developed by the New York City Department of Health.

grudge and carrying a stolen .38-cal. Smith and Wesson revolver . . . . 'This was a clear assassination,' said Chief John Hill, commander of the Brooklyn North Patrol Borough."[64] Other articles were more sympathetic, pointing out that Khalil was a fearful young man,[65] who acted in self-defense. The two poles—"unjustified assassin" and "fearful youth"—defined much of the debate that transpired at Khalil's trial.

From the first, his lawyer, John Russell, emphasized that Khalil had committed the crimes because he feared for his life. Russell was joined by noted civil rights lawyer, William Kunstler, and his associate, Ron Kuby. Together they organized a vigorous defense that focused on two points: (1) Khalil feared for his life and hence acted in self-defense and (2) he was suffering from mental illness related to stress and trauma from living in

East New York. The district attorney's office, represented by Ann Gutmann and Lance Ojiste, insisted that his actions were deliberate, considered, and without justification.

The trial opened in June 1993.[66] The prosecution presented evidence from the school and from counseling centers where he had been treated to make the case that Khalil had been disruptive and violent in the past. They also argued that his purported fear did not justify the shootings. Khalil, in their view, had not exhausted his options within the system of supports available at the school and in the community. It was his tendency toward violence that accounted for his actions, not his fear.

In support of their case, they called psychiatrist Robert H. Berger. Berger argued that Khalil was a manipulative young man, whose pattern was to seek quick gratification and avoid responsibility. Berger read from notes of a therapist who had treated Khalil, which stated, "Khalil appears to be fascinated by violence and hurting people who challenge him."[67] In Berger's view, Khalil was not acting under duress or the influence of extreme emotional disturbance. Rather, his actions were explained by his manipulative and antisocial personality.[68]

The defense team of William Kunstler, John Russell, and Ron Kuby argued that Khalil had acted in self-defense, based on the perception that he was going to be harmed and possibly killed. Kuby told us that Bernard Goetz, the 1984 "subway killer," had established the legitimacy of being afraid in New York City, and this was a cornerstone of the defense.[69]

Khalil testified about his perception of the danger he faced. He detailed the events leading up to the shooting, already described here. He also described that, on the day of the shooting, he saw Tyrone, Ian, and about eight to 10 other boys with them walking on the second floor. Khalil said that when he saw them he just kept walking, but then he saw Tyrone with his hand in his pocket and Ian with his hand "up under his jacket." At that moment, "I thought my life was in danger or that I was going to be seriously harmed."[70] He noted later in his testimony, "It was well known through Jefferson that Tyrone carried a razor. That they all—basically all their friends carried weapons."[71] Asked to explain why he shot Moore and Sinkler, he said, "At the time I was just—at the moment when I saw them reaching in their pockets, I was just real scared. The day before they had shot at me so there was no doubt in my mind that they would try to harm me today and I just—I just got petrified. I pulled out the gun and I shot twice."[72]

Forensic psychiatrist Stephen Teich testified for the defense as to Khalil's state of mind on the day of the shootings.[73] Teich testified, when asked about Khalil's mental state on February 26, 1992, "My conclusion was that at the time of this incident, Mr. Sumpter perceived that he was in danger and that he was acting in self-defense to prevent himself from what he perceived as potentially being killed. Now, I can't ascertain whether in fact . . . the threat was real or the threat was what he perceived as a cumulative result of the threats he had experienced the day before, his experience the day before and what happened in the period before. But it's clear in my mind that his perception of what was happening was that his life was threatened."[74]

In addition to a sense of impending harm, Teich also thought that Khalil was in a state of emotional distress, based on high levels of stress and past trauma. Teich reviewed with us the chronic stress and traumas that had contributed to Khalil's state of mental distress: his parents' divorce, his sister's angry departure from the family home, and the stabbing of a friend, which had led Sumpter to request a transfer from Graphic Arts High School to Jefferson.[75]

In summation, William Kunstler argued that, "the only issues that the defense is really raising is, one, whether he was acting in self-defense . . . or whether he was acting under this term *"extreme emotional disturbance,"* or essentially whether he intended to kill anyone under the murder statute."[76]

Ann Gutmann concluded that Khalil's actions were driven by anger, not by fear, "within the personality of a boy who felt that he was entitled to confront and punish those who dared to challenge him."[77]

The jury found that Khalil had acted under intense emotional distress and reduced the charges to two counts of first degree manslaughter. The defense team asked for leniency in the sentencing, arguing that "two young lives have already been tragically lost, let's not make matters worse by sending [Khalil] away."[78] On September 7, 1993, Judge Francis X. Egitto, who was known to favor long sentences, imposed the maximum sentence of 6 2/3 to 20 years in prison.[79] Khalil was released on October 23, 1998.

THE CONTEXT OF THE SHOOTINGS: THE NESTED SYSTEMS

Ecological Transition and Its Impact

The social milieu of East New York is, in no small measure, conditioned by physical processes of urban decay, exacerbated by cuts in essential municipal services in the early 1970s that were instituted well before New York City confronted any significant fiscal crisis.[80]

Firefighting services came under particularly careful scrutiny. Writing in 1969, the Rand Corporation saw East New York as the Brooklyn neighborhood most threatened by rising rates of fire and building abandonment:

> [Most] of the neighborhoods with the fastest increases [in alarms] cluster together somewhat to the east of the high-alarm area in Brownsville . . . .

> The neighborhood with the highest annual percentage increase is . . . in East New York. Its total number of alarms increased at an average annual rate of 44.3 percent in the period 1962–68, compared with a 10.8 percent increase for Brooklyn as a whole.[81]

The response of the city fathers was to eliminate about 10 percent of the city's fire companies, beginning in 1972 and 1974. The closings focused particularly on units located in the South Bronx, Harlem, and the threatened zone of Brooklyn, which included East New York. Figure 7-3 shows the result citywide when, over the course of a decade, these affected communities were particularly hard hit by a series of devastating

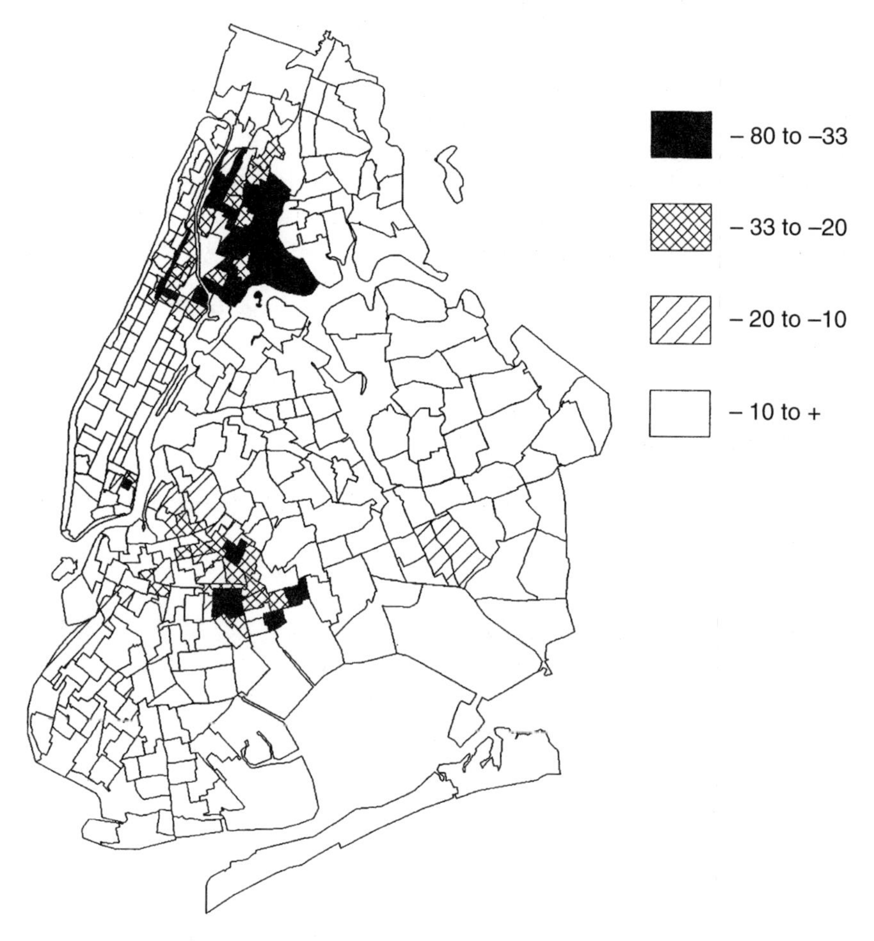

FIGURE 7-3 Percentage loss of housing units between 1970 and 1980. Figure represents most severe to least severe loss of housing units in each area.
SOURCE: Data from McCord and Freeman (1990).

fires. Some Health Areas[82] lost as much as 80 percent of housing units between 1970 and 1980. Note the concentration of heavy damage in the Brownsville-East New York section of Brooklyn.

Figure 7-4 shows the consequent, massive, forced displacement of community residents who lost their homes in these fires. Of particular interest are those health areas that lost or gained 2,500 or more in black population between 1970 and 1980. Again, while Harlem and the South Central Bronx were particularly affected, Brownsville-East New York also suffered greatly.

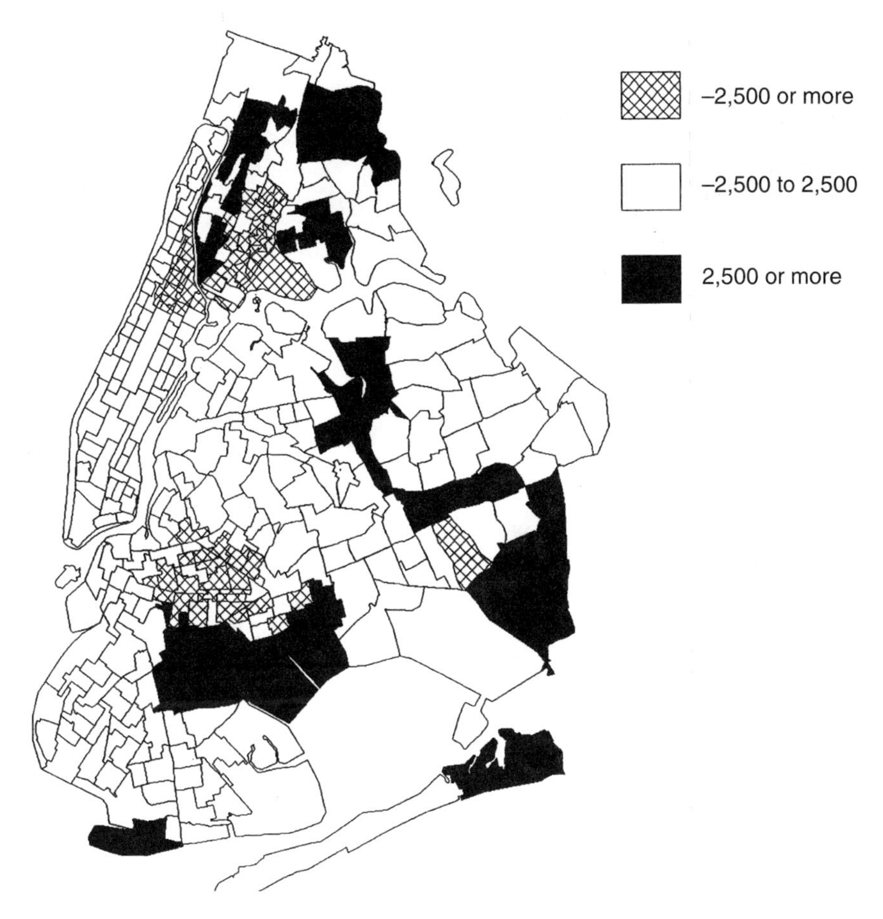

FIGURE 7-4 Absolute number of forced displacement of population from 1970 to 1980.
SOURCE: Data from McCord and Freeman (1990).

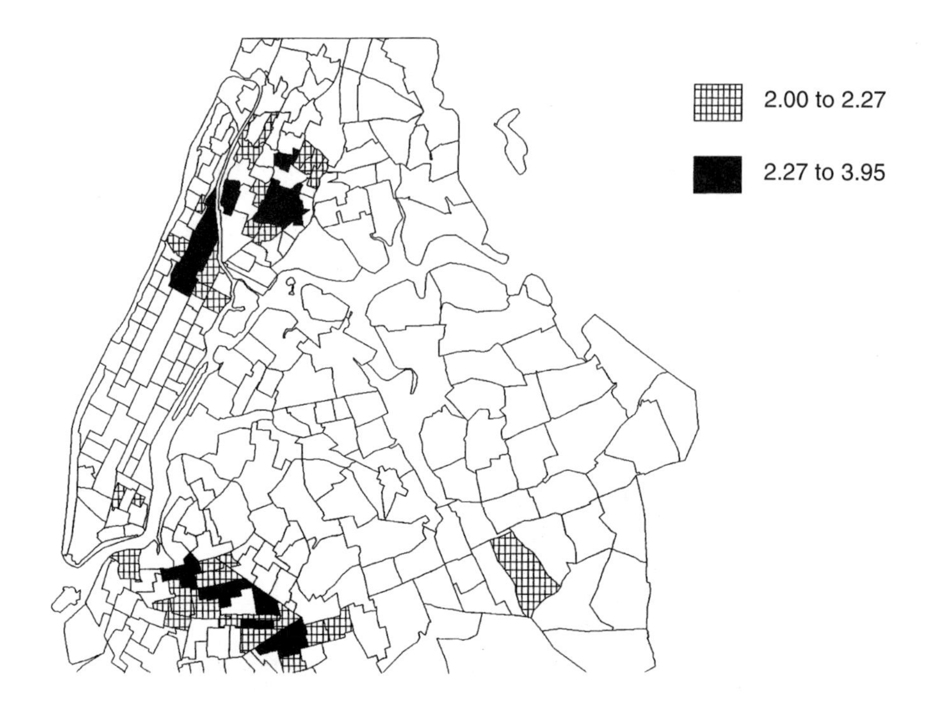

FIGURE 7-5 Population odds ratio for dying.
SOURCE: Data from McCord and Freeman (1990).

Figure 7-5 is taken from data developed by McCord and Freeman who demonstrated that, by 1980, adult males in Central Harlem had lower life expectancy than adult males in Bangladesh. Significantly, New York City Health Areas with standardized mortality rates that are two to nearly four times higher than those of the U.S. white population average coincide with those New York City districts that suffered the worst outbreak of contagious fire and abandonment. While the highest mortality rates were observed in Central Harlem, clearly the band across Northern Brooklyn, which includes East New York, was almost as badly affected.

Figure 7-1 shows the location of East New York's two zip codes, 11207 (marked as 7) and 11208 (marked as 8). In 1990 both zip codes reported similar proportions of adults not in the workforce (29.8 percent) and virtually identical proportions of students having completed high school (respectively, 19.5 and 19.7 percent). Zip code 11207 had 38.5 percent of families with children having two parents, and zip code 11208 had 48 percent.

TABLE 7-2 Average Annual Morbidity and Mortality for East New York Zip Codes and New York City, 1989–1991 (rate per 100,000)

Zip Code	Homicide	Infant Deaths[a]	Tuberculosis	AIDS	Syphilis
11207	73.8	97.4	56.6	93.7	479.8
11208	53.7	90.1	44.4	93.4	315.2
New York City	26.9	76.2	44.6	97.0	179.0

[a]Rate per 10,000.
SOURCE: New York City Department of Health.

Between 1980 and 1990, however, zip code 11207 lost 19.0 percent of housing and 13.8 percent of its population to continuing contagious urban decay, and zip code 11208 lost 5.4 percent of its housing but gained 6.2 percent in its population. As shown in Table 7-2, between 1989 and 1991 zip code 11207 reported far higher rates of homicide and syphilis that either its neighbor zip code, 11208, or the city as a whole.

Clearly, large sections of East New York began suffering contagious urban decay and depopulation earlier than the rest of the city and continued later than the rest of the city. Zip code 11207 seems the worst affected, and we attribute its markedly increased rates of homicide and syphilis and decreased rate of families with two parents to this pattern of decay and neglect. Thomas Jefferson High School was located in this particularly hard-hit area.

Drugs and Violence

Drugs and guns moved into East New York throughout the 1980s, reaching a peak between 1989 and 1993. All observers agree that guns were everywhere, and many people felt they needed to carry a gun. One Jefferson student was quoted in the paper as saying, "I'm not scared to go to school. I come here with a weapon, a 9-millimeter automatic. So it don't make no difference. I'd rather use my fists, but if it came down to that, I'm going to use it."[83]

Patrick Carroll, then head of the 75th Precinct, offered a number of ways of imagining the amount of violence that was going on:

- the number of "aided" cases—that is, cases in which officers offered aid to injured people—related to gunshots reached 600 in 1993, including those dead from their wounds;
- the sheer number of drug- and weapon-related arrests and the number of weapons confiscated each year; and

- two shootings of police officers, within a month of each other, right outside the precinct building, as the officers were on their way to the bodega across the street.[84]

In 1991, when people in the neighborhood came to Carroll for advice about the situation, he told them to go to Washington in support of the Brady Bill.[85] East New York was thus one of the first communities to mobilize in support of the bill. Carroll emphasized the effort the community made, finding the money and mobilizing to send 500 people in 12 buses to the demonstration.[86]

As in other New York City neighborhoods, crack cocaine contributed to a major decline in the area. Drug sales and consumption were everywhere. Because of the enormous profits, many entered the market, and battles for turf became an important part of the scene. The police attempted to impose order but were behind for a number of years. In the interim, violence, which had long been a part of the East New York scene, took on an even more important role in daily life. The 75th Precinct reported 109 homicides in 1990, 115 in 1991, 92 in 1992, and 129 in 1993, when it led the city.[87]

In addition to the close link between drug sales and violence, Carroll stressed that the drug dealers created the street culture that was so influential on the young. Drug dealers actively recruited young boys to work for them. Their cars, jewelry, and power made them important role models, outshining the model of honest but less remunerative hard work represented by the parents of many of the young people.

Community activists pointed out some other implications of the drugs and violence. One that they stressed was the civic paralysis that accompanied the widespread violence. One group explained that, because it was hard to get people to meetings, they started to escort people to and from meetings. When this became too dangerous, they switched to escorting people by car. After a while, the program lost steam because a huge effort had to be expended before even starting a meeting. Many similar stories were recounted to emphasize the multiple levels of disconnect triggered by the violence in the streets.

The Culture of Violence: The Street Ballet

Jane Jacobs, in her classic book *The Death and Life of Great American Cities* developed the image of the "street ballet" created by neighbors on an urban block:

> The stretch of Hudson Street where I live is each day the scene of an intricate sidewalk ballet. I make my own entrance into it a little after

> eight when I put out the garbage can, surely a prosaic occupation, but I enjoy my part, my little clang, as the droves of junior high school students walk by the center of the stage dropping candy wrappers. (How do they eat so much candy so early in the morning?) While I sweep up the wrappers I watch the other rituals of morning: Mr. Halpert unlocking the laundry's handcart from its mooring to a cellar door, Joe Cornacchia's son-in-law stacking out the empty crates from the delicatessen, the barber bringing out his sidewalk folding chair, Mr. Goldstein arranging the coils of wire which proclaim the hardware store is open, the wife of the tenement's superintendent depositing her chunky three-year-old with a toy mandolin on the stoop, the vantage point from which he is learning the English his mother cannot speak.[88]

The world envisioned by Jacobs is ordered by rituals that support the health and well-being of the neighborhood. The rituals provide for the safety of the block and form the community that can fight for the political rights of the area. This proved life-saving when Hudson Street was threatened by a highway, which was ultimately stopped by intensive local mobilization.

The concept of street ballet is essential for understanding the ways in which East New York functioned in the early 1990s. An experienced police officer named Lonnie Hayes gave the following description of the street ballet in Greg Donaldson's book:

> "Say you're walking down the street with your woman," Lonnie goes on, "and some guy around your building says, 'Hey I want your lady to suck my dick,' and you just grin." Lonnie displays a lame smile. "Then the guys get together and they start talking. You know, 'Hey, I told that guy's woman to suck my dick and he didn't do nothin'.' Then they start getting the idea that they should rob you." Lonnie acts out the robbery, with the victim raising his hand in a feeble gesture of protest. Then he assumes the role of the tough guy. "'Put your hand down. I told you not to move your fuckin' hand.' The next thing you know, they have your old lady up on the roof and they really *are* makin' her suck their dicks."[89]

The Culture of Violence: The Youth Culture

A key issue in understanding the violence at Thomas Jefferson High School is understanding the extent to which it fits into the pattern of the elaborate street ballet that was going on in East New York. Contrasting the street ballet of Jacobs with that of Donaldson, a whole series of differences emerges. Most obvious is that the first is prosaic, the second profane. In the second, the language of the story is assaultive and represents the threat of assault that hangs over the encounter.

Perhaps equally obvious is the extent to which the bystanders move into the space of the passerby in the second scenario. Greg Donaldson related a story that illustrates this point. He was walking down the street in East New York and passed some young boys playing with a ball. One threw the ball at Donaldson and hit him in the back.

The first point—the threat of assault—and the second—the move into personal space—are related in the following manner: the move tests the passerby's vulnerability to assault. If the passerby is found to be weak—note the "lame" smile Hayes referred to—then the others may take what he has. As robbery is a major economic activity among the young, largely deprived of other ways of getting goods they desire, one can conceptualize the difference between Hudson Street and East New York in economic terms. The business of Hudson Street is contained in stores. The business of East New York is in the streets. On Hudson Street, there is a flourishing market economy, while in East New York, the market economy has collapsed and the nascent capitalists have returned to primitive accumulation.

What is more subtle in all of this is that the "stuff" that can be taken includes not only material objects, but also reputation. When the passerby in Hayes's story smiles lamely, he establishes a reputation as "someone who can be taken," hence it is natural, in the language of that time, that people begin to think of robbing him, first of his things, then of exclusive sexual access to his woman. A reputation, once established, works as a kind of currency, an entitlement card that provides the cardholder with myriad privileges. But because the reputation of one person is "bought" at the expense of another person, the possibility that someone will want to take one's reputation always exists. This is highly reminiscent of what happened among gunslingers in the Wild West, who lived with the fear of the youthful challengers who would, one day, defeat them.

Part of the power of a street ballet lies in the fact that its symbols are understood by all who participate, and thus a given act triggers an appropriate next action. In this sense, the actions and next actions form a behavioral vocabulary that is easily read by people familiar with the scene.

Greg Donaldson, when hit by the ball, picked it up, put it in his pocket and kept on walking. In taking the boys' ball and walking on, he met their challenge and issued one of his own, one that they did not meet. Their failure to counterattack gave him reputation, and took reputation from them. Why, with so much at stake, did they not act to get their ball back? Part of the answer, Donaldson thought, lay in the fact that white men were outside of their system of thinking. A big, tall, unconcerned white guy in street clothes had to be an undercover cop, hence a man with a gun and a license to kill. Such were not attacked, Donaldson observed.

Returning to the story told by Lonnie Hayes, what was the threatened young man supposed to do? Lonnie did not answer that directly, but rather offered another story of challenge, to which he proposed two possible responses. "If you decide to take him out, you have to do it right, because if you fuck up, he'll kill you. You come up on him real fast with your head down and your hands in your pockets. Or you get your crew. That's why you need a crew. That's how you act when you live around here, and that is why so many brothers are in jail."[90]

In general, these kinds of challenges lead to "beef." East New York was often described as a tense, hostile atmosphere, in which people always had beef with each other. It is therefore critically important to investigate this term. In fact, one police officer we interviewed, aware that this report would go to Congress, suggested that, if Congress wanted to understand East New York, they should focus on beef.[91]

Beef is a term for interpersonal problems that are at a boiling point. Beef can start over any apparent mistreatment, including a wrong look, a disrespectful action, or a move into personal space. Handling beef was all-important as it contributed to the establishment of reputation, as suggested by Lonnie Hayes. Some aspects of beef had characterized rough Brooklyn neighborhoods for many generations: the critical difference in East New York, in 1991–1992, was that easy access to weapons meant that it was easy for the common thug to settle scores in a manner once reserved for major gangsters.

"Having beef," and responses required to that state, were truly a matter of life and death for Brooklyn teenagers at that time. We may infer that, for Jason and Khalil, their respective beefs had the weight of the world attached to them: they must have because they were so desperate to resolve the situations in which they found themselves.

Rules of the Street System

Two major rules in the vocabulary of violence can be discerned. First, a challenge had to be answered. There was really no getting around that fact. One young man told us, "I know it's a cliché and all, but where I come from there is no such thing as turn the other cheek. You gotta settle it right there and then. They'll come back with their friends. You have to assert yourself."[92] One interviewee told us his brothers would beat him all the time to make him tough.[93]

The corollary to rule one was that avoiding confrontation or resolving it with subtle maneuvers was possible, even advisable. An interviewee made the point that he thought people would always "catch beef." He thought the key was knowing what to do, how to fight, how to scare someone, how to get the message across that "you are not a punk."[94] Donaldson made much the same point to us, describing how adroitly Sharron Corley, the teen star of his book, handled dangerous fellows who often posed a tremendous threat to him.[95] The youths who lacked social skills, physical strength and agility, and intelligence were least likely to be able to employ these alternative strategies and most likely to be caught up in situations, either as victim or assailant.

Another corollary of the confrontation rule is that danger was highly localized, even in the East New York setting. Young people knew the geography well enough to know how to stay away from those places that were most violent. One Jefferson alumni pointed out that violence was a part of life, but you got into trouble only if you went someplace where you shouldn't have been. As an example, he explained that he did not go to graduation. It was known that he was moving shortly thereafter. Therefore, anybody who had beef with him would have to settle it then. Instead of taking the risk, he asked his mother to pick up his diploma.[96]

Second, reputation had to be established and protected. Some actions were highly detrimental to reputation, including doing well in school, while others were very supportive of establishing reputation, such as hanging with a respected crew. One young man described being a member of the LoLifes, a gang of sorts whose major objective was to shoplift Polo clothing. He wanted to be "down" with the LoLifes because it was an honor that came from everyone knowing that you were a part of one of the largest gangs in the city. He thought that most of East New York operated under that mentality: everyone did what they did to gain notoriety. People ran when they saw you coming and you never even had to do anything if you ran with the right crew.

Guns had a legitimate role in managing confrontation-reputation scenarios. One young man who was deeply involved in the street scene said that he would walk around with a gun and go to school with a gun. People didn't mess with him because they knew he was not to be messed with. He was willing to use the gun to remind them. He noted, "Once in a while I had to bust off a shot or two to tell niggas what's up."[97] Another young man, who was more ambivalently engaged with street life, explained that he took a gun to school most of the time. At first, the gun was for defense, but then it became an offensive tool. He used it to threaten people and to show off, thus gaining status as a "bad motherfucker."[98]

With the tension and pressure of violence omnipresent, some release was needed, and "wilding" appears to have provided a much-needed outlet from the rigors of daily life under a reign of terror. Wilding was a fundamental part of the world of boys. "Wilding" could be either a group or an individual behavior. From our interviews, we came to think of wilding as a state in which the urge to act out, most often in a violent and disruptive manner, overcame the young person's consciousness. It was a state of altered consciousness, described by our interviewees as a state in which "you have no idea what is going on. You of sort black out." Perhaps because of the lack of control people experienced in the wilding state, they often made conscious choices in setting the stage for their wilding, avoiding groups and places that were wilder than they wished to be.

At the heart of the act of wilding was the creation of fear in other people, which was experienced as very pleasurable. Interviewees tended to describe their own wilding with a certain amount of happy indulgence: "We would beat people up for fun" was a typical description. The wilding of others was viewed with more concern, edging toward a sense that they had entered a berserk state, that is, they were out of control in a deeper and more frightening way. Khalil Sumpter, some thought, may have "wilded out" when he killed Tyrone and Ian.

The language of the youth, including such complex terms as wilding, beef, and reputation, was highly developed and very specific (hence our effort to provide detailed definitions). The rules of behavior were well codified, and strictly enforced. In sum, it was impossible for a boy to grow up in East New York without participating, however obliquely, in the culture of violence. Although personal characteristics played a part in the unfolding of situations, an enormous amount of behavior was driven by the stringent rules around confrontation and reputation on one hand, and the single form of release, wilding, on the other.

School and Family

In the systems hierarchy guiding this study, school and family are placed subordinate to the larger systems of community collapse, the drug wars, and the street culture of violence. Carol Beck's impressive effort at reanimating Thomas Jefferson High School provides a case in point. She generally directed her efforts at creating a positive alternative to the streets. She suppressed violence in the schools. She attempted to be a positive role model and to bring in others who were also positive. This was a strategy that was highly admired by other adults, leading to her receiving the Readers' Digest American Hero in Education Award in 1991.[99] These actions, however, were distrusted by the youth, who had little faith that the world she described either existed or was attainable by them. They put much more faith in the immediacy of the violence that surrounded them and defined their minute-to-minute existence. One young man told *The New York Times*, "It's like this. If you were somebody's bigger brother and you tell them, if somebody do this to you, you do that to them, who you gon' listen to? Somebody who come to the school and tell you, 'Stay away from drugs and don't fight' or do you listen to your older brother?"[100]

Families, too, attempted to guide their children in positive paths. As Jason Bentley described his own family, his father often lectured him on the right way to act. Sadly, one can imagine that the lectures fell short of addressing the sights and sounds of East New York. As one example, Jason described, at approximately age 5, seeing an even littler boy stab his older brother with a broken bottle because he would not share his icee. Jason exclaimed, "Over an icee!" the chill of the moment still with him, nearly 20 years later. It is generally true that families have little training in the management of trauma, and it was certainly true that families were not prepared to deal with the magnitude and the ferocity of the traumas their children saw on a nearly daily basis.

Returning to the confrontation-reputation system, this system was largely lost on parents, teachers, and other adults. One adult, his voice dripping with sarcasm, told a story of a violent act, "triggered by something really important: somebody stepped on his sneakers."[101] A youth we interviewed told us, "If you by yourself unarmed and with a new pair of sneakers, don't even wear them sneakers if you ain't ready to fight for them."[102] Principal Carol Beck called Bentley's shooting "mindless,"[103] missing the very real fear that caused him to act as he did.

The chasm between the teenagers and the adults was evident in the gap in language. The young people interviewed were keenly aware of a breakdown in communication across the generations. By contrast, the adults usually focused on failing parents, rather than on the concept proposed by the young people that a whole generation had disconnected from another generation. Khalil Sumpter's mother reflected that her son's growing terror failed to make an impression on her because she couldn't imagine the kinds of goings-on that he was describing.[104]

The confrontation-reputation system and the disconnection from the adult world worked synergistically to create enormous isolation for the young people. Young people we interviewed were emphatic that they had to resolve problems on their own. This was perhaps least understood of all the issues surrounding the two shootings. A number of adults insisted that the youth had adults to whom they could have turned. The youth strongly rejected that idea. One young man said it was preposterous. There was no way that, given the reality of East New York at that time, the shooters could have talked to anyone. He insisted emphatically that they did not have that option.[105]

Growing Up in East New York

Seeing violence was a part of growing up in East New York. *The New York Times* conducted focus groups with 19 Thomas Jefferson students in the aftermath of the Sumpter shooting.[106] The reporters opened the conversation by asking the participants if they knew people who had been shot. These opening responses set the pattern:

> Mariana Bryant: Last night we was looking in my junior high school yearbook and my friend pointed out a number of people in that book that was dead.
>
> Q: About how many would you estimate?
>
> Mariana: Ten, all guys from eighth and ninth grade.
>
> Caroline Giffith: There was a shooting around my building over this girl. I was on my way to school. Two guys were going with one girl, so after one found out then he just pulled out his gun and shot him. . . . The one that got shot was 17.
>
> Sean Williams: One of the guys that died from this school was Wesley . . . He got shot in a barbershop right across the street on Pennsylvania. . . . I think he was 16.

Seeing violence was cited by a number of people as part of a developmental pathway that led young people to doing violence. Numbing, hypersensitivity to sound, and dissociative states were some of the mental states young people described that appear to have resulted from their exposure to violence. It is clear that, despite the remarkable levels of stress and trauma, there was little treatment available. Thus, the accommodations to injury were those available to young people functioning at a distance from adults: minimizing ("unless the gunshots were quite close, you just ignored them and kept on going"[107]); universalizing ("Ain't no place to go. You go to California and you can't even wear colored clothes out there. They've got that in Texas. You can't go nowhere"[108]); taking the offensive ("I was petrified. . . . I shot twice"[109]); and wilding ("we beat people up. . . . It was fun"[110]). These reactions were permissive of violence both directly, as in wilding, and indirectly, as in assuming that violence was the natural state of human existence.

In addition to facilitating the use of violence, seeing violence had other effects on development. Grief and anxiety interfered with learning. Chronic mourning interfered with the development of trust in the dependability of human beings. Fear limited freedom of assembly and freedom of speech, hence interfering with the development of civic life and civic skills. Not only was a childhood in East New York difficult to endure, it ill-prepared the young person for life in other parts of their country, not to mention the world.

IN RETROSPECT

This study was ending just at the approach of the tenth anniversary of the Bentley shooting and will be released around the time of the tenth anniversary of the Sumpter shooting. Much has changed in those 10 years, including in East New York. A hopeful article in *The New York Times* pointed to dramatic changes in the neighborhood, both in terms of the decline of the negative and the growth of the positive.[111] In the following paragraphs, we explore four issues: the connection between ecological transition and the evolution of the street ballet, the brief career of the violence epidemic, its lasting impact on a cohort that was in its teens, and the lessons for the future of schools.

Ecology and the Street Ballet

We have suggested that East New York was the product of contagious urban decay. The loss of housing, the displacement of large segments of the community, and the increasing isolation of East New York from the rest of the city all contributed to the conditions that resulted in the shootings at Thomas Jefferson High School.

In essence, East New York became a marginalized community in the period between 1970 and 1990. The withdrawal of many city services, ranging from firefighting services to sanitation and trash pickup, was among the many markers of this marginalization. During this period, many of the community's residents rarely ventured beyond its borders to visit other neighborhoods, other communities, or other boroughs of the city. Increases in joblessness and the concentration of poverty in East New York exacerbated the community's social isolation. Inevitably, the processes that created this marginalization would have an impact on families and on youth.

Childrearing practices are inevitably affected by the social structure of neighborhood life and by the opportunities that exist for adults and family members to participate fully in the economic and social life of the city. As East New York became increasingly disconnected from the rest of New York City and as the drug trade became more prevalent, community values shifted dramatically. As noted earlier, guns became more prevalent, shootings and gun-related homicides became commonplace, and community residents became more and more inured to the violence that surrounded them. More importantly, many accepted gun-related, violent behavior as a sad but inevitable aspect of community life.

The impact on the community's values was also dramatic. Many of the residents whom we interviewed described the value and importance of being perceived as tough, of having a reputation for being tough, of being prepared for danger, and of having survived one or more violent encounters. Being tough was an important means of not being a victim, and this was frequently reported as one of the major life priorities for adults and for adolescents during the 1980s and 1990s.

East New York was by no means the only community to suffer rapid, destructive changes in its social structure as a result of deurbanization. We suggest that an elaborate and characteristic behavioral repertory will develop within any community that is marginalized, as this community was.[112] Symbols of individual and group worth—which enable group members to carry out many social and individual objectives—will change and will inevitably begin to differ significantly from those of mainstream society. In such settings, actions become a form of communication, a language in which individuals communicate who they are, what they value, what they can do, and what they are prepared to do. They become aspects of one's personal identity—one's "rep"—that speaks as loudly (and often more effectively) than words.

The physical disintegration of the community, in other words, is likely to have an associated disintegration of the social "glue" that permits communities to remain cohesive and intact. We suggest that violent acts in particular may emerge as key behavioral symbols for the residents of communities that have suffered this degree of desolation and isolation. In the South Bronx, in Harlem, in the Central Ward of Newark, New Jersey, and in a host of other highly segregated, very poor communities, drug abuse, HIV/AIDS, high rates of tuberculosis and sexually transmitted diseases are all clustered closely together, along with high rates of homicide and interpersonal violence.

Communities function best, as Granovetter (1995) has shown, when weak ties (for example, those created by friendship and/or social networks) complement those defined by the strong ties created by membership in an ethnic group, an age cohort, one's occupation, family structure, place of residence, and so on.[113] In East New York, a principal effect of nearly three decades of contagious urban decay, from the early 1960s into 1990, was to disconnect the youth from the social structures that would ensure entry to the adult worlds of work and family-building. Put another way, not only was the East New York community marginalized from the larger city, but also the youth of the community were isolated from the adults.

Although we arrived at this idea from a theoretical perspective, the young people were well aware of it. Jason Bentley first pointed this out to us, noting that he thought the rift between the generations was a fundamental source of violence. The rift was accompanied by a shift in communication, such that adults and young people no longer communicated. He noted: "Kids trying to be adults and most of them not respecting the

BOX 7-2 Jason Bentley on the Rift Between the Generations

There are many factors that contribute to violence among young people. One of those factors is the breakdown of the relationship between the older and younger generations. The destruction of the family structure resulted in a loss of proper adult guidance for the youth. As things persisted, the rift between the generations grew farther apart, adding to the development of a hostile atmosphere in the community and violence amongst the youth.

Back in the 1980s "crack" hit the scene. Its impact was devastating to the communities it touched. It divided the community into three parts: those that sold the drugs, those that used the drugs, and those not included in the above. The division of the community was the catalyst for the rift that formed between the generations.

What the crack epidemic did was destroy the family structure. It snatched the father figure out of the home either by death, imprisonment, or it assassinated his respectability because he was strung out on crack. The same with the mother or any other authoritative figure in the household. Some families remained intact, but the majority were infected [affected] in some form or fashion.

Needless to say, without an authoritative figure in the home, guidance went out the window. Without proper guidance in the home, most young people turned to the streets for guidance. Some had guidance in the home but chose to ignore it.

Due to the lack of proper guidance young people started to take on the responsibilities of an adult before they were prepared for such a task. Most young people had to support themselves and sometimes their own families as well. Young females started having babies at young ages, becoming children without guidance trying to give guidance. Again, the father figure virtually nonexistent for various reasons. The destruction of the family structure caught in a continuing cycle.

Kids trying to be adults and most of them not respecting the adults in the community. The adults labeling every young person a hoodlum and distancing themselves from the youth. That's where the rift comes into play. The communication between the generations stops or is ineffective. The children aren't listening to the adults and the adults are not trying to hear what the youth has to say. Consequently, the relationship between the generations deteriorated.

So what you have is young people with a lack of respect and guidance and you have drugs. When you add guns to the equation, it equals violence. The violence produced the hostile atmosphere in the community. Fear gave the adults a legit excuse for distancing themselves from the youth and gave credence to the labeling of the youth. As a result the rift between the generations grew wider and wider and the scars of the community ran deeper and deeper.

The severance of the community, the destruction of the family structure, and the loss of proper guidance for the youth, all contributed to the breakdown of the relationship between the generations. The breakdown between the generations left the youth without moral, emotional, or practical support, guidance, direction, or discipline. All of these factors contributed to the development of violence amongst the youth.

adults in the community. The adults labeling every young person a hoodlum and distancing themselves from the youth. That's where the rift comes into play. The communication between the generations stops or is ineffective. The children aren't listening to the adults and the adults are not trying to hear what the youth has to say. Consequently, the relationship between the generations deteriorated."[114] See Box 7-2 for Jason's full analysis of this issue.

This rupture between the generations can be considered a "phase transition."[115] This major alteration in the state of the social system disrupted not only the social relationships but also the means of communication, creating a "noisy channel," that is, a social environment in which it was hard to be heard or understood. This drove the youth to create a "new" language, a language of behaviors that would enable the sending of messages of personal worth and efficacy. These new messages provided the individual with the ability to be known, to be respected, to be taken seriously in a community that had lost the capacity to give its residents a sense of their own worth and value using more conventional symbols, such as doing well in school, having a high-paying job, or doing well in sports. Extreme violent acts became an inevitable tool of this new behavioral language: the explosive noise of the gun could certainly be heard above the din of social disintegration.

The Brief Career of the Violence Epidemic

A remarkable aspect of the violence epidemic that hit East New York is the extent to which it has abated. Crimes of all kinds are down. In 1993 there were 129 homicides; as of June, only 10 homicides had been reported in 2001.[116] Guns have been taken off the streets. People have emerged from their houses to begin to live in a new way. There are gains in the number of housing units. The speed and the momentum of the shift have startled observers and caused many to wonder how to explain the changes.

Reverend Youngblood said of the Sumpter shooting that, as a man of God, "I have to look for the God factor. Maybe God allowed the violence to get out of hand so that we would finally pay attention to violence and young people." Certainly, the school shootings of 1991–1992 shook the neighborhood and the city, and the response was intense and widespread. A number of reforms were implemented that surely helped redirect interactions among people and alleviate the burdens on the young.

That said, it is important to consider the special effects of the violence epidemic on the cohort of young people that was reaching adulthood in those years. They saw enormous amounts of violence, and many directly experienced it. They lived in a social system that put enormous pressure on them to engage in violence, and to do so in relative isolation from helping adults. What are the ways in which these environmental forces may be expected to have shaped their well-being and their participation in adult life?

Khalil Sumpter's story is profoundly important as we consider these issues. Khalil, like many young men, committed a crime and was sent to prison. He was able to finish his high school education while there, but because of changes in the laws in the state of New York, access to higher education was denied him. He was able to learn a trade, becoming an accomplished cabinetmaker. On leaving prison he found himself in a new dilemma: prison, not skill, has been the deciding factor in his efforts to find employment. Although he is home in the technical sense of having been released from prison, he is not home in the more profound sense of being able to establish secure roots and a sense of belonging. Many young men of East New York are in a similar position. The current posture of society of permanently punishing those who have transgressed will multiply the effects of the violence epidemic for decades to come.

Those who avoided serious prison time still suffer from the peculiar conditioning to "stand on your own two." While most of America operates on the principle "It's not what you know, it's who you know," the youth of East New York operate on the basis of radical individualism. Much as they grew up in a fractured community, their sense of "I" operates to undermine the reformation of community, as well as the advancement of the self that can be accomplished only by interdependence.

Finally, an enormous number of these valiant young people were traumatized and like Khalil Sumpter suffer from trauma-related emotional disorders. One 1994 study of East New York junior high school students found that 50 percent had symptoms related to trauma.[117] The numbers would surely be higher for those born earlier who had lived through more of the violence epidemic. One man we encountered on the streets of the neighborhood talked about the violence he had experienced, lifting his shirt to show his scars. Tears formed in his eyes at the memory of the hostile and violent experiences he had endured. Nothing of any significance has been done to address the emotional distress that those young people are carrying with them as they begin their adult lives.

Issues for the Future

Everyone we interviewed was asked to compare the Jefferson shootings with other incidents of school violence in suburban and rural communities in the United States. Most took the widely publicized events at Columbine High School in Littleton, Colorado, as their reference point, concluding that the youth at Jefferson faced real danger, while those in Colorado did not. They inferred that the Jefferson shootings had some justification, while those elsewhere resulted from some form of mental instability. In essence, everyone aware of the situation in East New York emphasized that the shooting there occurred within the logic of the local culture, hence "made sense." Put another way, the East New York shootings were not "crazy."

Khalil gave a great deal of thought to the issue. When we first talked, he was keenly aware of differences but pointed out a possible similarity. He commented, "I don't think [school violence] is anything to do with the area where the kids are living, whether it's upper class areas or inner city ghettos. There's pent up anger and pain which gets vented at the cost of somebody's life. A kid should have no problems coming to an adult and asking how they would handle it. It's good if you have people like that. But that's not always the case. Kids get into a certain age group when the values instilled by the parents, all that is out the window. A kid will react according to the laws of the peers. I don't think the areas have a lot to do with the situation. Some situations might be related to jealousy over a pair of sneakers. It's the situation that the kid is trying to handle on his own that leads to the violence."

After reading a draft of the case study, he wrote us a letter. He felt that he had expressed himself poorly in the interview and wanted us to have a clearer version of his opinion on this important matter:

> My situation was based on my belief that my life was in danger. I really don't know all the details of the other incidents but I will say this. Communication and expression comes in infinite forms. Some people do so in the form of casual conversation, some in the form of music and art, and unfortunately, some people in my generation choose to express themselves in the form of violence. The result in some instances can be fatal. In East New York there was an unwritten set of rules or codes that many people abided by, young black men in particular. I didn't make any of these codes, for lack of a better term, but the way of thinking for most young men was either abide by these laws or become consumed by them. As stupid as these laws are, that was the norm at the time. I guess in a nutshell people in inner-city ghettoes respect violence, so as a result people become it. It doesn't always take the form of a violent act. But people become potentially violent by carrying guns or living by these laws. We live in an age where youth find comfort among youth instead of adults, and where peers have the greatest influence among peers, i.e., sex drugs and dialect. Until there's a comfort zone where adults can feel what youth go through or understand their concerns or insecurities, youth will continue to express themselves in a violent nature. I sincerely hope that people learn from my experience. My reality is very ugly but hopefully it can give people insight as to what's going on in the minds of America's youth. I wouldn't want anybody to face what I went through and what I'm going through. The mantra of many young men is that you have to "stand on your own two," as quoted in your report. But eventually you get tired of standing on your own two.

Consequently, when people eventually do get tired in some cases, it's too late. Let's do our best to reverse that trend.

Khalil's thoughts are worth considering. The shootings at Jefferson were among the first of the decade of the 1990s. Other shootings in other places take other forms, but reiterate a common theme: the use of the gun as the great equalizer in situations youth perceive to be untenable. Whether the threat is to life or to mental well-being is less the issue than the perception that the threat is intolerable. Khalil's emphasis on the need for an environment that fosters open communication is worth considering, not only in urban schools, but also in any school affected by this generation gap.

How might a different kind of environment have been created at Thomas Jefferson High School? Jefferson, in 1991–1992, was a massive high school, attempting to provide an education to 1,600 severely stressed young people. The school was not out of control at the time of the shootings, but it was not in control, either—in short, it was inadequate for the task it confronted. In a community where teenagers needed a massive amount of contact with adults, the school provided less than a minimum. This is not a fault of the principal, as the deficits lay far beyond anything she might have mustered from the Board of Education or private donors. It is an illustration of society's capacity to ignore the obvious. It took the desperation of Jason Bentley and Khalil Sumpter, trying to impose order on chaos, to bring some attention to the school. Even then, the attention focused on mechanical solutions, rather than human solutions. Principal Beck rated the retreats as contributing the most to an improved atmosphere in the school. That intervention cost $58,000, whereas the metal detectors that were installed everywhere cost $28 million.

If schools are to be functional places, then investment in the creation of nonviolent communities is in order. But it is not enough to say to teenagers, as did Mayor David Dinkins, "You must choose nonviolence." It is absolutely necessary that adults make it possible for teenagers to live by rules of conduct that permit cooperation, sharing, and mutual respect and trust. Furthermore, adults, inside and outside marginalized communities, must realize that youth will engage with adults only if they perceive that there is hope and a real promise for a future.

CONCLUSIONS

The shootings at Thomas Jefferson High School were among the first to capture national attention and contributed greatly to the public perception of school safety and youth violence. At the beginning of this case study we cited the Thomas theorem, which holds that what one perceives to be real is real in its consequences. School shootings are especially difficult phenomena to understand, and in searching for solutions one is often forced to distinguish between conflicting perceptions about what aspects of these tragic events are important and what aspects can or should be ignored.

For example, are the shootings at Thomas Jefferson fundamentally and conceptually different from those at, say, Columbine because they occurred in a poor, minority, inner-city community? In this case study we have presented a great deal of evidence that East New York was an isolated, impoverished community whose social ecology was dramatically influenced by the crack epidemic and the ready availability of guns. We have argued that elevated levels of interpersonal violence created an atmosphere of ever-present threat, particularly for adolescents. Young men growing up in this environment were forced to adopt attitudes and behaviors that permitted them to live with this high level of threat and to respond to it appropriately. With guns so readily available and with shootings so commonplace, it was inevitable that carrying weapons and using them would be perceived as the most effective strategy to use in interpersonal conflicts.

Schools occupy an almost sacred position in American culture. That the social ecology of the community would somehow penetrate the walls of a school and result in shooting fatalities is almost unimaginable to the average American adult. One can easily imagine the school as akin to the Catholic Church of the Middle Ages: a sanctum sanctorum that offered peace, security, safety, and refuge for all who entered. That schools would be the scene of violent death, perpetrated by adolescents with guns killing other adolescents, violates our most fundamental beliefs about schools as sacred places.

The truth, of course, is that schools are—to the contrary—intricately and intimately a part of the social ecology of the community. Students are not magically transformed when they step through the doors of the classroom. The conflicts that arise between young people at home, in the streets, and in the classroom always carry the potential for violent, even fatal resolution. For such conflicts to be resolved peacefully, a number of conditions must be present.

First and foremost, adults cannot be absent from the lives of the children in the community. Students in East New York during the period of the shootings believed that adults played no significant role in the resolution of interpersonal disputes and conflicts—of the beefs—that are so much a part of growing up. Our respondents were insistent on this critically important fact: when threatened with violence, adults were simply irrelevant.

It is this perception that is at the core of the problem of youth violence. The location of the violence is less important than the belief of so many young people that they are "on their own two" to create solutions. Thus, it is unrealistic to expect that schools would be spared the horrors of violent, at times fatal confrontations between young people, particularly in a nation with more than 200 million guns in the hands of private citizens. The miracle, in our view, is that there have not been more of them.

East New York was therefore, in many respects, the harbinger of events to come. Rather than an isolated example of ghetto violence, it is emblematic of the growing divide between adults and young people, and of the irrelevance of the schools as community institutions with the capacity to assist in the resolution of these problems.

It is also clear that schools must become more than buildings occupying space in their communities. They must become active participants in after-school efforts to bring adults and young people together. Thomas Jefferson High School was overwhelmed by the ambient violence in East New York in 1991–1992. Its teachers and administrators were not particularly active participants in any of the community efforts to deal with and prevent this violence. This is not to fault them in any way: Americans have always believed that the proper place for school staff is in the schools, and Jefferson's personnel were engaged in the business of education.

But as we write this report, we are also aware that times have changed. Many American adolescents perceive themselves to be in untenable situations for which violence may offer the only way out. Many have the means to obtain weapons yet do not have the maturity or the capacity to avoid using them if the threat is perceived to be unbearable.

In another age, adults had both the time and the ability to invest a considerable amount of time in assisting young people to mature into responsible, contributing members of the community.[118] This investment, which social scientist James S. Coleman termed "social capital," has gradually eroded as adults become more and more involved in working and making money and have assigned more and more responsibility to the schools for socializing their children. This rift between adults, schools, and students must be repaired.

There is a danger that the American people will demand that their legislators and their schools develop solutions to youth violence in general and to school violence in particular. The danger lies in the belief that bringing children to adulthood is the responsibility of social institutions. Schools do have a role to play, but that role must be viewed as a partnership between adults in the community and school staff, each of whom has a critical role to play in the lives of young people.

Finally, adults in the community, not just parents, need to be more present in the schools. The roles that they can play there can be defined and elaborated in each community where such efforts are undertaken. And school staff need to be more visible, more present in community activities and programs designed to affect the lives of young people. What must be avoided, in our view, is the belief that youth violence can be separated into two components: that occurring outside school and that occurring within it. It is this perception, more than any other, that appears to be especially fraught with peril. At Thomas Jefferson and at other schools that have experienced fatal shootings, a beef from the streets came into the school. The solution, in our view, is to extend the boundaries of the school.

There have been numerous examples of school-community collaborations that have been effective in bridging the gap between adult and adolescent roles. These efforts have been funded by public and private foundations and are too numerous to list here. It suffices to note that they exist, and because they exist, society's task is to disseminate and adapt these interventions more broadly.

Youth violence is not a new phenomenon. However, as it becomes increasingly prevalent in the nation's schools, and as it becomes increasingly the object of public concern, our hope is that the will to translate concern into action will be born.

ACKNOWLEDGMENTS

Without the help of many people, this challenging study could never have been completed in the time allotted. We are grateful, especially, to all the people who agreed to participate in interviews, many of whom went out of their way to provide us with documents, contacts, and other kinds of support. We owe special thanks to Theresa Hunt, at Community Board 5, who introduced us to the area, and to Reverend Johnny Ray Youngblood, who opened many doors for us. Greg Donaldson and Sam Freedman were generous with their time and insights. Greg also gave us permission to use the map from his book to illustrate this case study. John Dunn, first deputy commissioner of the New York City Police Department, and Lester Wright, deputy commissioner/chief medical officer of the New York State Department of Corrections, provided enormous assistance in negotiating their respective systems. We are indebted to Reverend Youngblood, Greg Donaldson, Stephen Teich, Lester Wright, Commissioner Patrick Carroll, Jason Bentley, and Khalil Sumpter for their feedback on earlier drafts.

At our own institution, John Oldham, Jack Gorman, and Allan Rosenfield provided essential administrative oversight and support. Paul Papagni and David Strauss helped us negotiate the institutional review boards. Richard Miller and Nancy Heim adapted Greg Donaldson's map for our use as an illustration and as the cover for the preliminary report.

Administrators too numerous to list here ensured that all the day-to-day problems were resolved. We are deeply appreciative to all of them.

Finally, we are grateful to Jason Bentley, Khalil Sumpter, and their families for talking to us about this most difficult period in their lives. Their generosity was fundamental to the success of our effort.

NOTES

[1]Schmitz, Stevens, Feldman, and Fullilove, unpublished manuscript.

[2]Thomas and Thomas, 1928.

[3]Engel, 1980.

[4]Donaldson G., 1993.

[5]Freedman S.G., 1993.

[6]Sullivan M., 1981.

[7]O'Connor S., 1997.

[8]Atlas.ti, Scolari, 1995.

[9]This version of the story is Bentley's. Greg Donaldson's version—the only other detailed accounting we found—offers a different account of the source of the beef but corroborates both the impression of Jermaine and the course of events Bentley described (pp. 227–232).

[10]Fieldnotes.

[11]Fieldnotes.

[12]Reports of the events vary, including in such basic facts as the spelling of names and the reporting of ages. Daryl, for example, was called "Darrell" in the original police report and in the first articles in *The New York Times*, but "Daryl" in later articles, as well as in Greg Donaldson's book. Other facts were inconsistently reported as well. The incident described here was referred to in a number of news articles as a "fight over a book bag." By contrast, the thrust of people's descriptions about life in East New York, the presence of guns, and the organization of the youth culture were highly consistent, despite variations in details of stories.

[13]Fieldnotes.

[14]*New York Newsday*, 11/26/91.

[15]*New York Newsday*, 11/26/91; Anderson returned to teach at Thomas Jefferson High School but was quickly given an unwanted transfer. When he asked why, he was told it was to avoid "the wrong kind of celebrity" (fieldnotes).

[16]*New York Newsday*, 11/26/91, cited "Since 1987, 50 of the school's students have been killed or wounded near the Pennsylvania Avenue campus" while The *Los Angeles Times*, 7/7/91, quoted Principal Beck as saying, "I have lost more than 30 children over four years—stabbed, primarily shot." Donaldson reported that 75 students had died violently in the previous four years (p. 322)

[17]*Los Angeles Times*, 11/27/91.

[18]*Time*, 3/9/92.

[19]Fieldnotes.

[20]*The New York Times*, 11/30/91.

[21]*The New York Times*, 11/30/91.

[22]*New York Newsday*, 12/7/91.

[23]Donaldson G., p. 237.

[24]*The New York Times*, 12/2/91.

[25]Fieldnotes; Donaldson. Reports agree on the sequence of events and the relationship between Khalil and Tyrone. Khalil's parents corroborated that there were phone threats,

but the identity of the caller(s) was challenged at trial. There was no independent corroboration that Tyrone fired shots at Khalil.

[26]Donaldson G., p. 144.

[27]*New York Newsday*, 2/28/92.

[28]There is some controversy about whether or not there were metal detectors in place at the school that day. Principal Beck had been hesitant to have them installed, even after the first shooting. According to several sources, there were no metal detectors at the school that day [*Los Angeles Times*, 2/27/91; NPR radio, 2/27/91; *Bergen County Record*, 3/1/92; fieldnotes]. There was general agreement that the presence of metal detectors would not have been an effective deterrent as there are 28 doors to the high school and no way to guard them all.

[29]*New York Newsday*, 2/27/92.

[30]Fieldnotes.

[31]*St. Louis Post-Dispatch*, 2/28/92.

[32]Fieldnotes.

[33]*New York Newsday*, 2/27/92.

[34]*New York Newsday*, 2/29/92.

[35]*New York Newsday*, 3/4/92.

[36]Fieldnotes.

[37]Ian had a nickname, E-Lo, signifying his membership in a gang called the LoLifes. He was memorialized in graffiti throughout the neighborhood which read, "R.I.P., E-Lo" (fieldnotes).

[38]*Will My Name Be Shouted Out?*, p. 290. Ian wrote this poem the year before he died on the occasion of his grandmother's death.

[39]*New York Newsday*, 3/4/92.

[40]Fieldnotes.

[41]Johnny Ray Youngblood's eulogy for Ian Moore.

[42]Johnny Ray Youngblood's eulogy for Ian Moore.

[43]Fieldnotes.

[44]Fieldnotes.

[45]Fieldnotes.

[46]Fieldnotes.

[47]Fieldnotes.

[48]*Morbidity and Mortality Weekly Report*, 10/15/93.

[49]Fieldnotes.

[50]*Morbidity and Mortality Weekly Report*, 10/15/93, p. 775.

[51]Fieldnotes.

[52]Fieldnotes.

[53]*New York Newsday*, 2/28/92.

[54]*New York Newsday*, 2/28/92; *Time*, 5/25/92; *The* [Manchester] *Guardian*, 9/7/92. It is not entirely clear what they meant by "immediately."

[55]*New York Newsday*, 3/2/92.

[56]*New York Newsday*, 2/28/92.

[57]*New York Newsday*, 2/28/92.

[58]Fieldnotes.

[59]*Time*, 5/25/92.

[60]Fieldnotes

[61]*The New York Times*, 3/3/92.

[62]Fieldnotes.

[63]*NewYork Newsday*, 2/25/93.

[64]*NewYork Newsday*, 2/27/92.

[65]*New York Newsday*, 2/28/92.

[66]*New York Newsday*, 6/30/93.

[67]Trial record, pp. 783.

[68]Trial record, p. 793.

[69]Fieldnotes. More recently, the perception of danger was the basis for the successful defense of the police officers who fired 41 shots into the African immigrant Amadou Diallo.

[70]Trial record, p. 407.

[71]Trial record, p. 408.

[72]Trial record, p. 416.

[73]Fieldnotes; *New York Newsday*, 7/1/93.

[74]Trial record, p. 573.

[75]Fieldnotes.

[76]Trial record, p. 880.

[77]Trial record, p. 922.

[78]*New York Newsday*, 9/4/93.

[79]*New York Newsday*, 9/8/92.

[80]Wallace D., Wallace R., 1998.

[81]Rand, 1969.

[82]A health area is a geographic unit designated by the Health Department for the collection of vital statistics.

[83]*The New York Times*, 12/2/91.

[84]Fieldnotes.

[85]*New York Newsday*, 4/23/91.

[86]Fieldnotes, *New York Newsday*, 4/23/91.

[87]New York Police Department statistics.

[88]Jacobs J., 1993.

[89]Donaldson G., 1993.

[90]Donaldson, p. 210.

[91]Notorious B.I.G., a rapper from Brooklyn, explained it in the song, "What's Beef?" This was one of the last songs he wrote before being murdered in what some thought was a beef-fueled feud.

[92]Fieldnotes.

[93]Fieldnotes.

[94]Fieldnotes.

[95]Fieldnotes.

[96]Fieldnotes.

[97]Fieldnotes.

[98]Fieldnotes.

[99]*Time*, 5/25/92.

[100]*The New York Times*, 3/3/92

[101]Fieldnotes.

[102]Fieldnotes.

[103]*New York Newsday*, 11/25/91.

[104]Fieldnotes.

[105]Fieldnotes.

[106]*The New York Times*, 3/7/92.

[107]Fieldnotes.

[108]*The New York Times*, 3/3/92.

[109]Trial record, p. 416.

[110]Fieldnotes.

[111]*The New York Times*, 6/10/01.

[112]Wallace, R ., Fullilove, M. and Flisher, A., 1996; Wallace, R., Flisher, A. and Fullilove, R., 1997; Wallace, R. and Fullilove, R., 1999.

[113]Granovetter, 1995; people in neighborhoods would not necessarily name the essential and supportive informal connections "weak" ties.

[114]Fieldnotes.

[115]Wallace R., Fullilove R., 1999.

[116]*The New York Times*, 6/10/01.

[117]Landsberg et al., 1999.

[118]Coleman, 1987.

REFERENCES

Atlas.ti
1995 Thousand Oaks, California: Scolari.

Board of Education of the City of New York
1992 *The Cohort Report: Four Year Results for the Class of 1992 and Follow-ups of the Classes of 1989, 1990, 1991 and 1991–92 Annual Dropout Rate.*
1992 *The Class of 1992: Final Longitudinal Report: A Three-Year Follow-up Study.* Board of Education of the City of New York.
1992 *The Class of 1993: Final Longitudinal Report: A Three-Year Follow-up* Study. Centers for Disease Control.
1993 Violence-Related Attitudes and Behaviors of High School Students—New York City, 1992. *MMWR* 42(40):773–777.

Champion, S.
1992 "Ego: Firepower," *The* [Manchester] *Guardian*, September 7.

Chicago Sun-Times
2000 *Deadly Lessons: School Shooters Tell Why*. Exclusive Report, October.

Chiles, N.
1992 "High School Slaying: Suspect's Parents Feared for Their Son on NY Streets," *New York Newsday*, February 27.

Chiles, N.
1992 "Circling the Safety Wagons: Detectors Wanted at High Schools," *New York Newsday*, February 28.

Coleman, J.S.
1987 Families and schools. *Educational Researcher* 11.32–38.

Collins, T.
1992 "Jefferson HS Boy Dies," *New York Newsday*, February 29.

Donaldson, G.
1993 *The Ville: Kids and Cops in Urban America*. New York: Ticknot & Fields.

Editorial
1992 "Are We Resigned to the Terror of Handguns?" *Bergen County Record*. March 1.

Engel, G.
1980 The clinical application of the biopsychosocial model. *American Journal of Psychiatry* 137:535–544.

Freedman, S.G.
1993 *Upon This Rock: The Miracles of a Black Church*. New York: Harper Collins.

Gambardello, J.A.
1992 "High School Slaying: Assassination: 2 Students Shot in Hall; Teen Charged," *New York Newsday*, February 27.

Goldman, J.
1991 "Lessons in Love Help Class Earn Diplomas in Survival Education," *The Los Angeles Times*, July 7.

Goldman, J.
1991 "For NY Students, It's a Time of Grieving Tragedy," *The Los Angeles Times*, November 27.

Goldman, J.
1992 "Tragedy Mars a School's Day in the Sun," *The Los Angeles Times*, February 27.

Granovetter, M.
1995 *Getting a Job: A Study of Contacts and Careers*. Chicago and London: The University of Chicago Press.

Hevesi, D.
2001 "East New York: A Neighborhood Reborn," *The New York Times*, June 10.

Hurtado, P.
1993 "HS Slay Suspect Says He Shot for Fear of His Life," *New York Newsday*, June 30.

Hurtado, P.
1993 "Teen Traumatized," *New York Newsday*, July 1.

Hurtado, P.
1993 "Leniency Sought for Teen Shooter of Two Schoolmates," *New York Newsday*, September 4.

Hurtado, P.
1993 "Teen gets Max in HS Slayings," *New York Newsday*, September 8.

International Association of Chiefs of Police.
1999 *Guide for Preventing and Responding to School Violence.*

Jacobs, J.
1993 *The Death and Life of Great American Cities*. New York: Modern Library.

Kalogerakis, M.
2001 *Adolescent Violence in America: A Historical Perspective.* The 33rd William Schonfeld Memorial Lecture, American Society for Adolescent Psychiatry Annual Meeting, Philadelphia, Pennsylvania, March 24.

Laboy, J.
1993 "School's Rage Gone Year After Killings," *New York Newsday*, February 25.

Landsberg, G., M. Spellman, and C. Devitt
1999 *The East New York: United for Safety Report (A Comprehensive Youth Violence Prevention Program)*. New York University, Ehrenkranz School of Social Work, Institute Against Violence. Funded by the CDC.

Lawson, H.
2001 *Reformulating the School Violence Problem: Implications for Research, Policy and Practice.* International Conference on School Violence and Public Policies, UNESCO, Paris, France, March 6.

McCord, C., and H.P. Freeman
1990 Excess mortality in Harlem. *New England Journal of Medicine* 322:173-177.

Menninger, W.W.
1995 *Uncontained Rage: A Psychoanalytic Perspective on Violence.* Sixth Annual Earl J. Simburg Lecture, March 23.

Morrow, L.
1992 "Childhood's End," *Time*, March 9.

Muir, E., J. Devine, and P. Lucas
2001 Violence in Schools: A Dialogue on Security and Discipline. [Web Page]. Available: www.nyu.edu/education/metrocenter/violence/diaSD.htm. [Accessed March 15, 2001].

National Public Radio
1992 Reported by Alex Chadwick on February 27, 19992. "Teen Held in Brooklyn School Shootings."

Negron, E.
1991 "At Memorial, Students Aim to Halt Violence," *New York Newsday*, December 7.
Negron, E.
1992 "400 Students Want Out of Jefferson," *New York Newsday*, February 28.
Newkirk, P. and P. Hurtado
1992 "Students, Neighbors: Suspect a Timid Kid," *New York Newsday*, February 28.
Newman, M.
1991 "Two Who Lost Their Way in Urban Wilds," *The New York Times*, December 2.
New York City Police Department
1992 The Police Department Statistical Report on Complaints and Arrests.
O'Connor, S.
1997 *Will My Name Be Shouted Out? Reaching Inner City Students Through the Power of Writing*. New York: Simon and Schuster.
Perez-Rivas, M.
1991 "In Despair, They Unite," *New York Newsday*, April 23.
Perez-Rivas, M.
1991 "Death Takes No Recess: Gunfire Kills HS Student, Hits Teacher," *New York Newsday*, November 26.
Perez-Rivas, M.
1992 "Mourn Slain Teen: Dinkins: We Must Stop these Deaths of Children," *New York Newsday*, March 4.
Pierre-Pierre, G.
1994 "Fewer Killings Tallied in '93 in New York," *New York Times*, January 2.
Rand.
1969 Appendix 2, Incidence of Fire Alarms, 7/27/73 Plaintiff's Exhibit 221 for Identification, Uniformed Firefighters Association lawsuit on the civil rights of fire company closings.
Schmitz, S., N. Feldman, and M. Fullilove
High-risk Situations and Public Health Practice. Unpublished manuscript.
St. Louis Dispatch
1992 "Two Students Gunned Down in Brooklyn School Hallway," February 28,
Sullivan, M.
1981 *Getting Paid*. Ithaca: Cornell University Press.
Supreme Court of the State of New York, County of Kings
1993 Criminal Term: Part: 30. The People of the State of New York against Khalil Sumpter. Indictment No. 2511/92. June 29.
Tabor, M.
1991 "Mourners of Slain Student Ask, 'When Does it Stop?'" *The New York Times*, November 30.
Taylor, C.L.
1991 "Dealing with Violence Around Them," *New York Newsday*, November 26.
Thomas, W., and D. Thomas
1928 *The Child in America*. New York: AA Knopf.
Time
1992 "Thomas Jefferson High School, Brooklyn, New York," May 25.
Vossekuil, B., M. Reddy, R. Fein, R. Borum, and W. Modzeleski
2000 *Safe School Initiative: An Interim Report on the Prevention of Targeted Violence in Schools*. U.S. Secret Service National Threat Assessment Center. Washington, DC: U.S. Department of the Treasury.
Wallace, D., and R. Wallace
1998 *A Plague on Your Houses*. London and New York: Verso Press.

Wallace, R.
1985 *A Preliminary Study of the Brooklyn Fire/Abandonment Epidemic, 1971–1983: The Crisis That Was, the Disaster to Come*. Public Interest Scientific Consulting Services.

Wallace, R., M. Fullilove, and A. Flisher
1996 AIDS, Violence and Behavioral Coding: Information Theory, Risk Behavior and Dynamic Process on Core-Group Sociogeographic Networks. *Social Science and Medicine* (43)339–348.

Wallace, R., A. Flisher, and R. Fullilove
1997 Marginalization, information and infection: The correlation of ghetto risk behaviors and the spread of disease to majority populations. *Environment and Planning* (A 29)1629–1943.

Wallace, R., and R. Fullilove
1999 Why simple regression models work so well describing 'Risk Behaviors' in the US. *Environment and Planning* A, 31.719–727.

Wright, C., and K. Friefeld
1992 "Dinkins Tackles Schools' Violence," *New York Newsday*, March 2.

Youngblood, Reverend Johnny Ray
Eulogy for Ian Moore, St. Paul Community Baptist Church, undated typescript.

8

A Cross-Case Analysis

Because of the variability among the cases presented in Chapters 2–7, the committee conducted a cross-case analysis. It is valuable to compare the cases systematically to see what this small sample of incidents can tell us in terms of factors that are present in all the cases, those that are present in some cases and not others, and those that are entirely absent. Through such activities we can parlay this small number of complex cases into a more coherent pattern of understanding. Later, in Chapter 11, we consider the cases as though they were a small dataset that can be used to influence our judgments about likely causes and plausibly effective interventions. We begin by describing the committee's analytic framework for creating and analyzing the cases and the method of inference used. We then test the plausibility of different substantive claims that are made about the important causes of these events.

ANALYTIC FRAMEWORK AND METHOD OF INFERENCE

As described in Chapter 10, violence can best be understood through processes operating at multiple levels of explanation, taking account of the dynamics that seem necessary, if not sufficient, to produce a violent incident (National Research Council, 1993). These levels include factors operating at the social, community, group or subcultural, family, and individual levels of society. They often change—sometimes rapidly. In addition, sequences of events, powered and guided by microsocial processes, are often important in transforming the potential for violence into the concrete reality of an event. The ideas and cognitions that operate in the heads of offenders are important as they become agents of violent events. The aim here is to bridge these different levels of analysis.

Our interests go beyond what might have caused the individual shooting and what might have prevented it. We are also interested in the incident as a cause of both a community and a policy response. Table 8-1 is a representation of the committee's analysis. Although lacking precision, the table is meant to convey the extreme rarity of school rampage violence within the larger picture of lethal school violence, other serious youth violence, youth violence in general, and violence in general.

In reviewing the cases, the case authors and committee members developed multiple hypotheses about what might be driving the behavior and checked them against the data collected in the six cases. For example, in one case an offender had a more successful older sibling; in another the family of the offender had recently moved into town; in another the offender knew a great deal about a previous school rampage shooting, and those attending school were organized in a large number of subcultures with few overlapping members and little interaction with one another. We then considered whether there was a plausible causal connection between the feature observed and the likelihood that the violent incident would have occurred or would have taken a particular shape. If the factor showed up in any of the other cases and seemed to be working in the same way, then there was a plausible hypothesis about an important causal phenomenon. If little such evidence was found, then the hypothesis seemed less likely to be true, or less important even if true, because its impact could not be seen across the cases examined.

We also considered some common hypotheses—for example, that the offenders had been bullied and the attack was an attempt to get back at those who bullied them, or that the schools in which the incidents had occurred had allowed cliques to form and to skirmish with one another in the school—and looked at the cases for any examples. If none of the cases was characterized by such a feature, that cast significant doubt on either the truth or the importance of that factor as a contributing cause of the violence.

These methods, although not systematic enough to confirm or refute alternative causal explanations, could be viewed as an effort to undertake the preliminary scientific task of generating causal hypotheses and doing the rough work of casual inspection that helps make some hypotheses seem stronger than others. In this approach, part of the task is to reduce confidence in any particular claim by imagining the variety of things that could be true and then ordering the claims according to some rough sense of plausibility.

SIMILARITIES AND DIFFERENCES

Table 8-2, at the end of this chapter, lays out the details of the case studies for ease of comparison.

All seven of these incidents involved young people arming themselves, in several cases with semiautomatic weapons, and opening fire on their schoolmates and teachers, killing or seriously injuring them (mean fatalities: 2.2; mean injuries: 4.5; all but 4 killed or injured were other students). The incidents occurred in places that were supposed to be safe—the hallways and common areas of schools and at a school-sponsored event. The offenders were all young themselves—most were age 13 or 14 (range: 11–15). The offenders in six of the seven incidents were convicted of homicide in adult courts, and most were sentenced to long prison terms.

These events also took place in a society that seems to encourage or condone violence. The popular culture is tolerant of violence. The United States has the world's largest supply of privately owned weapons (used in four of these incidents). And recently there was a dramatic increase in violence occasioned at least in part by an epidemic of cocaine use fueled and supported by violent illegal drug markets. In the committee's view, levels of violence are influenced by such structural factors, and we cannot exclude their importance in producing the events examined, although it is hard to find direct evidence of these factors in the cases.

All of these events occurred in the *sturm* and *drang* of adolescent development. The youth who committed the offenses were young men trying to become grownups. They were intensely concerned with their status and power, with their masculinity, and with their relationships with members of the opposite sex. They were all vulnerable to exaggerated hopes and fears and to the perceptions and judgments of peers and adults. At this time of life, it is easy to imagine that one is under threat, or that there are certain things one has to do to gain attention or standing in the world. It is also a time of life when adult authority and norms are being contested. These processes were at work in both the inner city and in the suburban and rural areas.

An important difference between the inner-city and suburban and rural contexts lies in the coherence of the events. In each of the three incidents in the two inner-city schools, the offender had a specific reason to shoot when he did, where he did, and at whom he did. Moreover, the reasons given were comprehensible to their peers and even to the adults and officials who investigated and responded to the events. The shootings were occasioned by a very specific relationship between the shooter and some other student at school, which ripened into an unresolved dispute, in which there was a continuing, seemingly credible threat of violence

TABLE 8-1 Stages in the Development and Aftermath of the Incident

Factors Leading To Incidents	Durable Conditons That Create Potential For Violent Incident	Micro/Social, Situational Processes That Transform Potential Into Act
Society Wide	Poverty Economic inequality Racial discrimination Culture of violence Weapon availability	
Community	Economic isolation Rapid mobility Inter-group conflict	
School	Inadequate resources Ineffective management Ineffective discipline Weak faculty/student relations Poor physical security	Arbitrary disciplinary actions Unexpected negative feedback
Local Youth Culture	Violent norms/scripts Inter-group conflict Non-inclusive cliques Bullying Status hierarchies	Status threats Physical threats Audience for violence
Family	Broken homes Emotional distance Inattentive parenting Sibling Competition	Parental crises Parental rebukes
Individual	Psychology Low cognitive functioning	Acute feelings of inadequacy Fear and rage

against the person who ended up doing the shooting. There were even intimate, local encouragements for the shootings to occur, in the sense that there was a youthful audience that knew about the dispute, understood the rules by which such disputes could be settled, and would have viewed the reluctance of the shooter to act as an invitation to degrade his status and make him a victim because of his reluctance to use force to defend his status and, in one case, the status of his brother.

The four suburban and rural incidents lacked this coherence and social clarity. That is not to say that these shooters didn't have their reasons to shoot. Like the inner-city shooters, the suburban and rural shooters

The Incident	Immediate, First Round Consequences of Incident	Social Interpretation and Response to Incident	Longer Term, Second Round Consequences and Response to Incident
	Notoriety Media coverage	Commentary in media Commentary in expert communities	Changes in federal policy Research studies
Location Weapons	Shock Self-defense	Local political discussion	Civil court processing
Number of victims Seriousness of injuries	Security Decisions	Broad discussion of cases and prevention Grief counseling	Installation of new security Creation of new relations w/students
Relationship of victims to offender	Shock Self-defense	Hidden discussion among youth?	Altered relations Willingness to report
How ended: –Suicide –Legal interven. –Citizen interven. –Flight	Humiliation Sadness	Soul searching	Getting on with life
	Criminal justice processing	Trial	Civil court processing

were intensely interested in defending or elevating their social standing. They believed that they were being ignored, or were under attack, or had been unjustly treated. That provided much of the energy they used in preparing for and executing the attack.

Yet in the suburban and rural shootings, the shooters' perceptions seemed to have little basis in reality, or, if they were real, they were not widely understood or shared by others. The acts were more the product of aberrant thought than of requirements of social circumstances—as reflected in the occasion, the victims, and the targets of the shooting. The suburban shooters were not being threatened with physical violence at the time they shot. While there was often some relationship between the suburban shooters and those at whom they shot or who were in the vicinity of the shooting, they mostly did not have specific targets or hit those whom they might have been targeting. Their grievances were more diffuse, and their shooting more indiscriminate. An important finding of the committee is that events of this kind did not seem to have occurred in the inner-city schools.

TESTING THE PLAUSIBILITY OF VARIOUS EXPLANATIONS OF VIOLENCE

The Role of Economic and Social Structure

A well-established tradition in sociology and criminology is to begin the search for the causes of violence in the structural characteristics of society. The idea is that violence occurs among those in social positions who are disadvantaged relative to others in society, or who are excluded and alienated from the dominant culture, or who are victims of oppression and injustice. This tradition is reflected in the inner-city cases. The case writers locate the violence exactly where this view would expect to find it—in communities that have been socially, economically, and politically marginalized. It is hard not to draw the conclusion that these factors played an important role in causing the violence observed.

What is startling, however, is that some portion of the lethal violence observed happens not in economically marginalized communities, but in ones that are relatively well off economically, socially, and politically. Paducah is not a thriving community; it has its pockets of poverty. But the other three suburban or rural communities studied were relatively affluent, and homogeneously so. Jonesboro, Edinboro, and Rockdale County were all thriving, affluent areas in which few were left behind. It would be hard to attribute the shootings in these areas to economic or social disadvantage.

The role of economic and social status cannot be wholly excluded. In one of the suburban cases, the social position of the shooter was relatively low in his community, which may suggest to some that class did play an important role in shaping the character and motivations of a future offender. And there are plenty of reasons to think that great tensions lurk beneath the surface of the apparently successful communities, that great gulfs divide adults from youth in these places, and that the places themselves are a bit unstable due to recent economic growth. An important similarity across the urban and rural and suburban environments was the presence of rapid social change in five of the six communities, which can produce instability even when the changes are positive ones.

The Role of Culture

A second idea with widespread currency is that lethal school violence is caused by a societal and popular culture that tolerates, encourages, or even demands violence. There are different versions of this argument.

In one version, a claim is made that inner-city violence seeped into pop culture, embodied in the mass media, and tended to glorify and encourage violence. The focus of the mass media on violence is seen not only as a reflection of a general enthusiasm for violence, but also as an influence that barrages American society with graphic images of violence. The important role of the mass media is thought to be particularly attractive and dangerous to youth who can spend hours viewing TV, listening to music with violent lyrics, and playing video games that feature violence.

It is hard to exclude this hypothesis on the basis of the evidence. All the incidents occurred against the backdrop of a popular culture that seems to tolerate if not encourage violence. However, it should be noted that the offenders in the cases did not seem to be more obsessed with these materials than the millions of kids who did not go on a shooting spree.

A second version of a claim about cultural influences emphasizes the gap between adult culture and youth culture. The argument is that if adults are not much present in the lives of youth, they will not be able to guide them toward such adult values as self-reliance, self-discipline, civility, mutual respect, patience, generosity, and empathy toward others. Without adult influence, youth culture might turn out to be particularly vulnerable to dangerous influences from the media or from peers. One extreme version of this is the emergence of a gang culture in which youth gangs perform the protecting, explaining, and socializing roles that would ordinarily be performed by families.

From the outset of our work, the committee was much attracted to this hypothesis. The inner-city cases were profoundly influenced not only by the general culture of violence in their neighborhoods, but also more particularly by youth gangs whose interactions created an important part of the social circumstances that animated and authorized the shootings—at least in the minds of the boys who fired the weapons. The suburban and rural cases also showed strong evidence that the world of youth was not very well understood by adults, especially youth who were forming their own culture supporting some forms of antisocial behavior.

In Rockdale County, a syphilis outbreak among teens caused by widespread sexual promiscuity of a particular group of kids who gathered at an unmonitored house to watch pornography on TV and imitate the acts they saw offered strong evidence of the absence of adult and parental guidance. In Paducah as well, the case writer observed that the "social dynamics of adolescence were almost entirely hidden from adult view."

So there is in these cases a gap between the adult and the youth culture. Communication did not flow easily across these boundaries. The adults at school did not seem to know the kids very well, or to be much present in their lives other than as administrators and teachers. These adolescents did lots of things that the adults in the communities would view as dangerous. We can also see that this gap matters, because it allows gangs, cliques, and rivalries to grow, and it lets festering disputes and grievances go unnoticed and unresolved. When information became available that should alert adults to the likelihood of a fight or an assault by one youth against another, the information often did not cross the boundary that divides adults and officials from the adolescents.

It is unclear whether the gap between adults and youth in these particular communities exists in other communities that have not experienced these tragedies. In our cases, this gap was evident in both the poor inner-city neighborhoods and the more well-to-do suburban sites—both in terms of the quality and intensity of their engagement and in terms of the substantive values they embraced. The committee discussed ways of closing this gap. But it is important to recognize that this gap can never be fully closed and probably should not be. Successful human development, and the development of society as a whole, depend on new generations being able to separate themselves to some degree from their parents and the traditions they embody.

A third version of a cultural explanation for lethal violence in schools and school rampages focuses on the pervasive presence of guns in the United States. It was not difficult for the shooters to obtain weapons, getting them from friends or stealing them, unnoticed, from family members or neighbors. And many of the shooters had some experience with guns. Three of them had gone hunting or shooting with an adult prior to the time of the shooting. At least one other had practiced shooting by himself and so had some experience with how to use the weapon. Again, it seems obvious that easy access to guns facilitated the lethal school violence and school rampages.

In sum, we cannot rule out the big cultural explanations: the distinctive American tradition of violence, the impact of the mass media, the gap between adult and youth culture, and the role of guns in U.S. society. But for purposes of scientific explanation, the potential impact of smaller, faster-moving aspects of culture and their carriers are also of interest: the special role of violent rap music and video games, the role of gangs in spreading a culture of violence among kids, and the impact of the press coverage given to the rural and suburban shootings themselves. These things are a possible link between the inner-city violence and the rural and suburban violence.

The Role of the School

The fact that the violence took place in schools tends to magnify its importance and social consequences. Because people expect schools to be safe havens for adolescents, when violence happens in them, the consequences are particularly severe.

Another important implication is that the schools can be seen as a platform for launching interventions to prevent and control violence. Finding ways to engage schools in efforts to accomplish this goal would seem to be an urgent task.

A third implication is that conditions inside the school may be a potential cause of the violence. In one version of this idea, one could see the failure of the schools to put in place preventive measures as an important cause of the lethal violence and school rampages—an error of omission rather than commission. Or one could see the schools themselves as "criminogenic"—the social relationships and norms existing within the schools might have actually caused the violence to occur.

A fourth implication is that the school, as the largest and most common social setting for adolescents, is the most likely setting for interpersonal violence among them. And because it is the most common social setting, it also serves as the primary public arena for acting out. In most of the rural and suburban cases examined, the school served not only as a convenient place to commit violence, but also as a public stage on which to perform and to be seen.

In considering these hypotheses in light of the cases, the committee sees evidence for the first claim—that the location in school makes the social consequences of the violence worse, regardless of where the school is located. There is also evidence that the school was considered an important place to launch efforts to prevent future incidents—not only in the places that experienced the violence, but also in communities across the country.

It is much harder to determine whether the schools were an important cause of the violence—through either omissions or commissions. The schools successfully constructed a protective boundary that separates them to some degree from conditions in the community; generally speaking, kids are safer in schools than in other locations. The difficulty is that the boundary is not impermeable. The violence of inner-city communities can reach into the schools, and the impulses associated with rampage shootings like those we have seen in workplaces can come into schools as well. In hindsight, it seems that there might have been some things that the schools could have done to prevent and deal more effectively with the violence when it occurred.

While evidence is scarce that the schools somehow generated the violence as a consequence of the way they were structured and administered, the sense of community between youth and adults in these schools, which research has shown is protective against crime, was lacking. In the worst example, the school allowed a school newspaper to print an article that humiliated one of the students who became a shooter. The adults involved may have been too distant from the students to prevent some social processes leading to the potential for violence or resulting in an intolerable humiliation for some particularly vulnerable youth.

In order to prevent violence from occurring, the adult culture and the school's administration and faculty may have to find ways to successfully engage the youth culture, to make every student feel valued, and to keep the youth culture in the school from becoming lethally dangerous. In hindsight, the schools in the cases may have been insufficiently preventive. But they do not seem to have been criminogenic in themselves.

Characteristics of the Offenders

Another tradition in criminology focuses attention on the more or less stable individual characteristics that make some individuals more likely to offend than others. In these respects, there may be important differences between the cases of violence in the inner city and the rural and suburban areas. Many social pressures—large and small, durable and transient—led toward the incidents of lethal violence observed in the inner-city schools. In the suburban and rural schools, the social pressures leading to violence seem much less visible. This may leave more to be explained by individual-level factors.

The shooters had some characteristics that, based on evidence from research, would place them at high risk for serious offending: being male, having mental health and, in one case, substance abuse problems, and having previous minor behavior problems. Most had recently begun hanging out with delinquent or more troubled friends and had a recent drop in their grades at school. All had easy access to guns (see Farrington and Loeber, 1999).

Of special note is how young all the shooters were. None was older than 15, and the youngest was 11. At least two important aspects of this age period have implications for what occurred. First, it is during early adolescence when peer relations and finding one's place in the social order of the school become most important. Second, during this period, cognitive abilities, including perspective taking, are developing. Young adolescents' ability to make accurate judgments, especially about social relationships, may be lacking.

Most of the shooters did not have either records or reputations that would link them to violent crime, but there are some important differences between the inner-city and the suburban and rural incidents. Two of the three offenders in the inner-city cases had previous arrests for serious crimes and were known to the police. One of the inner-city shooters and all of the suburban shooters had no previous arrests for serious crimes and were largely unknown to the police. However, we cannot conclude from these facts that the underlying level and seriousness of criminal offending was different in the inner city and the suburban cases, or that the inner-city kids had more experience in committing crimes than the rural and suburban kids. One had committed an undetected felony-level theft, one had stolen and sold his father's gun, and one had molested a two-year-old child. We can say, however, that the urban teens were more likely to have previous records of offending than the suburban kids, who essentially had none.

The shooters' records of school performance were similar for five of the boys. In the two inner-city cases, the offenders had good records until 8th grade, when their performance began to slip; that shift was attributed either to the fact that receiving good grades was seen as a sign of weakness or that the offender became involved with a gang culture. In the committee's view, this shows the influence of local subcultures in the schools that were powerful influences on young, impressionable adolescents beginning their high school careers and their developing hostility to "good" performance. In the rural and suburban cases, two of the offenders struggled in school. For the others, their grades ranged from average to good until around 8th grade.

Disciplinary records in school were similarly varied. At least one of the inner-city offenders and two of the suburban offenders had disciplinary records; the others didn't. The offenders could not be said to be notorious "bad actors" in their schools. They seemed to be adolescents struggling to make their way academically and socially in the competitive environment of gangs and cliques that characterize most junior and senior high schools.

In considering the causes of these offenses and potential points of intervention, a crucially important question is the extent to which the shooters can be considered mentally ill. For events that seem largely unaccountable (as in the suburban cases), and when there are no apparent powerful social factors shaping conduct (as in the suburban/rural cases), it is tempting to seek the explanation not just in individual characteristics, but in mental illness. Such an analytic move should be suspect, even though it may seem objective and logical. One reason is that such diagnoses are difficult to make objectively after the fact in cases like these. Moreover, in young adolescents, it can be hard for nonprofessionals to distinguish the early stages of mental illness from the ordinary confusion that kids tend to have about the world.

Only two of the shooters (both suburban/rural) had any diagnosis or clear sign of mental health problems prior to age 12. One of these had been diagnosed with attention deficit hyperactivity disorder in grade 4, and the other, the 11-year old shooter, had shown signs of conduct disorder, noted by the case study authors. After the events occurred, three other offenders (one-inner-city, two suburban) were diagnosed with mental illness. Another, one of the youngest shooters, had molested a two-year-old child. In addition, two of the offenders made an explicit suicidal gesture during the shooting, and one of these (later diagnosed with schizophrenia) repeatedly tried to commit suicide once he was in custody. Suicidal thinking was a prominent feature in all of the suburban and rural shooters studied.

A common legal standard for finding people not guilty by reason of insanity is that they are not aware of the consequences of their actions at the time they took them. None of these cases rises to that standard. In fact, there is evidence of both premeditation and rational calculation. In all the cases, the offenders made preparations to commit the offenses: they acquired weapons. They made plans for the shooting. They warned their peers that something big would happen. All this adds up to strong evidence of a kind of rationality that exposes the shooters to harsh judgments of culpability.

At the same time, in the cases of the rural and suburban shooters, the worries that prompted the shootings appear to be exaggerated, as were the hopes that attended the shootings. The circumstances were not well judged, the purposes were obscure, and the means were inappropriate to the ends. This sort of thinking does not provide a legal excuse for the action, but it is important to recognize it in adolescents in making a just and effective response.

The information gathered on the family background of the shooters included not only the basic structural conditions, but also something about the family interactions and the ways in which they were changing over time. We expected to find a significant amount of family pathology and a low degree of parental involvement in the lives of their children. There was some evidence of this, but it was by no means a universal pattern across the offenders. Two of the three inner-city shooters and two of the five rural and suburban shooters lived in intact, stable families at the time of the shooting. In two rural cases, the case writers found evidence of parental conflict. In one case, a videotape of the interaction between the mother of the offender and the offender suggested to the case writer an astonishing degree of distance in the relationship. There was no evidence either of family violence or child abuse and neglect in any of the cases examined. The parents seemed to remain involved with their children and vigilant of their conduct. Relationships within the family did not change markedly in the period leading up to the shooting.

The case writers also gathered information about the social standing of the offenders and the quality of relationships they had with their peers. Although the expectation was that the shooters would be isolated loners, this was true of only one suburban shooter—the one diagnosed with attention deficit disorder before the shooting and who attempted suicide shortly after the incident occurred. All the others had a status that could be described as marginal. The urban youth were members of marginal groups, and the rural youth seemed to be on the periphery of many groups rather than firmly at the center of any single group. One theme was that the offenders were the kinds of youth who sought to draw attention to themselves through practical jokes or joking around. Often, their humor seemed inappropriate to others. This may suggest the importance to these youth of their standing in their peer community, and that they were struggling with a significant gap between their desired status and the status they actually experienced.

More interesting and important is that, in six of the seven shootings, the offenders had recently changed their relationship with their peers. The only case in which this was not true was the case involving the loner. Also of importance is that in the three suburban-rural cases, the shooters had a recent experience of being rejected by a girl. This supports the idea that one of the factors fueling these shootings was a struggle to find status, understand masculinity, and develop relations with members of the opposite sex. When these desires were frustrated and there seemed to be nowhere to turn, a dramatic, violent act may have provided an attractive avenue of expression.

Contagion Mechanisms

The case study method allowed the committee to explore whether important contagion mechanisms were operating to spread and elevate the violence. While we cannot say much about whether such mechanisms were at work in spreading lethal violence in inner-city schools, or the extent to which inner-city violence seemed to have leapt out of the inner city and touched off the increase in lethal violence in the suburban and rural areas, we can address the extent to which contagion mechanisms seemed to elevate and spread school rampage shootings.

It turns out that only some of the suburban shooters were aware of other shootings, and in only one case (the one involving the loner) were the other shootings accorded a significant role. In that case, the offender was powerfully influenced by the shooting at Columbine High School in Littleton, Colorado. He knew about it, studied it, and took inspiration from it. In the committee's view, that counts as a copycat incident. Although that offender might have done something else hazardous to himself and others if the Columbine incident had not occurred, it seems clear that Columbine had a very large effect on the shape his actions took. In another case, the shooter showed a high interest in the Jonesboro shooting, which occurred exactly one month previously. In that case the shooting incident itself was very different, so it is not clear whether he was copying the Jonesboro incident; his general behavior might have been inspired by it, however. Given that these are rare events, even one copycat shooting makes contagion an important contributor to the kinds of shooting sprees that occurred in the rural and suburban cases.

Community Responses

The committee analyzed the communities' responses by looking at how they learned about and formed interpretations of the events; the role of the local and national media; the role of community leaders; and the policy responses made by the communities: what they did to deal with the grief and anxiety that spread in the aftermath of the events, what they did to improve security in schools, how they handled the offenders, and what lessons they drew from the experience. There was significant variability in the responses as well as some common elements.

Media Coverage

Media attention created enormous difficulties for the communities in which these events occurred. In the cases for which information was collected, the media coverage of the event was considered to be inaccurate by the community and turned out to be so inaccurate that the case writers could not rely on it. It is also clear that the media coverage was experienced as destructive and unhelpful to the communities and the schools. This was particularly true for the suburban and rural schools that experienced shooting sprees, which attracted huge, sustained national media coverage. The reports from Paducah, Jonesboro, and Rockdale County indicate significant local hostility to the national media and their negative impact on the local communities. In Chicago, the media coverage was so intense and so inaccurate that it caused the case writer to conclude that the case had been decided in the press before the trial was conducted, distorting the facts and limiting the dispositional options.

Community Forums

The communities varied a great deal with respect to the creation of local forums for discussing the meaning of the events and the appropriate responses. All the communities responded in the immediate aftermath of the shootings with more or less elaborately organized efforts to provide grief counseling to those who were victims or eyewitnesses or who were swept up in the emotions following the events. While this was a common response, there was an interesting difference between the inner-city neighborhoods and the suburban neighborhoods beyond the immediate reaction.

Both inner-city communities responded with antiviolence marches and rallies organized by informal and formal community leaders. They developed new programs for strengthening the relationships among students of the school, adults in the school, and adults in the surrounding communities. Both communities succeeded in engaging the interest and commitment of elected leaders and the appointed heads of the school system. In short, the inner-city communities reacted through widespread mobilization.

In contrast, the response of the rural and suburban communities was more sporadic, ad hoc, and less political. Only in Jonesboro and Edinboro was there anything resembling community rallies and meetings. In Jonesboro, a ministerial alliance was formed, but there was little evidence of its impact on the community. In Edinboro, a series of meetings were held, but they did not involve many citizens or have much impact on the policy and programmatic responses of governmental organizations. In Paducah and Rockdale County, there was essentially no community or political response; the actions taken were all official ones, as the schools made professional responses to the problem.

Community Understanding

The processes of reacting to, trying to understand, and making a just and effective response to the events led the communities to two quite different interpretations. In both inner-city cases, the community appeared to arrive at a diagnosis that attributed the lethal school violence to a generally violent atmosphere, to the presence of gangs in the schools, and to a wide communication gap between adults and youth about the danger in general, but particularly the youth involved in a gang culture. In two of the rural cases (Paducah and Jonesboro), the communities join the inner-city communities in attributing the violence to a lack of communication between adults and youth. They saw this as both a general contributing cause of the violence and as a specific problem that prevented threats made by the shooters in advance of the events from being heard and taken seriously by adults. In the other two suburban cases (Edinboro and Rockdale County), the problem was not attributed to a general community problem but instead was seen as the result of a single troubled kid growing up mentally ill in a troubled family. None of the suburban communities attributed the violence in their communities to violence in American society, perhaps because they felt insulated from the violence that was happening elsewhere. Nor did they attribute the violence to gangs, because they did not see the cliques that formed in their schools as gangs that created occasions for or supported violent acts. The adults and youth in these communities may have been influenced in some ways by the violence of the wider society, but if they were being so influenced, they were not much aware of it.

Criminal Justice Response

These different interpretations of the events did not lead to different criminal justice responses to the shootings. All except two of the shooters were tried as adults on the most serious charge that the evidence would support, usually first- or second-degree murder. All those charged with murder in adult court eventually pleaded guilty to lesser offenses that recognized mitigating circumstances. In two of the inner-city cases, the mitigating circumstance was either "acting under extreme emotional distress" or "acting in self-defense." In two of the rural/suburban cases (Paducah and Rockdale County), the mitigating circumstance was mental illness. Four of the shooters received very long sentences: 45 years for the Chicago offender; life without parole for 25 years for the Paducah shooter; 30–60 years in prison, eligible for parole at age 45, for the Edinboro youth; and 60 years of custody, including 40 years of probation, for the mentally disturbed adolescent in Rockdale County. The two New York shooters received somewhat more lenient sentences: 3–9 years in prison, part served in a youth facility for one, and 6–20 years in prison for the other (of which only 5 years were served, but the offender remains under supervision in the community for the rest of his sentence). The juvenile offenders from Jonesboro got the maximum sentence allowed under the juvenile law—an indeterminate sentence to age 21. The fact that the Jonesboro offenders will be getting out of jail relatively soon has created great concern in that community.

Only one victim of the inner-city shootings chose to pursue civil litigation against the school board. In three of the four suburban shootings, civil litigation was also initiated by the victims against the shooters, the shooters' families, and (in two cases) against those who were the source of weapons for the shooting. In Paducah, the suit also named high school officials and the producers of the video games and pornography that were found on the computer of the shooter. It is not clear that these legal processes have done justice or brought an emotional sense of closure to these communities. They seem to come too late and to be too disconnected from the community to be of use in helping restore a sense of self-confidence or peace.

Policy Responses

The communities all made what might be considered localized, school-level responses to prevent the occurrence of lethal violence in the future. Not all were well considered, however. All the communities except Rockdale County (the community in which a mentally disturbed loner was the shooter) made some kind of technical or security response to the school shootings. The two inner-city schools turned to heavy reliance on metal detectors in the school. The use of metal detectors, long a policy in New York, was intensified in the wake of the school shootings there. In Chicago, Mayor Richard Daley seized on the school shooting to institute a citywide policy mandating the use of metal detectors; the case authors note there have been no subsequent lethal shootings in Chicago schools since the adoption of this policy.

The rural and suburban schools did not go for metal detectors but rather built fences around the schools and instituted the use of nametags to identify faculty and students. Such measures are designed only to keep those who are not members of the school community out of the schools. Yet in all the cases in which these measures were installed, the shooter was a member of the school community and would not have been kept out of the school by them, so such preventive measures may be of questionable effectiveness.

Four communities supplemented the technical security arrangements with the redeployment or hiring of new security personnel. New York instituted a policy of more intensive police patrols inside and outside the schools. Chicago hired new security officials from the local communities to enforce a zero tolerance policy for weapons carrying and fighting in the schools. Both Paducah and Jonesboro also hired community resource officers to provide a regular presence in the schools and to educate students and enforce rules about weapons, fighting, and violence. Edinboro and Rockdale County did not put additional security officials in schools.

Finally, all the communities also made some sort of programmatic effort to strengthen relationships between adults and youth in the schools and among the kids themselves. They increased the adult presence in the schools and encouraged the adults to engage more closely with the students. Much of this took the form of closer supervision and monitoring by school officials, but additional help was provided to students who seemed to be in trouble, and some more general programs focusing on character education and leadership were provided to encourage the students to assume more responsibility for establishing an appropriate normative environment or moral order in the schools. In East New York, a special program of retreats involving adults and youth was initiated by a powerful city councilwoman. There is little evidence from the cases about the impact of these programs on the character of school life, the degree of security that individuals felt, and so on. But undertaking such initiatives was clearly important to these communities, at least as an expression of concern if not something that was instrumentally effective in protecting them from what was a very low-probability event.

TABLE 8-2 FOLLOWS ON PAGES 266-283

TABLE 8-2 Comparison of Cross-Case Variables

Variable		NY—Jason	NY—Khalil	Chicago
1.0 The Incident				
1.1	Date	11/25/1991	2/26/1992	11/19/1992
1.2	Time of day	Midmorning	Morning	No specific info but looks like morning
1.3	Victims:			
1.3.1	Number killed	1	2	1
	Number wounded	1	0	2
1.3.2	Victim status	Student and teacher	Students	Students
1.3.3	Relationship to offender	Bystanders	Rivals	Rivals
1.3.4	Targeted by offender?	No: uninjured other person was target, but done in self-defense	Yes but claimed self-defense	No but self-defense was element as a result of a gambling dispute
1.4	Location of incident	School hallway	School hallway	School hallway/ stairway
1.5	Perpetrator at time of attack:			
1.5.1	Number arrested	1	1	1
1.5.2	Demographics: age; grade; family structure and employment status	Black male; age 14; 9th grade; two-parent stable family, both parents working	Black male; age 15; 10th grade; parental conflict/divorce; sister also withdrew from family, mother works	Black male; age 15; 10th grade; stable two-parent family, middle/working class
1.5.3	Suspected co-conspirators	No, but aiding brother	Friend supplied gun	Gang affiliation
1.5.4	Under influence of alcohol or illegal drugs	No	No	No
1.6	Apparent motivation:			

Paducah	Jonesboro	Edinboro	Rockdale County
12/1/1997 7:42 am	3/24/1998 12:35 pm	4/24/1998 9:40 pm	5/20/1999 Around 8:am
3 5 (2 serious)	5 (1 teacher) 10 (1 teacher)	1 3 (1 serious)	0 6 (1 serious)
Students	Students and two teachers	Teacher killed; teacher and two students injured	Students
One victim was girl he had unreturned crush on (friends, but never dated). Other members of prayer group. Open question whether he shot them because they were prayer group or whether they were the convenient group in an open space.	Ex-girlfriends of both boys; one cousin of Andrew, other students, including neighbors of Andrew and two teachers, student bystanders	None	None
Not clear	Not clear	No	No
School lobby	Right outside school	School dinner dance in building outside of school	School's common area
1	2	1	1
White male; age 14; 9th grade; affluent, two-parent family, well educated	White males; one age 13, 7th grade; one age 11, 6th grade. 13-year-old broken/troubled family, working class; 11-year-old intact, established family, middle class	White male; age 14, 8th grade; middle-class; two-parent family, older father who owns business parental conflict	White male; age 15, 10th grade; parental divorce age 4, mother remarried age 7, affluent family
Yes, but no proof; may have thought Goth youths would help him take over the school	Two offenders; report of third older kid but never corroborated	No	No
No	No	No	No

TABLE 8-2 continued

Variable		NY—Jason	NY—Khalil	Chicago
1.6.1	Resolve interpersonal dispute	Yes: on behalf of brother—defending brother, dispute understood by others	Yes: others understood there was a dispute	Yes: others understood
1.6.2	Self-defense	Yes: defending himself and his brother	Yes: believed his life was in danger and his family threatened	Yes: feared for his life in gang-related gambling dispute
1.6.3	Revenge	No	No	No
1.6.4	Live up to script or code	Yes	Yes	Yes
1. 6.5	Copycat	No	No	No
1.7	Weapons:			
1.7.1	Used in attack	Gun	Gun	Gun
1.7.2	In hands of perpetrators at time of attack	Hand Gun	Hand Gun	Small semi-automatic pistol
1.8	How attack ended	Shooter ran outside	Shooter ran outside	Shooter ran across street
1.9	How perpetrators apprehended	Arrested by police right outside school	Arrested by police right outside school	Arrested by police across street from school
2.0	**Community Setting**			
2.1	Type of community	Urban		Urban
2.2	Community socioeconomic status	Isolated, advanced urban decay, hypersegregated, no economic structure, no jobs. Adults commute outside to work. History of economic disinvestment		Urban decay, economic disinvestment, and rapid social change regarding race/ethnicity of residents. Much low-income housing and housing projects

Paducah	Jonesboro	Edinboro	Rockdale County
No	Not clear	Not clear	No
No	No	No	No
Only in a very general sense. The people he shot at were not the ones who tormented him. Was important to show something to the bullies, earn respect from entire school community.	Yes: Mitchell claims intended only to scare people, claimed "anger" was his motive. Andrew was "mad at a teacher." Both felt as though they were "put upon" by fellow students.	Maybe: conflicting reports over whether particular students were to be targeted, although none of the victims appears to have been. Made general statements that he wanted to "kill nine people" he hated and then kill himself	No
No	No	No	No
No	Likely not	Maybe: showed high interest in Jonesboro shooting	Yes: Columbine a factor, may have intended suicide
Gun .22 pistol	Gun 30.06 semi-automatic rifle w/scope; rifle	Gun .25 caliber semi-automatic pistol	Gun .22 caliber rifle; .357 Magnum handgun
Shooter surrendered to school principal	Stopped shooting probably in response to construction workers shouts and fled toward van	Fled outside to back of banquet hall	Chased outside by other students. Knelt down and put gun in his mouth
Surrendered to principal	Arrested by police near van	Apprehended at rifle point by banquet hall owner	Surrendered to school official
Rural	Rural	Rural	Suburban
River town, higher than national average unemployment rate in 1997; wide range of economic backgrounds from trailer parks to mansions	Jonesboro itself has thriving and diverse manufacturing, service and retail economy. Shootings happened outside Jonesboro in a largely rural area with practically no industry or service sector. Working- and	University town, thriving economy, middle- to upper-middle-class area. Education, manufacturing, and retail businesses are cornerstones of economy	Affluent area with a median household income higher than surrounding counties or the U.S. 75% home ownership

TABLE 8-2 continued

Variable	NY—Jason	NY—Khalil	Chicago
2.3 Social relations in community:			
2.3.1 Community heterogeneity	Hypersegregated, poor, black		Racially and ethnically heterogeneous: 32% white, 46% black, but Joseph's tract and others around it hypersegregated
2.3.2 Community conflict	Drug war and epidemic of youth violence early 1990s		Racial tension and conflict, gang violence
2.3.3 Community change	Community decay since 1960s but entry of crack in 1980s accompanied by high homicide rates and violence		Rapid change in racial composition, lowered housing values, low income housing and racially homogeneous gangs emerged. Layoffs in 1990s
2.4 Level of violence and crime in the community:			
2.4.1 All violence	Neighborhood characterized by violent drug wars. Led city in homicides in 1993		Overall violence on South Side high at the time. Mostly gangs
2.4.2 Youth violence	Homicide leading cause of death		Gang-related fights in school hallways daily
2.5 Youth alienation from adults:			
2.5.1 Separate youth culture	Yes: kids did not tell parents or other adults how bad the violent environment was		Yes: parents felt they had little influence over kids, especially related to gangs. Police did not assert control in neighborhood, did not respond to calls. Kids thought they had no one to turn to
2.5.2 Prevalence/density of youth gangs	Crews important to survival, but not much about criminal gangs in these cases		Criminal gangs had significant presence in neighborhood

Paducah	Jonesboro	Edinboro	Rockdale County
	middle-class families. Low crime rates, considered "great place to live" by residents.		
Considerable black population in area, but school almost all white. Considerable class heterogeneity.	Christian. Rural areas are virtually all white, although Jonesboro has a small but significant black population and a growing Hispanic population.	Mostly white (92%) area. Vintage small-town America: community service clubs, lots of community sports and cultural events	Over 75% white area, small inner city where blacks and poor whites with ties to rural past live
None	None	None	None
Rapid change from rural to industrial and service-based economy	Small but growing generally cohesive community. Influx of poorer residents and a significant problem with crystal meth	Rural, with little change; steady economic growth	Rapid social change from rural/exurban to suburban. Population tripled in a short period of time
Low violent crime	Low crime rates	Low crime rate	Rare
School fights rare, some off-grounds fighting	Youth violence low, but bullying is an issue. Youth gangs are in area and drug use by teens at the high school a significant problem	No	Rare
Yes: profound disconnect between adults experiences and small group of disaffected teens. "Social dynamics of adolescence almost entirely hidden from adult view"	Yes and no: kids knew right away that it was Mitchell and Andrew, but adults shocked. Some kids had reported problem behaviors of Andrew to adults	Yes: students did not tell adults about Andrew's threats, adults shocked. Parents largely unaware of problem behavior among youth	Evidence that adults don't know about or monitor youth behavior (syphilis outbreak)
No	Gangs were supposedly a problem in the town of Jonesboro, but crime rates were low, so not clear if these were criminal gangs	No	No

TABLE 8-2 continued

Variable		NY—Jason	NY—Khalil	Chicago
3.0 School Setting				
3.1	School type	High school		High school
3.2	Economic and social status of students	Poor/working class		Working class/poor
3.3	Level of school violence	Not high		High: daily fights in halls but killings almost never occurred
3.4	Social relations in school	Kids members of crews, embedded in peer relationships/conflicts; needed protection from seeming weak, couldn't do things seen as weak, like get good grades		Difficult to avoid gang entanglements, school was recruitment site for gangs, daily fights in halls of school
3.4.1	Divisions among students	Different crews, usually made up of friends/relatives. "Beef" a feature of social interaction		Lots of racial and gang conflicts
3.4.2	Separation of teachers from students	Yes		Yes
3.4.3	Separation of school administrators from students	N/a		N/a
4.0 Background and character of perpetrators				
4.1	Prior offending/police contact:			
4.1.1	Offender known to police?	No	Yes	Yes
4.1.2	Offending by shooter	Gun carrying	Robbery arrest	Burglary, gang activity, possessing stolen property
4.1.3	Victims' offending	No	Robbery arrest	Robbery and drug arrests for two surviving victims
4.2	Prior arrests for serious offenses:			
4.2.1	Offender	No	Yes	Yes
4.2.2	Victims	No	Yes	Yes

Paducah	Jonesboro	Edinboro	Rockdale County
High school Mixed	Middle school Middle class	Middle school Middle class	High school Upper middle class
Low	Low	Low	Low
Both vertical and horizontal differentiation, some students move easily between groups, some antagonism between groups	Bullying a significant problem at Westside but not clear that the problem is any bigger than other middle schools	Generally good	Generally amicable, open to newcomers, fluid cliques
Cliques: preps, prayer group, Goths, jocks	Some bullying. cliques: athletes and cheerleaders, band kids, druggies, cliques generally not seen as impermeable	Cliques, but no disruptive divisions	Lots of cliques: rednecks, jocks, preps, Christians, wiggers, "straight-edge mafia or vegans"
Yes	Yes	Yes	Yes
N/a	N/a	N/a	N/a
No	No	No	No
Increasing stealing: CDs, guns, $100 bills from dad, fax machine and murder weapon from neighbor	Mitchell: threats to teacher and other kids, minor vandalism, molested a two-year-old Andrew: reports that he tortured and killed animals	Alcohol and other drug use, illegal discharge of a firearm	Small amount of minor delinquency, stole stepfather's gun and sold it, experimental soft drug and alcohol use
No	No	No	No
No	Mitchell was brought before a juvenile court for the molestation issue	No	No
No	No	No	No

TABLE 8-2 continued

Variable		NY—Jason	NY—Khalil	Chicago
4.3	School record of offender:			
4.3.1	Academic achievement	Good grades until 7th grade for both boys, then no. Getting good grades seen as sign of weakness at T.J. High		Good grades until 8th grade, then slipped, cut class a lot
4.3.2	Disciplinary record at school	Not included in cases studies		Suspended for gambling in bathroom
4.3.3	Changes in school status	Grades slumped after 8th grade		Grades slumped after 8th grade
4.4	Mental illness of offender:			
4.4.1	Formal diagnosis prior to event	No	No, but believed to have emotional problems	No
4.4.2	Formal diagnosis after event	No	Yes: defense diagnosis was mental illness relating to stress and trauma from living in East New York; prosecutor claimed antisocial personality	No
4.4.3	Suicide attempt during incident	No	No	No
4.4.4	Previous suicide attempt/threat	No	No	No
4.4.5	Suicide attempt after incident	No	No	No
4.4.6	Psychological counseling prior to incident	No	Yes	No
4.4.7	Recent changes in mood	No	Fearful	Fearful
4.5	Family background of offender:			
4.5.1	Family structure	Intact two-parent family, siblings	Divorced parents, sister withdrew from family	Intact two-parent family

Paducah	Jonesboro	Edinboro	Rockdale County
Grades slumped in 8th grade, then improved freshman year. IQ of 120	Mitchell: As and Bs, Andrew: average student who needed extra help in elementary school	Andrew struggled with grades which got steadily worse until he had mostly D's and F's	Trouble with grades after age 8
Five disciplinary infractions for minor behavior problems	Mitchell: three in-school suspensions Andrew: nothing out of the ordinary	None	None
Grades slumped in 8th grade then improved	No for both cases	Grades slumped in 8th grade	Grades slumped in months previous to attack
No	None that we know of	No	Yes: attention deficit disorder, took Ritalin
Yes: diagnosed with "dysthymia and schizotypal personality disorder" and "dysthymia and traits of schizotypal personality disorder with borderline and paranoid features" by defense psychiatrist Currently has schizophrenia	No conclusive evidence for either boy	Yes: preschizophrenic ideation diagnosed by defense psychiatrist	Yes: clinical depression, defense characterized as major depressive disorder with psychotic features, prosecution as mild dysthymia.
Possibly: reportedly asked another student to kill him	No	No	Yes: immediately after the incident
Suicidal thoughts but no attempt	Andrew may have threatened self in the months prior to the shooting.	Suicidal thinking and threats. Left suicide note	Suicidal thinking and threat
Yes, several	No	Unknown	Yes, almost died in prison
Yes	Mitchell: yes	No	Yes
Not clear, but depression may have magnified fears and insecurities and affected judgment	No, but both reported to be angry. One teacher thought Mitchell became withdrawn before shooting.	Evidence of depressed mood	Increasing social withdrawal
Intact two-parent family, sister	Mitchell: divorced parents, mother remarried, little contact with father, one brother and two half-sisters	Intact two-parent family, but conflict between parents; two brothers; two stepbrothers from previous marriage of	Parents divorced, mother remarried, no contact with dad. T.J. close to stepfather; sister, and older stepbrother

TABLE 8-2 continued

	Variable	NY—Jason	NY—Khalil	Chicago
4.5.2	History of family violence	No	No	No
4.5.3	History of abuse and neglect	No	Possible	No
4.5.6	Recent changes in family relations	No	Yes: sister withdrew from family	No
4.6	Status of perpetrators in school/community:			
4.6.1	Social standing in community	Marginal group member, had friends	Marginal group member, had friends	Marginal group member, had friends
4.6.2	Member of youth gang	No	No	Yes
4.6.3	Interest in violence media	Don't know	Don't know	Don't know
4.6.4	Involvement in gun culture	Guns part of daily life	Guns part of daily life	Don't know
4.6.5	Victim of bullying	Yes	Yes	Don't know
4.6.6	Bullying others	No	No	No
4.6.7	Recent changes in peer relations	Yes	Yes	Yes
4.6.8	Recent peer rejection	No	No	No
5.0 Preparatory Actions of Perpetrators				
5.1	Evidence of planning	Carried gun but otherwise no	Got gun from friend to defend himself from previous threats	Got gun from friend, loaded and test-fired the gun the night before

Paducah	Jonesboro	Edinboro	Rockdale County
	Andrew: intact well-established family, two half-siblings lived elsewhere most of the time	father, but not part of Andrew's life	
No	Mitchell's dad had explosive temper	No	No
No	Mitchell: possibly emotional by father, molested by neighbor. No evidence of abuse for Andrew	No	No
No	Mitchell's father threatened he might have to live with him.	Recent family conflict	No
Marginal member of many groups, had friends but difficulty forming friendships with others and socially insecure in relationships, active in school band, class clown	Mitchell: had conventional friends, close to brother, involved in school and community activities, sports, and choir Andrew: had friends but not popular, played trumpet, class clown	Conventional friends (small group of long-time friends that included boys and girls) but began hanging out with more troubled group	Loner by middle school—increasingly passive and withdrawn, did not join groups or play sports
No	Mitchell: wannabee (Westside Bloods)	No	No
Yes	Mitchell: yes Andrew: yes	Yes	Yes
No	Mitchell: limited use of guns Andrew: yes	No	Yes: one of his most important sources of social identity
Yes	Mitchell: yes Andrew: probably	No	No
Yes, but was not physically imposing	Mitchell: yes Andrew: yes	Yes	No
Yes: trying to impress Goths	Yes	Yes: started hanging out with kids with behavior problems	Yes: began hanging with kids considered to be "fringe" but mostly a loner
Yes: had recently dumped girlfriend in favor of other girl who did not return his attentions.	Mitchell: had just lost a girlfriend	Yes: had recently lost a girlfriend, was rejected by another girl he asked to the dance	No
Yes: elaborate planning but not clear if alone or with others	Yes: elaborate planning	Planning involved but not clear if alone or with others	Yes

TABLE 8-2 continued

Variable		NY—Jason	NY—Khalil	Chicago
5.2	Source of weapons	Carried a gun	Friend	Bought from neighborhood boy
5.3	Threats/warning	No	Responded to threat	Responded to threat
5.3.2	Warnings to adults	No		Not really, but mother knew he had to defend himself but not that he had a gun
5.4	Reason for particular place and day:			
5.4.1	Reaction to immediate events	Yes	No	No
5.4.2	Preventive response to threat	Yes	Yes	Yes and no
5.4.3	Proactive	No	No	A little
6.0 Community Response				
6.1	Impact of media coverage on community:			
6.1.1	Accuracy of media coverage	Slight inaccuracies at first but quickly corrected and followed by high-quality investigation		Not accurate
6.1.2	Impact of media on community deliberation and welfare	Not discussed		Case decided in press before trial: jury pool tainted, no room for plea bargain
6.2	Community interpretation/understanding of events:			
6.2.1	Community forums: existence, nature, impact	Antiviolence marches and rallies attended by celebrities, diverse community organizing under the auspices of Centers for Disease Control and the New York City Health Department to educate the community about violence, retreat for students in Catskills with local politician		Antiviolence rallies and a new Rites of Passage program for the students and teachers to teach antiviolence and leadership skills, leadership from principals, and the local school councils

Paducah	Jonesboro	Edinboro	Rockdale County
Stolen from neighbor	Stolen from father and grandparents	Stolen from father	Stolen from parents
Hinted something big would happen the week before	Gave many hints, said something big was going to happen tomorrow	Gave many hints that no one took seriously	Only a few oblique remarks to peers in weeks before shooting
No	Adults at school told of Andrew's threats by at least one student and student's father	Teacher knew of "will" given to another student, teacher reported threatening conversation between Andrew and another student to school administrator	No
No	No	No	No
No	No	No	No
Yes	Yes	Yes	Yes
Not accurate in the residents' views, and in many instances our own	Mixed: local media generally more accurate	Not accurate	Not accurate
Community united against presence of media, media too aggressive, interviewed students without parent's consent	Community hates national media, media too aggressive as in Paducah	Community considered media presence to be intrusive	Community despises media as too aggressive, distorting their area, making it difficult to move forward
No community forums but more school professional days were added to identify students with serious problems, school counselor added to help freshmen transition to high school, outside therapist visits to school one day per week to talk with students	Jonesboro Ministerial Alliance formed (religion-based) to resolve community problems, no evidence it played any role, teacher debriefing held the day after shooting to discuss how to deal with students, additional social workers and full-time counselors added, federal assistance and help from National Organization for Victims of Crime (NOVA) sought	Town government launched no major initiatives. Series of meetings over the summer with community members to develop recommendations for the schools mostly focused on security, character education program developed for students, more training in conflict resolution instituted for teachers, Christian group sponsored youth center	Community-wide invitational meetings, church meetings, invited youth from Littleton to a retreat, school system increased number of psychologists and social workers in the system, introduced parent education program, held series of meetings to discuss school security needs, and made changes in behavior and dress codes

TABLE 8-2 continued

Variable		NY—Jason	NY—Khalil	Chicago
6.2.2	Engagement of political representatives	Councilwoman Wooten developed year-long retreat program, Mayor Dinkins called for an antiviolence movement, Chancellor Fernandez established 40 new smaller schools		Mayor Daly used incident to get metal detectors installed system-wide, local school council's representatives engaged in public debate over security measures
6.2.3	Community understanding of event	Violent atmosphere, gulf in communication between kids and adults		Gangs, gulf in communication between kids and adults
6.2.4	Grief counseling	Yes		Yes
6.2.5	Impact on community climate	More attention to community violence and danger		More attention to gang problem
6.3	Criminal justice response:			
6.3.1	Charges filed	2nd degree murder	1st degree manslaughter	1st degree murder
6.3.2	Prosecuted as juvenile or adult	Adult	Adult	Adult
6.3.3	Defense offered	Plea to lesser charge of 1st degree manslaughter	Acting under influence of extreme emotional distress	Act of self-defense
6.3.4	Disposition/ sentence	3–9 years in prison, part in youth part in adult facility	6 2/3 to 20 years in prison, served 5 years	45 years in Illinois State Prison

Paducah	Jonesboro	Edinboro	Rockdale County
Political representatives were not engaged, mostly by choice of the school	No information	Town government launched no initiatives	Participation in community meetings described above, judge ordered tough laws posted in schools
Freak, inexplicable event, could not be predicted or explained	Generally cannot understand why this happened but when pressed cite several factors including lack of communication between kids and adults, decline in religiosity, media influence, availability of guns	Troubled kid from troubled family happened to live there	Mental illness and family problems
Yes	Yes	Yes	Yes
Potential negative impact on community's ability to attract high-quality professionals to jobs, many students and teachers still in treatment, civil suits unresolved	Community angry about sentences for boys, some felt there is better cohesion in the community, shooting no longer widely discussed, civil suits unresolved	Community felt this could have happened anywhere—troubled boy, many people blamed parents, general sense that community is ready to move on	Sense of shame among some young people to say where they are from, but relates as much to syphilis outbreak as shooting. Community did not feel different from any other community
3 counts of murder, 5 counts attempted murder, 1 count burglary	Delinquency was actual charge, but based on 5 counts capital murder and 10 counts first degree battery for each boy	Criminal homicide/1st degree murder	18 counts of aggravated assault, six counts of cruelty to children, and 5 counts of illegal possession and use of firearms
Adult	Juvenile	Adult	Adult
Pleaded guilty but mentally ill under an Alford plea	Mitchell: pleaded guilty Andrew: pleaded not guilty, judge ruled that in juvenile proceeding insanity and incompetency defense could not be raised	Pleaded guilty to 3rd degree murder	Pleaded guilty but mentally ill to above charges
Life without parole for 25 years	Indeterminate sentence to age 21 for both boys	30–60 years in prison, eligible for parole at age 45	Sentenced to 60 years of custody including 40 years of probation

TABLE 8-2 continued

Variable	NY—Jason	NY—Khalil	Chicago
6.4 Civil litigation	None	None	Civil suit against the school board by one of the victims' parents
6.5 School security responses:			
6.5.1 School security: hardware	Metal detectors installed		Metal detectors installed system wide
6.5.2 Efforts made to strengthen relationships in school	Yes: through retreats and violence education programs		Yes: through Rites of Passage program, zero tolerance policy instituted
6.5.3 Other security	Daily police patrols inside and outside school, supplemental programs dealing with violence prevention and crime at 40 schools, additional after-school programs proposed, replace very large schools like T.J. High with 50 smaller schools		Hired community security volunteers, instituted zero tolerance policy including suspensions, for some offenses and automatic police referrals for a variety of offenses from serious assault to minor drug violations

Paducah	Jonesboro	Edinboro	Rockdale County
			supervision. Eligible for parole after 18 years
Victims' families filed suits against Carneals, neighbor from whom gun was stolen, students who knew or may have been involved, teachers and principals of Heath High and Middle schools, producers of Basketball Diaries, makers of point and shoot video games, porn Internet sites offender visited	Civil suits brought by victims' families against shooters and their parents, Andrew's grandfather, and gun manufacturers for failure to install trigger locks	Civil suit by teacher's widow against Wurst family	3 of 6 victims filed civil suits against offender and his parents
Fences built around school, identification tags required	Wooden slat fence built around school	Use of metal detector wands failed but metal detectors used for prom, restricted building access, name badges for staff	School added additional surveillance cameras
Extended freshman orientation period, teachers search student bags in morning, more professional days for teachers to identify problem students, part-time guidance counselor and therapist added to staff	Adult monitors on school buses, increased attention to isolated or troubled kids, therapist added, more teacher training, school resource officer added	More teacher training in conflict resolution, new character education program, but most changes security oriented	Increased number of psychologists and social workers, introduced a new parent education program
Hired school resource officers (police) to interact with students and maintain security	Instituted common sense zero tolerance policy, hired school resource officers (police) to maintain safety and security and educate students about them	None	New dress code, strict weapons in school policy involving automatic referrals to police and harsh sentences

Part II

Understanding and Preventing Lethal School Violence

The cases presented in Part I present very specific, concrete images of lethal school violence in America in the 1990s. The images are tragic and compelling. They force one to ask why. But even with the benefit of hindsight, the answers seem exceedingly mired in complexity. In each case, there seem to be so many factors that give impetus to the events; so many things that if they had not been present might have lessened the likelihood that the events would have occurred; so many moments when a particular small intervention might have averted tragedy.

However, for purposes of stimulating the imagination about the factors that might be important causes, and the kinds of interventions that might be effective in preventing and controlling such tragic events, the very richness of the cases is their value. One's mind is opened to a variety of possibilities. Stereotypes and confident assumptions with which one began the inquiry are undermined. One emerges from these cases less confident that one knows anything for sure, at least partly because many more things now seem potentially important. Even though it sounds paradoxical, learning that one knows less than one thought, that there are many more possibilities than one imagined, that one's favorite theory is not particularly well supported by a confrontation with the detailed facts of these cases, represents a gain in knowledge.

But there is another use of the cases as imperfect pieces of evidence that allow distinctions among different kinds of violence, as well as the development of some hypotheses that may seem a bit more likely than others. In order to use cases in this way, we had to find a way to put them in an aggregate context, to see where these particular incidents of violence might fit in a broader overall pattern, and to decide how the different kinds of violence that appeared in the cases might be related to one another.

It also is important to put these cases in the context of theoretical literatures that help sort different kinds of violence into different classes distinguished from one another in terms of their character, causes, and effective modes of prevention and control.

In this part of the report, we put the cases in context, first, by looking at where they fit in the aggregate patterns of violence that have beset the country, and, second, by seeing where they seem to fit in the academic literature on violence, and what that literature says about the causes and effective means of controlling such incidents.

Part II ends with the committee's observations about potentially important strategies for preventing or controlling the violence and recommendations for future research. In offering these, we have stretched to the breaking point the evidentiary power of the sources at our command. We do so because we think the nation needs some considered observations about the nature of this problem, and that it cannot necessarily wait to get more information before it acts. Already, communities have begun to take steps to guard themselves from these incidents. Costs are being incurred; consequences—intended and unintended, good and bad—are beginning to accumulate. We therefore recommend an agenda for additional research, and that it be done quickly to ensure that the nation is better off a year from now, and five years from now, in terms of the ability to understand and control these events.

9

Lethal School Violence in Statistical Context

This chapter presents information about levels and trends in the form of lethal school violence under study, placing them in the con- text of other forms of violence. Such statistical information helps us interpret the information from the cases by revealing the size of the phenomenon under study relative to other forms of violence. It also enables an investigation of the likely causes of the particular form of violence that interests us and, more particularly, whether it seems to move in concert with other forms of violence or in its own independent patterns.

EXTENDING THE DEFINITION

To do any kind of statistical analysis, it is necessary to develop an operational definition of the phenomenon to be studied. As described in Chapter 1, the operational definition we embraced was "lethal school violence," which includes the following elements:

- *Lethal* violence
- That took place in *schools*
- Was committed by *students* of the school and
- Resulted in *multiple victimizations.*

The committee's application of this definition was fluid enough to include at least one case in which no one died.

In fact, it was the idea that the subject of study was a series of incidents defined not only in terms of the amount of carnage, but also in terms of the character of the attacks and the motivation of the offenders that generated a second concept beyond the operational definition of lethal school violence that guided most of our work. After we reviewed the cases and the various literatures that seemed related to them (including the emerging literature on rampage shootings), we began referring to some of the school cases as "school rampage shootings."

In one sense this is hardly controversial. The label "school rampages" seems a reasonable descriptor of events in which a student of a school shot at and killed members of his school community.

Yet it is worth noting that the idea of a "rampage shooting" has connotations and meanings that go beyond the operational definition we began with. The idea of a rampage, for example, suggests that an important cause of the event was the agitated, confused mental state of the perpetrator. When we refer to such events as rampages, we are already making claims about the likely causes of these events as well as their consequences.

It is also worth noting that this assigns the school shootings to a class of violence (rampages) about which there is both theoretical and empirical information (see, e.g., Fox and Levin, 1998). We could equally well have assigned these incidents to a different specialized class of violence. We could, for example, have characterized the incidents as "school mass murders" (see, e.g., McGee and DeBernardo, 1999). This would have had somewhat different connotations than school rampages. The point is that, as soon as we start using concepts that are more abstract and evocative than a specific listing of the attributes that provide the operational definition of the phenomenon we seek to measure, we are at risk of subtly biasing both our own and our audience's understanding of the events we are trying to analyze.

To examine the choices of how to think about these relationships, the committee used a series of Venn diagrams that reveal the relationship among more or less inclusive ideas. Figure 9-1 shows the relationship between three distinct but partially overlapping sets: Set A is the idea of "lethal school violence in incidents involving more than one victim" (this was the core operational definition for most of our work); Set B is the idea of "lethal school violence" (which is the phrase we often used, but as an analytic matter would include cases in which only one person was killed); and Set C is the idea of "school rampages" (which includes cases in which more than one person was injured, and there was a significant potential for lethal violence as well as those in which people were actually killed).

As a practical matter, the violence we investigated included the union of sets A and C. The portion of Set B that involved only single victims (whether fatal or not) was excluded from the analysis. This put the pri-

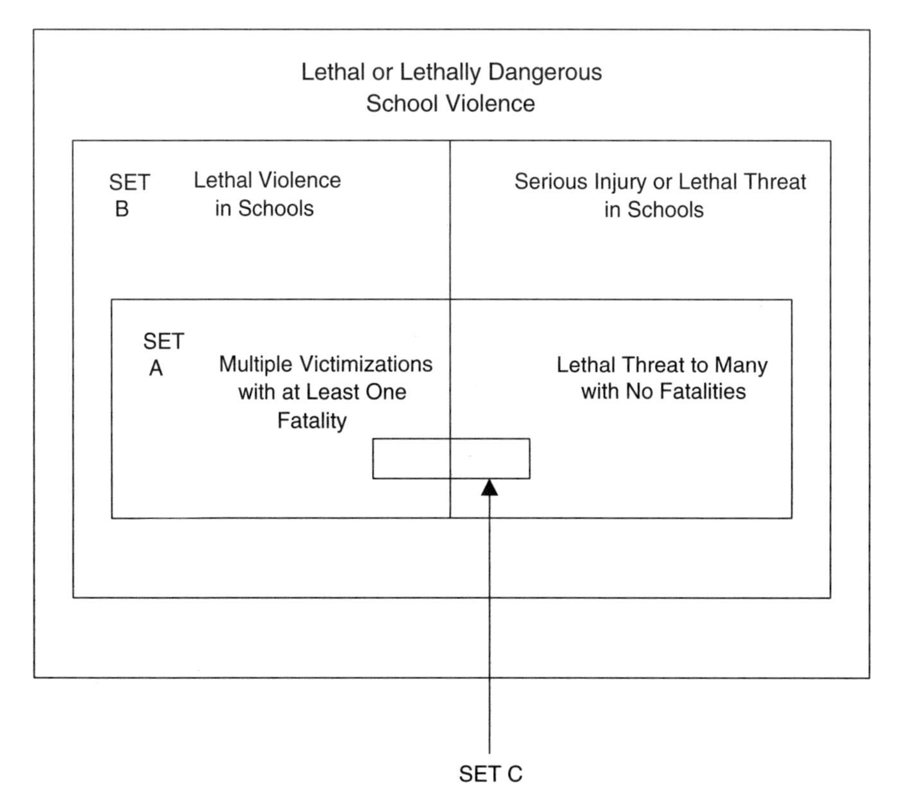

FIGURE 9-1 Lethal school violence and school rampages as adjunct sets.

mary emphasis on a very small number of cases. Moreover, it turned out that suburban and rural schools were much more significantly represented in these cases than inner-city schools, where many of the incidents of school violence involved single victims. That was not because we thought these cases were unimportant; the committee held the opinion that the special attention given to the relatively few incidents involving multiple victimizations might distort society's picture of lethal school violence in general. It was, instead, that we had been directed by Congress to give special, but not exclusive, attention to this apparently new form of violence. We also recognized that a significant amount of public and scholarly attention had been devoted to lethal youth violence in urban areas, and that less had been focused on trying to understand this new and unexpected phenomenon.

A Venn diagram is also helpful in showing the relationship between the kind of violence that was our primary focus and the other categories of violence with which it was being compared. Figure 9-2 embeds Figure

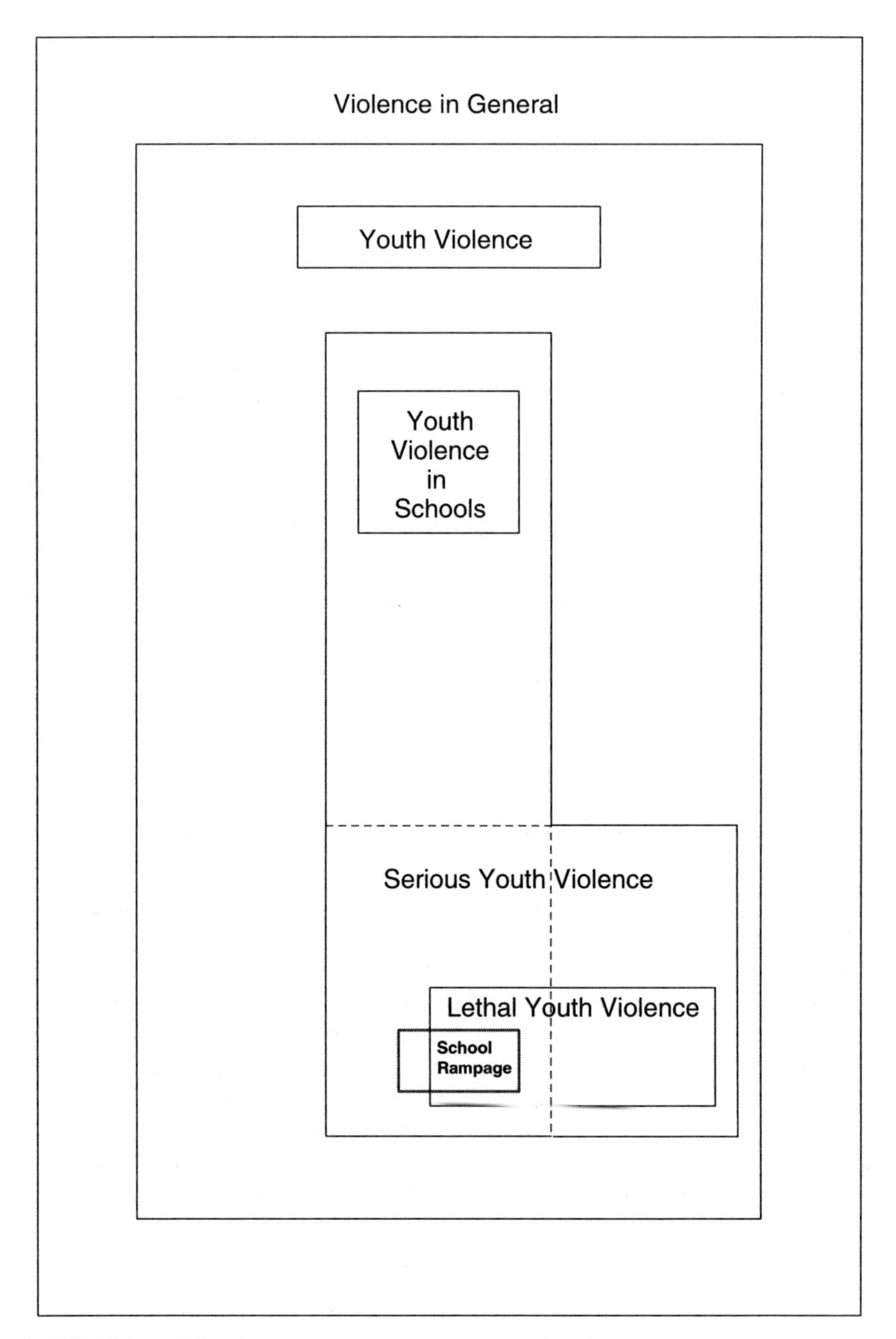

FIGURE 9-2 School rampages as a component of violence.

9-1 in the larger frame of more general definitions of violence and more common forms of violence. The largest set is "all violence." An important subset of all violence relevant to this study is all violence committed by or experienced by youth. Two important subsets of youth violence relevant to our subject are "serious youth violence" on one hand and "all school violence" (regardless of the seriousness of the victimization) on the other. These subsets overlap, but neither is a subset of the other. There is a lot of serious youth violence that takes place outside of school. There is a great deal of school violence that is not serious. Within the category of "serious youth violence," we can create a subset of "lethal school violence." Some of this happens in schools, and some happens outside schools. Within the category of "lethal school violence," we can create two subsets that define lethal incidents in which only one person was killed and distinguish that from other incidents in which many were killed or injured in the same incident. Finally, we can describe a subset of serious (but not lethal) school violence in which many were victimized but no one died.

As noted above, the core of the study focused on incidents of lethal school violence in which more than one person was injured ("lethal school violence" for shorthand purposes) and incidents in which many were victimized regardless of whether anyone was killed ("school rampages.") In the rest of this chapter, we present data on levels and trends in forms of violence organized in terms of these different sets. We look first at levels and trends in the kind of violence that was at the core of our investigation: lethal violence, in schools, by students, resulting in multiple victimizations, and other situations in which multiple victimizations occurred in a way that suggested the fact that no one died was simply a matter of chance. We then look at how this violence fits within the larger patterns of youth and school violence.

TRENDS IN LETHAL SCHOOL VIOLENCE

The principal dataset available in assessing the level and trends of lethal school violence (with or without multiple victims) was produced and maintained by the National School Safety Center (2001). It is a census of school-associated violent deaths that occurred between 1992 and 2001, based on a systematic collection of newspaper accounts of the incidents. We used this dataset to separate from all lethal violence in schools that portion that meets our core definition. This excludes incidents in which only one person was killed or injured, as well as incidents in which a student started shooting, but no one died. Figure 9-3 presents a simple count of these incidents over the time period from 1992 to 2001. It makes clear that incidents of lethal violence including multiple victims while serious, are indeed very rare events. Overall, there have been 13 of these

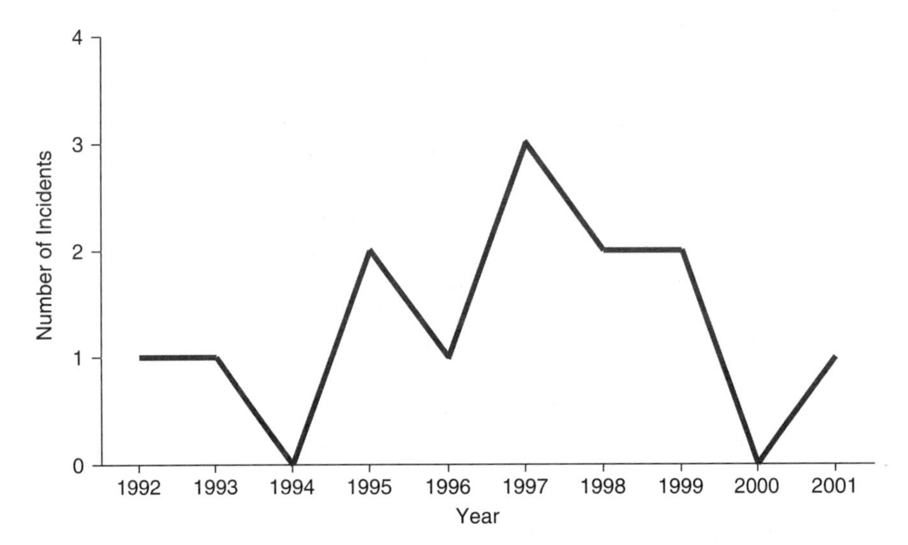

FIGURE 9-3 U.S. multiple-fatality, student perpetrated incidents, 1992–2001.
SOURCE: Data from the National School Safety Center.

incidents in this period. In the peak year, 1997, there were only three of these incidents.

There are major concerns about these data for our purposes: (1) they go back only to 1992, (2) they focus only on events in which there was at least one fatality (thereby excluding incidents in which many were victimized, but no one killed), and (3) the sample was limited to the United States. Each of these presented a problem for the committee as it sought to put the issues of multiple-victim lethal school violence and school rampages in a broader context. Consequently, we sought to expand and improve on this database.

In constructing our own database to complement this one, we relied on the same basic source—namely, newspaper accounts. But we looked in newspapers *prior to 1992,* and in *countries other than the United States.* We also had access to the United States Secret Service study that looked at this form of violence over a longer period than 1992–2001. Consistent with our broader operational definition that includes incidents in which many were victimized but none killed, we included incidents that resulted in multiple serious injuries but no fatalities. Figure 9-4 presents the results of that inquiry for the United States: our best estimate of the number of U.S. multiple-victimization, student-perpetrated school violence incidents between 1974 and 2001. Table 9-1 provides some details on these incidents.

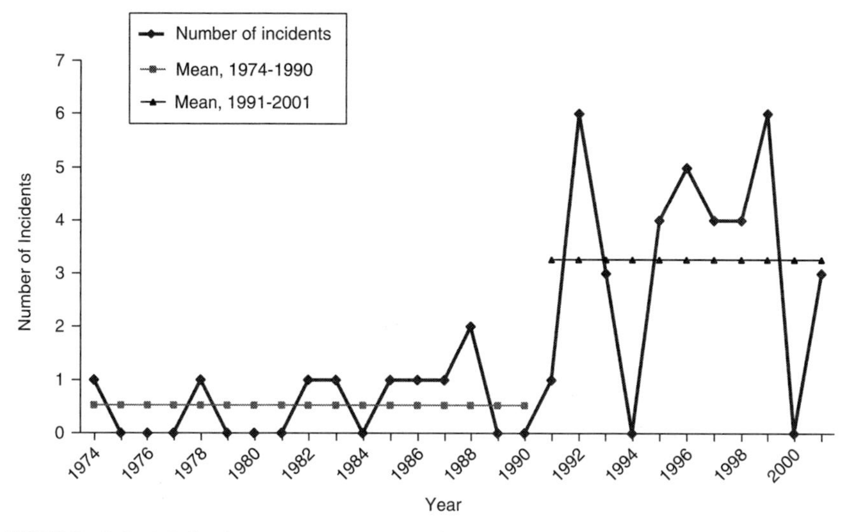

FIGURE 9-4 Multiple-victimization, student-perpetrated incidents, 1974–2001.

Figure 9-4 shows that student-perpetrated school rampages (with or without a fatality) are not entirely new phenomena. There were two such incidents in the 1970s and six in the 1980s. And yet it also seems clear that the frequency of student-perpetrated school rampages resulting in multiple victimizations increased dramatically after 1994. The difference is highlighted in the figure by lines showing the mean number of such incidents per year in the 17-year period from 1974 to 1990 and the 11-year period from 1991 to 2001. The mean number of student-perpetrated ram-

TABLE 9-1 Multiple-Victimization, Student-Perpetrated School Violence in the United States

Date of Incident	Location	Name of Offender	Fatalities	Injuries
12/30/1974	Olean, NY	Anthony Barbaro	3	11
2/22/1978	Lansing, MI	Roger Needham	1	1
3/19/1982	Las Vegas, NV	Patrick Lizotte	2[a]	2
1/20/1983	Manchester, MO	David F. Lawler	2[b]	1
1/21/1985	Goddard, KS	James Alan Kearbey	1	3
12/4/1986	Lewistown, MT	Kristofer Hans	1	3
3/2/1987	DeKalb, MO	Nathan Faris	2[b]	0
2/11/1988	Pinellas Park, NJ	Jason Harless and "companion"	1	2

(continued on next page)

TABLE 9-1 continued

Date of Incident	Location	Name of Offender	Fatalities	Injuries
12/16/1988	Virginia Beach, VA	Nicholas Elliott	1	1
11/25/1991	Brooklyn, NY	Jason Bentley	1	1
2/26/1992	Brooklyn, NY	Khalil Sumpter	2	0
5/14/1992	Napa, CA	John McMahan	0	2
11/4/1992	Detroit, MI	Renard Merkerson, Montrice Coleman	0	6
11/4/1992	Detroit, MI	Unknown	0	2
11/19/1992	Chicago, IL	Joseph White	1	2
12/4/1992	Great Barrington, MA	Wayne Lo	2	4
1/18/1993	Grayson, KY	Scott Pennington	2	
2/1/1993	Amityville, NY	Shem McCoy	1	1
3/18/1993	Harlem, GA	Edward Bryant Gillom	1	1
1/23/1995	Redlands, CA	John Sirola	1[b]	1
10/12/1995	Blackville, SC	Toby R. Sincino	2[b]	1
10/30/1995	Richmond, VA	Edward Earl Spellman	0	4
11/15/1995	Lynnville, TN	Jamie Rouse	2	1
2/2/1996	Moses Lake, WA	Barry Loukitas	3	1
2/8/1996	Palo Alto, CA	Douglas Bradley	1[b]	3
3/19/1996	Las Vegas, NV	An unidentified 8th grade student	0	2
5/14/1996	Taylorsville, UT	Justin Allgood	1[b]	1
7/26/1996	Los Angeles, CA	Yohao Albert Rivas	0	2
2/19/1997	Bethel, AK	Evan Ramsey	2	2
10/1/1997	Pearl, MS	Luke Woodham	2	7
12/1/1997	West Paducah, KY	Michael Carneal	3	5
12/15/1997	Stamps, AR	Joseph "Colt" Todd	0	2
3/24/1998	Jonesboro, AR	Andrew Golden, Mitchell Johnson	5	10
4/25/1998	Edinboro, PA	Andrew Wurst	1	2
5/21/1998	Springfield, OR	Kipland Kinkle	2	21
9/29/1998	Miami, FL	Felly Petit-Frere	0	3
1/8/1999	Carollton, GA	Jeff Miller	2[b]	0
1/14/1999	New York, NY	Camillo Douglas	0	2
4/20/1999	Littleton, CO	Eric Harris, Dylan Klebold	15[b]	23
5/20/1999	Rockdale County, GA	Anthony B. "T.J." Solomon, Jr.	0	6
10/11/1999	Las Vegas, NV	Maynor Villanueva	0	2
12/6/1999	Fort Gibson, OK	Seth Trickey	0	4
2/22/2001[d]	Portland, OR	14-year-old student	0	3
3/5/2001	Santee, GA	Charles Andrew Williams	2	13
3/22/2001	El Cajon, CA	Jason Hoffman	0	5[c]

[a]Includes offender killed by police.
[b]Includes offender suicide.
[c]Includes offender injured by police.
[d]Stabbing incident; all others are shootings.
SOURCE: List of incidents was compiled using data from the National School Safety Center (2001), the U.S. Secret Service (2000), and a separate media search using online national and local newspaper archives.

pages increased from an average of 0.53 incidents per year to an average of 3.27 incidents per year.[1]

It is important to note that these are very small numbers. It is also important to note that the increase observed in the 1990s could be explained at least in part by a reporting phenomenon. It seems likely that the media would cover fatalities in schools, and particularly fatalities that occurred with multiple victimization, with a high degree of consistency and reliability over the entire period from 1974 to 2001. What we cannot be sure of, however, is whether the media would have covered incidents involving multiple victimizations without a fatality as consistently or reliably over this period. While it seems likely that multiple victimizations in a school setting would be newsworthy throughout this period, we cannot be entirely sure that the media weren't particularly sensitized to the issue of school rampage shootings in the late 1990s, and therefore began covering these more assiduously (even when they did not involve fatalities) than had previously been true. If the media were sensitized to these events, part of the increase could be accounted for by the increased likelihood of news accounts of such events, not by an increase in the real underlying rate of these events. Still, the difference in the rate of these events is impressive and would easily be rejected as a chance occurrence if the reporting were accurate, even though the numbers are very small.

Our media search also uncovered five student-perpetrated school rampages in other countries (Table 9-2). While these results may be biased by the less certain coverage of international events, it seems note-

TABLE 9-2 International Multiple-Victimization, Student-Perpetrated School Violence

Date of Incident	Location	Name of Offender	Fatalities	Injuries	Method	Data Source
5/28/1975	Brampton, Ontario, Canada	Michael Slobodian	2[a]	13	Shooting	Media
12/7/1999	Veghel, Netherlands	17-year-old student Father and sister (age 15) were charged as accessories	0	5	Shooting	Media
3/16/2000	Rosenhem, Germany	16-year-old student	1	1[b]	Shooting	Media
4/20/2000	Gloucester, Ontario, Canada	15-year-old student	0	6[b]	Stabbing	Media
3/26/2001	Machakos, Kenya	Felix Mambo Ngumbao, Davies Otieno Onyango	67	19	Arson	Media

[a]Includes offender suicide.
[b]Includes offender self-injury.

worthy that only one incident occurred in 1975 and no additional shootings occurred until 1999. The 1999 shooting was followed by three other rampages involving different means of inflicting harm on others (arson, stabbing, and shooting). This suggests that school rampages are not unique to the United States and, since no international school rampages were evident until 1999, rampages in other countries may have been somehow influenced by the U.S. epidemic in the 1990s.

One final point: a December 2001 article in the *Boston Globe* reported that since the April 1999 Columbine tragedy, 12 U.S. school rampage shootings have been discovered and thwarted before they came to fruition. Ideally, we could put these events on Figure 9-4 as a further indication of the trends in time of these school rampage shootings. There are three problems in doing so, however. First, it is quite likely that, given the public concern about the school rampages, the newspapers would be much more likely to report on thwarted incidents in this period than they would have in earlier periods. Second, given efforts to mobilize students to report these events and law enforcement to take them seriously, it is quite likely that the police would find more such events and that they would treat each event as a serious plot that was really to be carried out rather than mere fantasizing by the kids involved. Third, in any case, Figure 9-4 records events that actually occurred. Presumably, for every act that actually occurred, there were some others in which some preparations were made, but for a variety of reasons, the act never occurred. Consequently, we would have to assume that there were even more attempts to be found than completions. What we are observing in the thwarted events, then, are some incidents that might never have occurred even if the police had not found them in time.

For all these reasons, it is inappropriate to put these thwarted shootings in the same figure as the other data. Still, the fact that these thwarted events were planned during this period is consistent both with the idea that planning for such events increased in the latter half of the 1990s, and that society and the police got a bit better at learning about and thwarting the events. But the data cannot prove this claim.

While the data depicted in Figure 9-4 are weak by scientific standards, they are still important to include in the effort to understand multiple-victim lethal school violence. What they suggest is that school rampage shootings are not a recent phenomenon, nor are they uniquely a U.S. phenomenon. It seems likely that the United States has experienced an epidemic of these incidents in the latter half of the 1990s—that is, an unexpected increase in their number. There may also have been some contagion mechanisms at work—that is, some kind of copycat influence.

If the international and thwarted incidents are included in the basic time trend of observed school rampages, then copycat mechanisms seem likely. But there also seems to have been a small previous increase in these incidents in the late 1980s that no one much noticed. The lack of notice may have prevented the escalation of these shootings through the copycat phenomenon. But this is largely speculation, not a scientific claim. It seems unlikely that this phenomenon is either entirely new or entirely unique to the United States. It may have gotten worse recently and—even more speculatively—that may be in part the result of a kind of contagion. But the problem has endemic and international aspects as well as epidemic and U.S. ones.

TRENDS IN RELATED FORMS OF VIOLENCE

The comparison of school rampage violence with related forms of violence in scale and over time is important for policy as well as scientific reasons. On the policy side, we are interested in its size and trend relative to others to help in making judgments about the relative importance of the different kinds of violence. What proportion of all violence, or all violence involving youths, or all violence that occurs in schools do rampages account for? The question of how fast this component of the violence problem is growing is also of interest. If it is increasing rapidly and other forms are fading, then there is more urgency about efforts to keep it from getting out of hand, even if it is not a major piece of the problem. Comparing school rampages with other forms of violence indicates what priority is appropriate to give to this kind of violence.

There are also scientific reasons to be interested in the relationship of school rampage shootings to other forms of violence. One possibility is that all forms of violence spring from the same basic causes that operate at a structural level in society. Another is that one kind of violence tends to cause other kinds of violence somewhat independently of the structural causes. For example, in this view, one might say that the epidemic of youth violence in the inner city from 1985 to 1995 set the stage for or caused the outbreak of school shootings in 1995–2000.

We place our estimate of the trends in school rampages against the backdrop of trends in the following other kinds of violence: (1) intentional lethal violence in general, (2) intentional lethal violence among youth, and (3) intentional violence in schools.

Figure 9-5 presents the yearly number of all homicide victims in the United States between 1976 and 1999 as measured by the Federal Bureau of Investigation's Supplementary Homicide Reports (SHR) (Fox, 2001).[2] There are two distinct peaks: homicides increased to a peak of 22,953 victims in 1980, decreased through the early to mid-1980s, increased to a second peak of 24,711 victims in 1991, and then decreased to 15,483 victims in 1999. As the figure shows, homicide victims ages 12–18 represent

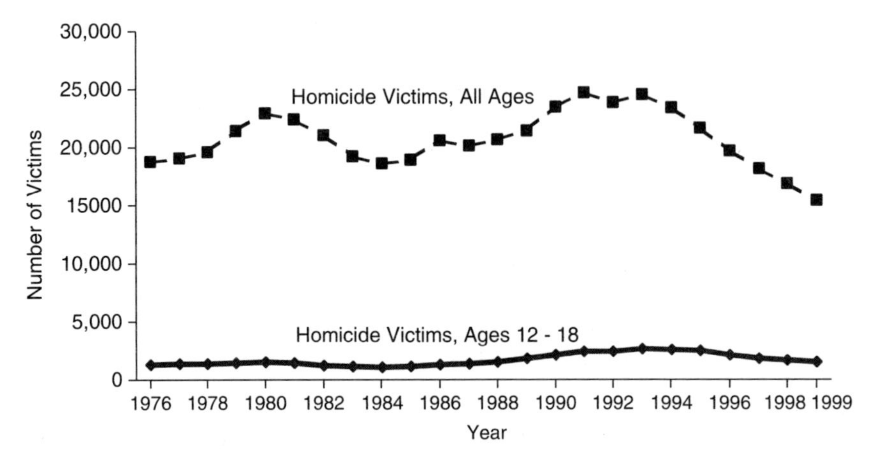

FIGURE 9-5 Homicide victimization in the United States.
SOURCE: U.S. Federal Bureau of Investigation (2001).

a relatively small portion of the yearly body count. At the peak in 1993, there were 2,634 homicide victims ages 12–18, which represents slightly less than 11 percent of the 24,571 total homicide victims that year.

Figure 9-6 presents the yearly counts of homicide victims age 12–18 and homicide offenders age 12–18.[3] The number of youth homicide victims increased dramatically between 1985 and 1993, then decreased significantly from 2,476 victims in 1995 to 1,495 victims in 1999. More directly relevant to the focus on rampage shooters, the yearly number of

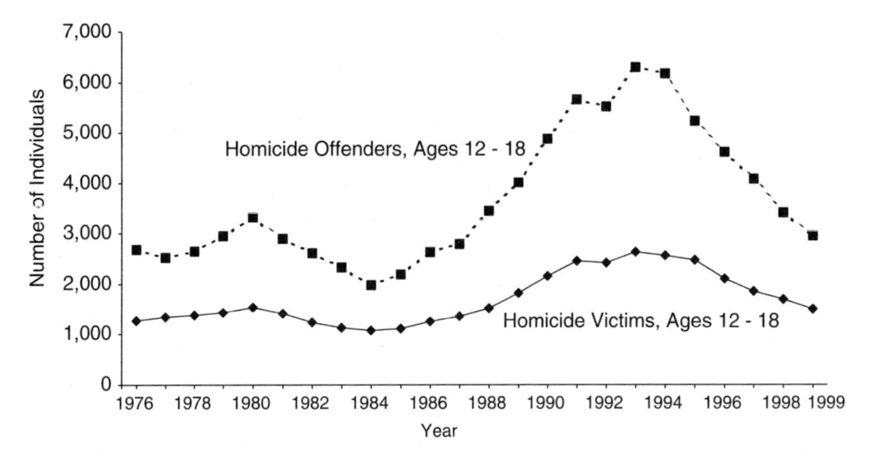

FIGURE 9-6 U.S. homicide victims and offenders ages 12–18.
SOURCE: U.S. Federal Bureau of Investigation (2001).

homicide offenders age 12–18 followed a similar trajectory. The number peaked in 1993 at 6,285 offenders and decreased significantly to 2,959 offenders in 1999.

These patterns of intentional lethal violence in general and intentional lethal violence among youth stand in stark contrast to the epidemic of student-perpetrated school rampage shootings. The school rampages were increasing when broader forms of intentional lethal violence were decreasing. The pattern also contrasts with trends in reported school violence. According to National Crime Victimization Survey estimates, the number of reported violent and serious violent crimes against students decreased between 1992 and 1999 (Table 9-3). Using data collected from media databases, state and local agencies, and police and school officials, Anderson et al. (2001) found that the rate of single-victim student homicides decreased significantly between 1994 and 1999, while the rate of multiple-victim student homicides increased significantly during the same period. Between 1994 and 1999, Anderson et al. (2001) identified 220 school-associated events resulting in 253 violent deaths.[4] The vast majority of the events involved single deaths, while only 8.2 percent (18 of 220) involved multiple deaths. While these data do not distinguish school rampage shootings, it is clear that multiple violent deaths in schools represent a small portion of all lethal violence in schools.

Student-perpetrated school rampage shootings could be thought of as a subset of multiple killings or mass murders committed by adolescent offenders. Figure 9-7 presents the yearly number of homicide offenders age 12–18 who killed two or more individuals in a single incident. The time series follows the same general trajectory as the trends depicted in Figure 9-6. This is probably because the youth homicide epidemic was

TABLE 9-3 Number of Nonfatal Crimes Against Students Ages 12–18 at School per 1,000 Students by Type of Crime, 1992–1999

	1992	1993	1994	1995	1996	1997	1998	1999
Violent	48	59	56	50	43	40	43	33
Serious Violent	10	12	13	9	9	8	9	7
Total	144	155	150	135	121	102	101	92

NOTE: Serious violent crimes include rape, sexual assault, and aggravated assault. Violent crimes include serious violent crimes and simple assault. Total crimes include violent crimes and theft. "At school" includes going to and from school.

SOURCE: National Crime Victimization Survey, 1992–1999 from P. Kaufman, X. Chen, K. Peter, S. Ruddy, A. Miller, J. Fleury, K. Chandler, M. Planty, and M. Rand. 2001. *Indicators of School Crime and Safety, 2001.* Washington, DC: U.S. Departments of Education and Justice.

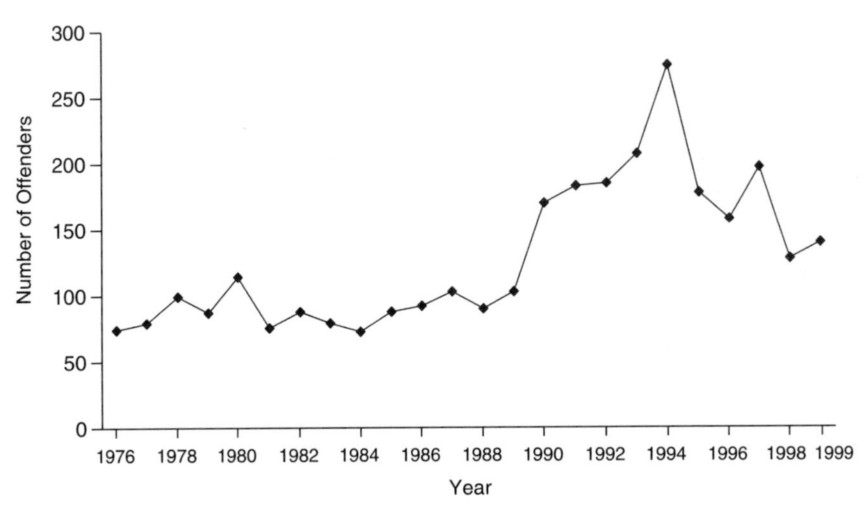

FIGURE 9-7 U.S. homicide offenders, ages 12–18, multiple victims.
SOURCE: U.S. Federal Bureau of Investigation (2001).

driven by gang-related and drug-related violence; there is no reason to believe that a majority of multiple homicides would not share similar circumstances. However, when these types of homicides are parsed from the data, the same pattern is evident.

CONCLUSIONS

The analyses reported here suggest a number of broad conclusions. First, school rampage shootings, while extremely serious, account for a very small component of all the lethal violence in the United States. Even if we take the smallest subset of violence—lethal violence in schools—we find that school rampages are a small component of this sort of violence. Although it has been growing, it remains small.

This point can be emphasized still further by showing deaths in school rampages as a component of all traumatic death for young people and the relative risk of dying from different causes that adolescents face. These comparisons are not meant to minimize the pain and suffering and shock that the school rampage shootings generate. But they underscore the other kinds of risks to which children are exposed in addition to school rampages.

Second, the trends in school rampages seem to be somewhat unrelated in time to the other forms of violence. They do not rise and fall together. Nor do they seem to rise and fall in some other relationship to one another, with one rising after the other rises, or one falling while some other component rises. It is quite possible that this form of violence moves independently of other forms of violence, just as the level of domestic homicide or gangland murders seems to move somewhat independently of homicides in general.

NOTES

[1]The difference between the numbers of events in these two periods would be significant in any probability model constructed to test whether the number of student school rampages increased. However, the issue of statistical significance is not directly relevant here, as we have a census of the events. As a descriptive matter, there is little doubt that the number of student school rampages increased over time.

[2]SHR victim data were weighted to match national FBI Uniform Crime Reports estimates of homicide victimization.

[3]SHR offender data are limited by missing offender data associated with unsolved homicides. The estimates reported here use an imputation algorithm made available with the SHR computer file that provides estimates for these unsolved homicides based on the characteristics (age, race, and sex) of offenders associated with solved homicides (see Fox, 2001).

[4]These events included homicides, suicides, legal intervention, or unintentional firearm-related related death of a student or nonstudent in which the fatal injury occurred (1) on the campus of a public or private elementary or secondary school, (2) while the victim was on the way to or from such a school, or (3) while the victim was attending or traveling to or from an official school-sponsored event (Anderson et al., 2001).

10

Literature Review

Reviewing the existing research on school rampages and related forms of violence was important to the panel's work for two reasons: first, to look for substantive findings about the causes of such events and, second, to consider whether theories or suggested analytic frameworks could help in understanding the panel's detailed case studies.

This chapter presents evidence from several different bodies of research. First is the research that focuses narrowly on school rampage shootings. Given the apparent newness of the phenomenon and the public interest the incidents have generated, the sources of information we needed to tap lay beyond the usual research procedures and standards. In addition to academic research published in peer-reviewed journals, we examined journalistic and professional practitioner studies.

We also looked at the research on violence, on youth violence, and on school violence—including the role of bullying in provoking retaliatory violence in adolescents.

In exploring the incidents of lethal violence in schools and the school rampage shootings, we discovered some more specialized bodies of research that had grown up around incidents similar to those under study. Thus we looked at the research on mass murder, on public rampages in workplaces and other public places, and on shootings followed by suicides. Finally, concerned about issues of contagion, we looked at research on copycat violence and suicide clusters.

RESEARCH ON SCHOOL RAMPAGE SHOOTINGS

A small number of very recent studies have focused on incidents in which students at a school opened fire on other members of their school community, killing or seriously injuring more than one person. Four studies of youth school rampage shootings are reviewed here: two from the academic literature, two from the professional practitioner literature. While there are some noteworthy discrepancies, the studies paint broadly similar pictures of this rare phenomenon.

Adolescent Mass Murderers

In the wake of the Columbine High School mass murders in April 1999, a group of researchers focused their attention on adolescent mass murders—many of which occurred in school settings and resembled the shootings at Columbine (Meloy et al., 2001). The researchers defined adolescent mass murder as the intentional killing of at least three victims (other than the perpetrator) in a single incident by an individual age 19 or younger. They identified cases through a search of medical, social science, and criminal computer databases. A case was included only if there was sufficient credible information.

The researchers collected evidence from primary data sources—for example, courtroom testimony, scientific articles, interviews with law enforcement personnel involved in the case, and video interviews with the perpetrators, survivors, family members, and witnesses—as well as secondary data sources—for example, newspaper articles. They produced a dataset that covers 27 incidents of mass murder involving 34 perpetrators. Eight of the perpetrators committed their attacks at school. The 27 mass murders accounted for 126 people killed and 84 people injured. All the mass murders occurred between 1958 and 1999, with more than half occurring between 1995 and 1999.

The adolescent mass murderers had the following characteristics: all were male, 80 percent were white, 70 percent were described as "loners," 43 percent had been bullied by others, 37 percent came from separated or divorced families, 44 percent were described as "fantasizers" (daily preoccupation with fantasy games, book, or hobbies), 42 percent had a history of violence, 46 percent had an arrest history, 62 percent had a substance abuse history, and 23 percent had a documented psychiatric history.

The researchers also noted that 48 percent of the mass murderers were preoccupied with war or weapons. This measure included such behaviors as the acquisition of a large number of weapons, war and weapons-related media, military uniforms, frequent trips to the shooting range, infatuation with street gangs, preoccupation with martial arts, idealization of fictional and nonfictional violent characters, and taking a nickname associated with a violent figure or violent theme.

Precipitating events or triggers, such as personal loss or status threat, were documented in 59 percent of the attacks. These triggers usually preceded the violent event by only a few hours or days.

Meloy and his colleagues (2001) developed a typology of adolescent mass murderers that includes the following categories: family annihilator, classroom avenger, criminal opportunist (who commits mass murder during the commission of a crime, such as eliminating witnesses to a robbery), bifurcated killers (who combine family annihilation and classroom revenge), and a miscellaneous group with diverse motives ranging from sensation-seeking to occult beliefs. Family annihilators and classroom avengers were much more likely to consciously ponder mass murder and premeditate an attack plan. Classroom avengers were more likely to be the victims of bullying and preoccupied with fantasies compared with family annihilators and criminal opportunists. However, criminal opportunists were more likely to have a preoccupation with weapons. Classroom avengers were more likely than family annihilators and criminal opportunists to experience clinical depression, while a history of antisocial behavior predominated among the latter groups.

Classroom Avengers

Two researchers took an approach that focused even more closely on our subject. McGee and DeBernardo (1999) examined 12 shooting incidents that occurred in American middle and high schools between 1993 and 1998. These incidents were selected for study by the authors because they considered them "nontraditional" school shootings. In their view, traditional school shooting incidents involved juvenile gangs, inner-city problems, minority or ethnic status, turf warfare, drugs or other criminal activity, like armed robbery or extortion. And nontraditional incidents involved multiple rather than single victims and were more similar to episodes of adult workplace violence, described as workplace vengeance, than to incidents of violence associated with gangs, drugs, and street crime.

These authors developed a behavioral profile of the 12 shooters in these incidents through a subjective analysis of available data, including unconfirmed anecdotal accounts from official police reports and popular media. The authors caution that their behavioral profile is not a definitive portrait and might well change as more complete information becomes available.

The picture of the "classroom avenger" that emerged from these sources is one of a physically healthy, working-class or middle-class white male who lived in a rural area or a small city. Family background and relationships were often quite dysfunctional; parents were often divorced or separated, and parental discipline has been meted out in ways that are both harsh and inconsistent. In terms of peer relations, the authors found that classroom avengers had problems with bonding and making social attachments. Although they did not show any overt sign of mental disorders, their mood was significantly depressed. Their depression was not readily apparent, however, because they did not complain or show physical signs, such as sudden weight loss or lack of energy. Instead, their depression was manifested through sullen, angry irritability and seclusiveness as well as "action equivalents" of depression, such as vandalism, temper outbursts, and comments such as "my life sucks." They blamed their personal failures on others and were easily frustrated by the slightest adversity. Self-esteem was unstable and vacillated between feelings of worthlessness and self-reproach on one hand and narcissistic self-aggrandizement and superiority on the other.

Although their physical appearance was unremarkable, they usually had a negative body image, viewed themselves as unattractive, and were frequently perceived by peers as "nerds" or "geeks." McGee and DeBernardo (1999) suggest that these youth were friendless, immature, and socially inadequate loners who prefer the company of younger children and inappropriately continued to play with soldiers and "G.I. Joe" games. Their associates were also outsiders who often shared a highly eccentric or nihilistic view of the world. Academic performance was normal to somewhat above average, but it declined in the weeks before their violent outburst. Extensive histories of delinquency and police involvement were rare, but covert vandalism and cunning dishonesty were common. The classroom avengers were not interested in typical teen preoccupations, such as dating, cars, and sports. Rather, guns, bomb making, and violent media fascinated them. The violent events always involved firearms, which were readily available in the home.

In the immediate weeks prior to the shooting incident, the researchers found that the classroom avengers had been exposed to psychosocial stressors that seemed to act as triggering events for the shootings. Specific triggering events include reprimand or discipline by parents or school authorities; some form of public ridicule; treatment perceived as unfair or demeaning; loss of a real or imagined relationship, particularly with a female love object; and hostile rejection or taunting, teasing, or bullying by peers.

The classroom avenger styled incidents after actual and fictional events. The authors suggest certain events, especially ones in close geographic proximity, can precipitate an attack via a copycat response. Shooting sprees were often preceded by journal or letter writing in which the details of the looming outburst were spelled out. Classroom avengers often verbalized their impending attack in the form of threats, boasts, assertions of intent, or warnings.

The school shooting sprees were premeditated and motivated by vengeance (McGee and DeBernardo, 1999). The shooters fantasized about revenge and retaliatory triumphs over their adversaries, and mental rehearsals of their acts of violence began well in advance of the actual attack. In fantasies, they selected victims and witnesses, time, location, means, and course of action. They planned predatory aggression that was selective, calculating, and premeditated. These acts were sophisticated and creative, unlike purely impulsive acts of suddenly erupting rage. The authors observe that all elements of the "Menninger triad" (1938) were present: (1) they wished to die (suicide), (2) they wished to kill (homicide), and (3) they wished to be killed (victim-precipitated homicide). The authors suggest that the psychiatric diagnoses of a classroom avenger are atypical depression and mixed personality disorder with paranoid, antisocial, and narcissistic features.

Targeted School Violence

Since the early 1990s, researchers at the United States Secret Service have been working on ways to improve the agency's assessment of threats in order to better protect dignitaries from targeted violence. To many in the Secret Service and at the U.S. Department of Education, the problem of protecting dignitaries from violent attacks seemed similar to the problem facing those interested in preventing school rampages.

With the support of the secretary of education, the Secret Service's National Threat Assessment Center studied 37 school shooting incidents involving 41 attackers (Vossekuil et al., 2000).

The incidents examined were limited to shootings in which the attacker chose the school for a particular purpose (not simply as a site of opportunity) and that were not related either to gang or drug activity, or an interpersonal or relationship dispute that happened to occur on school grounds. For each incident, teams of researchers and investigators reviewed primary source materials, such as investigative, school, court, and mental health records. In addition, center personnel conducted interviews with 10 of the attackers.

All of the incidents were committed by boys or young men. In more than two-thirds of them, one or more students, faculty, or others at the school were killed. Firearms were the primary weapons used. In over half of the incidents, the attacker selected at least one school administrator, faculty, or staff member as a victim.

Vossekuil and his colleagues (2000) reported that the incidents of targeted school violence were rarely impulsive. In almost all of them, the attacker developed the idea to harm the target before the attack; in well over three-fourths of the incidents, the attacker planned the violent event.

Insofar as these events were premeditated and deliberate, they can be seen as rational. What seem less rational, however, are the attacker's perceptions of the circumstances that prompted and focused the action and the normative rules that seemed either to require or allow the attacks to take place.

More than half of the attackers had as their motive a general kind of revenge against an undifferentiated target, and over two-thirds had multiple reasons for their attack. Yet most witnesses to the events leading up to these incidents did not see the same reasons for anger and vengeance that the attackers saw, nor did they think the concerns rose to a level that would justify such reckless attacks. If the idea of rationality includes some sense of objectivity in assessing threats to one's status and welfare and some commitment to protecting rather than attacking the welfare of one's fellows, then these events were less than rational.

Prior to most of the incidents, the shooter told someone about his idea or plan (Vossekuil et al., 2000). In more than three-quarters of the cases, the attacker told someone, almost always a friend or peer, about his interest in launching an attack at school. In less than one-quarter of the cases, the attacker directly communicated a threat to his target before the outburst. In almost every incident, the attacker engaged in some type of behavior—such as attempting to get a gun, writing disturbing essays or poetry, inappropriate humor—that caused others, such as school officials, police, and fellow students, to be concerned about him.

In contrast to the McGee and DeBernardo (1999) study, the National Threat Assessment Center report concludes that there is no accurate or useful profile of the school rampage shooters. They were much more impressed by the diverse characteristics of the shooters than their similarities. The school shooters were described as coming from a wide variety of racial and ethnic backgrounds (in nearly one-quarter of the cases, the attackers were not white); coming from a wide range of family situations, ranging from intact families with numerous ties to the community to foster homes with histories of neglect; having academic performances that ranged from excellent to failing; having a range of friendship patterns, ranging from popular to socially isolated; having varied behavioral histories, ranging from no observed problems to multiple problem behaviors; and most having little change in academic performance, friend status, disciplinary problems, or drug or alcohol abuse prior to the attack. Although access to firearms was common among the attackers, the report also differed from other research on the importance of weapons to the shooters. In most cases, the shooters did not express a fascination with weapons or explosives.

Although the attacker acted alone in at least two-thirds of the cases, they were influenced or encouraged by other students in almost half of the attacks. In more than three-quarters of the attacks, other students knew about the attack before it happened. Some knew exactly what was planned, while others only knew that something "big" or "bad" was going to happen.

In more than two-thirds of the cases, bullying played a key role in the attack. In these instances, the shooter felt persecuted, threatened, attacked, or injured by others before the outburst. Some of the attackers had experienced intense bullying and harassment for a very long period of time.

In more than three-quarters of the incidents, the attackers had difficulty coping with a major change to a relationship or a loss of status (e.g., a personal failure) prior to their school attack. More than half of the attackers had a history of feeling extreme depression and nearly three-quarters of the attackers had threatened or attempted to commit suicide prior to the incident.

School Shooter Threat Assessment

The National Center for the Analysis of Violent Crime of the Federal Bureau of Investigation (FBI) has also developed a model of threat assessment for school shooters by analyzing 18 school shooting cases (Federal Bureau of Investigation, 2000). The incidents involved both single and multiple offenders. Actual shootings occurred at 14 schools; in the remaining 4 cases, the student or students had planned and made significant preparations but were detected by law enforcement before the shooting took place. These 18 cases were supplemented by an unidentified number of cases in which the center was already preparing a threat assessment.

The FBI study described a process they call "leakage," in which a student intentionally or unintentionally reveals clues to feelings, thoughts, attitudes, or intentions that may signal an impending act (Federal Bureau of Investigation, 2000). These clues can take the form of subtle threats, inappropriate jokes, boasts, or innuendos and can be either spoken or conveyed in stories, journal entries, essays, poems, songs, or drawings. Leakage is considered to be a cry for help and the most important clue that precedes the violent act.

The FBI's findings focus on the personality traits of school shooters; their relationship with parents and the quality of parenting; social relationships at school, including relationships with deviant peers and the school culture; and the influence of violent media. The authors recommended that any sudden changes in outside interests or drug and alcohol use should be closely monitored. They also noted a copycat aspect to many of these events, as school shooters seem to be influenced by other shooting events that generate intense media scrutiny. The FBI suggests that school administrators, parents, and law enforcement officials should be more vigilant in monitoring disturbing student behavior in the months following a well-publicized incident elsewhere in the country.

GENERAL RESEARCH ON VIOLENCE

The research reviewed above tends to focus on the characteristics of the offenders. It is as though an implicit assumption is being made that the character, motivation, and circumstances of the offender are the principal causes of these events; furthermore, that being able to identify such offenders before they commit these crimes would be the most obvious and most direct means of dealing with the problem. Yet other broad classes of factors may turn out to be important either as significant causes of the events, as important targets for intervention, or both. In the general research on the causes of violence, the emphasis on the offender's individual character and motivations and the role of mental illness is often reduced in favor of other explanatory factors.

One approach emphasizes broad social factors, such as poverty, racism, and a culture of violence, that are expressed in the conduct of particular individuals. Another points to more idiosyncratic situational factors, such as unfortunate combinations of acute problems in an individual's life with the ready availability of weapons. A reading of the general research on violence suggests a broad range of variables that may be contributing to school rampages and provides some insight on the effectiveness of interventions focused less on the stable characteristics of individuals and more on either broad social factors or situational factors.

Violence in General

In 1993, the National Research Council (NRC) published *Understanding and Preventing Violence.* This work is a comprehensive attempt to catalogue what is known about possible causes of violence (National Research Council, 1993–1994, 4 volumes). One of the important results of that work was the development of an analytic framework for identifying the many different possible causes of violence, shown here as Table 10-1. For our purposes, there are several important things to understand about this framework.

TABLE 10-1 Matrix for Organizing Risk Factors for Violent Behavior

Units of Observation and Explanation	Proximity to Violent Events and Their Consequences		
	Predisposing	Situational	Activating
Social			
Macrosocial	Concentration of poverty Opportunity structures Decline of social capital Oppositional cultures Sex-role socialization	Physical structure Routine activities Access: Weapons, emergency medical services	Catalytic social event
Microsocial	Community organizations Illegal markets Gangs Family disorganization Preexisting structures	Proximity of responsible monitors Participants' social relationships Bystanders' activities Temporary communication impairments Weapons: carrying, displaying	Participants' communication exchange
Individual			
Psychosocial	Temperament Learned social responses Perceptions of rewards/penalties for violence Violent deviant sexual preferences Cognitive ability Social, communication skills Self-identification in social hierarchy	Accumulated emotion Alcohol/drug consumption Sexual arousal Premeditation	Impulse Opportunity recognition
Biological	Neurobiologic[a] "traits" Genetically mediated traits Chronic use of psychoactive substances or exposure to neurotoxins	Transient neurobiologic[a] "states" Acute effects of psychoactive substances	Sensory signal-processing errors Interictal events

[a]Includes neuroanatomical, neurophysiological, neurochemical, and neuroendocrine. "Traits" describes capacity as determined by status at birth, trauma, and aging processes such as puberty. "States" describes temporary conditions associated with emotions, external stressors, etc.
SOURCE: NAS, 1993.

First, the NRC panel on violence recognized that one could search for the causes of violence along two quite different dimensions. One dimension involves what could be considered the "structural level of analysis"—that is, one could try to find the explanation for violence either in the aggregate structures of society or in the characteristics of individuals. Furthermore, in looking at the aggregate structures of society, one could look at characteristics of very large aggregates of people and places in society, which presumably do not change very much or very fast, or at characteristics of smaller aggregates, which presumably have wider variation across society and change more quickly. In looking at the characteristics of individuals, one could look at characteristics at either the psychological level or the biological level.[1]

The other dimension involves the way that the potentials for violence that are contained within social structures and individual characteristics are transformed from a latent potential to the actual production of a violent event. In this dimension, the dynamics of time, of situations, of chance combinations of factors that lead toward violent acts are introduced into the understanding of violence. The report explains the importance of addressing these more dynamic, situational processes (National Research Council, 1993:298–299, emphases added):

> A violent event requires the conjunction of a *person* with some (high or low) predisposing potential for violent behavior, a *situation* with elements that create some risk of violent events, and usually a *triggering event*. Development of an individual's potential for violence may have begun before birth: perhaps with conception involving an alcoholic father, or through abnormal prenatal neural development. It may have begun during early childhood in a violent household, or though school failure, or through frequent exposure to violence in the neighborhood or from the media.
>
> A hazardous *situation* for violence could involve a dispute, perhaps aggravated by a miscommunication in a bar because of loud background noise, which was misinterpreted as an insult because of intoxication and escalated because participants were afraid of losing face in bystanders' eyes. The *surrounding community* could be gang turf, the site of illegal drug or gun markets, or a neighborhood where large numbers of unsupervised teenagers reside. It may be the scene of recent aggravating events such as police brutality, or of frequent brawls between members of different ethnic groups. The neighborhood may be experiencing social disruption as stable families move to the suburbs, as businesses close, and as public services decline.

The significance of bringing the dynamic, situational factors into view in the explanation of violence has at least two important implications for our work. First, it increases the importance of thickly descriptive narrative case studies as an important way of understanding the causes of violence. Only case studies can pick up the detail and the narrative flow of events that convert a mere potential for violence into the real thing. The structural factors can identify parts of society in which violence may be more or less likely to occur, and in doing so, highlight high-risk situations and individuals. But they cannot necessarily show everything that went into the creation of a particular event, or the variety of things that could have been done to prevent that particular event (as well as some others more or less like it) from occurring.

Second, it raises the important question of whether there is any reason to prefer interventions that focus on the more or less stable characteristics of social aggregates or the more or less stable characteristics of individuals, over interventions that focus on interrupting small, somewhat idiosyncratic microprocesses and the things that sustain them as they carry a potential for violence into the real thing.

Of course, the benefits of preventing something from occurring rather than reacting to it after the fact are obvious. And there is benefit to eliminating the potential of something bad from occurring as opposed to remaining constantly vigilant, and then scurrying around to try to stop it once it appears. But an important question is whether one kind of intervention should be preferred over another from the outset.

The NRC panel on violence provides some guidance on this matter: "A major problem in *understanding* violence is to describe the probability distributions of predisposing factors, situational elements, and triggering events at the biological, psychosocial, microsocial, and macrosocial levels. The problem in *controlling* violence is to choose among possible interventions" (p. 299). The report goes on to say: "We do not assume that any single level is more fundamental than the others in explaining a particular type of violence. Rather, . . . violent events and community violence levels arise out of interactions across the levels, and these interactive processes differ from one type of violence to another" (p. 296). Finally, the panel observes (p. 300):

> In most violent events, contributing situational elements are most visible in the microsocial encounter that precedes the event. These elements include the dynamics of communications among participants, such as disputes, threats and counterthreats, exchanges of insults, robbery and resistance, and the urgings of bystanders. Both the nature and interpretation of these exchanges may be conditioned by preexisting social relationships among participants: an intimate relationship, a power or status hierarchy . . . or a culturally defined relationship. . . . Because situational elements from all levels contribute to the outcome, the possibility exists that even without full causal understanding, altering one link in a chain of events might have prevented a violent event or prevented an assault from becoming a homicide.

Youth Violence

Much research has been done on the development of serious and violent juvenile offending careers and the risk and protective factors that influence them. In an analysis of this research, the Study Group on Serious and Violent Juvenile Offenders at the U.S. Department of Justice found that the majority of such offenders are male, tend to have problems with substance abuse, mental illness, and school performance, typically display early minor behavior problems that lead to more serious acts, and have disproportionately been victims of violence themselves (Loeber and Farrington, 1998). Serious and violent juvenile offenders differ substantially from juveniles involved in more typical, minor acts of delinquency: they tend to have earlier onset of delinquency and longer offending careers, tend to be chronic offenders, and are overrepresented in inner cities. The study group concluded that violent behavior is a result of the interactions of individual, family, school, peer, and neighborhood factors; joining a gang and becoming a drug dealer were more proximal risk factors (Loeber and Farrington, 1998).

Findings from another study indicate that gun ownership by adolescents is related to a wide range of delinquent behaviors, including gun carrying, gun crime, gang membership, and drug selling (Lizotte and Sheppard, 2001). In this Rochester Youth Development Study, 5 to 10 percent of the boys studied carried illegal guns, and about 6 percent owned a gun for protection by age 15. Joining a gang increased the likelihood of owning a gun for protection, carrying a gun, and having peers who owned guns for protection.

The most important lines of research on the causes of youth violence have focused on what the NRC panel on violence would have called "macrosocial structures." They point to such problems as decaying communities with limited economic and social opportunities, the consequences of which include ineffective socializing efforts of family, school, religion, and neighborhoods, the absence of parental supervision, and the diminished role of the family (Bennett et al., 1996). Neighborhood structural disadvantage and social disorganization have been found to concentrate deviant behavior, such as child abuse, low birthweight, cognitive problems, and later delinquency and violence, in some communities (Sampson, 1997).

However, what was anomalous in the major epidemic of lethal youth violence that peaked in the 1990s was not its social location, but its scale and its dynamics. The epidemic produced a level of youth homicides that was well beyond anything that would have been predicted from the demographic or social characteristics of the period. Moreover, the time profile of the epidemic had features often associated with the spread of infectious diseases: it did not go up steadily, but zoomed up in a dramatic way and then suddenly dropped. Accounting for this pattern with social demographics is difficult for the simple reason that they did not change so fast or in the same way that the levels of youth violence did. Social conditions may have been setting the stage for the epidemic and channeling its location, but the particular ferocity of the epidemic could not be explained simply by structural characteristics.

Some criminologists argue that the increase in youth homicide in the 1990s was due to increasing propensities for offending in each age cohort and a demographic increase in the number of adolescents (DiIulio, 1995; Fox, 1996; Wilson, 1995). However, Cook and Laub (1998) showed that there was not an increasing propensity for violence among youth because the increase in rates of offending occurred in all cohorts, not just one, and in 1985 the rates of offending and victimization within these same cohorts were historically unremarkable. So it could not be that the violence was caused by the sudden appearance of a new, particularly violent group of young "predators" who differed from their older brothers. Their older brothers changed, too.

To explain what occurred, analysts turned to explanations that give a prominent role to a set of factors that would be considered microsocial processes, to situational factors, and to the kind of fast-moving cultural trends that are associated with fads. Blumstein (1995) hypothesized that the increase in youth homicide was a result of the nature of the crack cocaine markets. The low price of crack increased the number of transactions, creating a need for drug sellers to recruit a large number of new sellers. The resultant recruitment of adolescents into the drug market led young people to arm themselves for protection, which in turn caused violent encounters to become more deadly.

In a later study, Blumstein and Rosenfeld (1998) argued that the subsequent decrease in youth homicide could be due to increased stability in the crack market. The number of new crack users diminished, and those involved in the drug trade had the opportunity to develop dispute-resolution mechanisms other than violence. As a result, the need to keep recruiting youth as sellers abated, fewer territorial disputes erupted, and the need for youths to carry guns for protection decreased. In addition to changes in the drug markets, Blumstein and Rosenfeld (1998) identified other factors that could have contributed to the decline in youth homicide: economic expansion that created more opportunities for legitimate jobs, and police and community efforts to limit opportunities for the drug trade, remove guns from kids, and reduce conflicts among youth.

If the cocaine epidemic explains the inner-city epidemic of lethal youth violence, then one would have to conclude that the rural and suburban epidemics were of a different kind, for there is no evidence that the school rampages were associated with crack markets. The inner-city epidemic still could have had an influence on the suburban and rural shootings, through the images of violence and spread of a culture that made it imaginable that children would carry guns to school, from the fear and despair that these shootings might exacerbate among other adolescents who were having a difficult time adjusting to adult lives. Such a "second mechanism" might relate inner-city to suburban and rural violence. These influences of the inner-city epidemic on the suburban and rural epidemic of rampages are a different claim than the hypothesis that the two different epidemics were caused by the same external factors.

One could also explain the youth homicide epidemic of the 1990s by giving more emphasis to cultural and subcultural factors. Fagan and Wilkinson (1998) point out that there is no direct evidence of a causal link between adolescents' involvement in the drug trade and homicides committed by adolescents. They argue that the epidemic occurred as guns became an important part of social interactions among urban youth. The possession and use of a gun had become a symbol of power and control, a way to gain status and identity, and a means to enhance feelings of safety and personal efficacy among teenagers. The increased youth demand for guns, the available supply, and the culture that teaches kids lethal ways to use guns had a large and complex impact on the overall level and seriousness of youth violence.

Another explanation lies in the particularly fast-moving aspects of culture that are viewed as fads. Rock music and video games may have helped to spread a culture of violence among kids. The spread of gangs across the country may have provided a medium for the spread of commitments to violence and knowledge about how to use weapons. Gang culture may have spread even more quickly and widely than gang infrastructure, as suburban and rural kids who had never had contact with a real gang member took up the stance and behavior of the gang that they learned from movies. Klein (2002) suggests that, although many gang members migrate to other cities and bring their gang experiences and culture with them, the majority of street gangs are homegrown. The diffusion of gang culture—and youth culture—lends the appearance (and often reality) of similarity among such groups.

In the inner-city case studies in Part I, there is much evidence of the influence of gang culture, and in the suburban and rural case studies little evidence of it. There is some evidence of the influence of violent games and media in both the urban and the suburban and rural cases, but given the general popularity of these things among all adolescents, it is hard to know whether the communities and youth we examined are unusual in their enthusiasm for this sort of media. The shooters in the suburban cases may have been influenced by violent media, but not many of the other kids were, and the subcultures one sees in the communities in which the rampages occurred were very different from the adolescent subcultures of the inner cities at the height of the crack epidemic.

Research on youth violence also points to the importance of "social information processing deficits" among adolescents (as summarized by Gottfredson, 2001). Impulsiveness and self-control are linked with problem behavior through cognitive processes. Antisocial adolescents tend to misinterpret social cues. They attribute hostile intentions to peers when none may exist. They have difficulty evaluating the likely consequences of their actions and considering alternatives. They also have trouble regulating behaviors in communication, including using appropriate eye contact and tone of voice. Several studies have linked these cognitive and behavioral deficits with peer rejection (e.g., Dodge et al., 1986; McFall, 1982; Perry et al., 1986).

Finally, contagion mechanisms may have played a role in producing the unexpected dynamics of the inner-city violence, and these mechanisms may have played a role as well in spreading the inner-city violence to the suburban and rural areas, or in producing the spate of school rampages independently of the inner-city violence. The important focus here could be imitative behaviors spread through an interested and open youth subculture by the media. It may be that the school rampage shooters took their inspiration from the youth violence of the inner city. Or it could be that one school rampage shooter took his inspiration from an earlier school rampage shooting with little connection to inner-city violence.

School Violence and Bullying

Many of the youth school shooters were reported to experience bullying. Bullying is not particular to schools; it also goes on outside schools and, like other forms of victimization, is imported. Although there is no universal definition of bullying, there is widespread agreement that it includes several key elements: physical, verbal, or psychological attack or intimidation intended to cause fear, distress, or harm to the victim; an imbalance of power, with the more powerful child oppressing the less powerful one; absence of provocation by the victim; and repeated incidents between the same children over a prolonged period (Farrington, 1993).

Bullying is surprisingly common. In his review of the research, Farrington (1993) suggests that over half of children have been victimized and over half have been bullies. Data from the National Crime Victimization Survey suggest that the prevalence of bullying is lower than what Farrington reports, but it still represents a large problem. In 1999, the percentage of students ages 12–18 who reported being bullied at school during the previous 6 months was 10 percent for students in grades 6–7, 12 percent for students in grades 8–9, and 7 percent for students in grades 10–11. In a recent national survey of youth in grades 6 through 10, 30 percent reported some type of involvement in moderate or frequent bullying, as a bully (13 percent), a target of bullying (11 percent), or both (6 percent) (Nansel et al., 2001).

The prevalence of bullying is of great concern, as it causes immediate harm and distress to the victim and has negative long-term consequences for the victim's mental health. It also has negative consequences for the bully, since the bully may become more likely to engage in other aggressive behavior. Understanding bullying is important because it is related to crime, criminal violence, and other types of aggressive antisocial behavior (Farrington, 1993).

Farrington (1993) observes that bullies tend to be aggressive in different settings and over many years. Adolescent bullies tend to become adult bullies and then tend to have children who are bullies (Farrington, 1993). Like offenders, bullies tend to come disproportionately from families with lower socioeconomic status and poor childrearing techniques, tend to be impulsive, are more likely to be involved in other problem behaviors such as drinking and smoking, and tend to be unsuccessful in school (Farrington, 1993; Nansel et al., 2001). Olweus (1992) reported that individuals with a history of bullying had a fourfold increase in criminal behavior by the time they reached their mid-20s. Victims of bullying tend to be unpopular and rejected by peers and tend to have low school attainment, low self-esteem, and poor social skills (Farrington, 1993). There is evidence that social isolation and victimization tend to persist from childhood to adulthood, and that victimized people tend to have children who are victimized (Farrington, 1993; Nansel et al., 2001). Males who are bullied tend to be physically weaker than males in general (Olweus, 1978). Boys are bullies more than girls, but girls and boys are equally victimized (Farrington, 1993). Boys are overwhelmingly bullied by boys, and girls are bullied equally by boys and girls.

In general, bullying incidents occur when adult supervision or surveillance is low (e.g., playgrounds during recess). The most common location is the playground. Not all bullying incidents come to the attention of teachers, and teachers and other children do not always intervene to prevent bullying (Mellor, 1990; Whitney and Smith, 1991; Ziegler and Rosenstein-Manner, 1990).

Successful bullying prevention programs generally aim to alter the school environment to make norms against bullying more salient (Farrington, 1993). These programs provide information to the school community about the definition, level, and consequences of bullying. Prevention efforts seek to establish clear rules against and consequences for bullying and to consistently enforce them. Attempts to create more communal social organizations are likely to be effective for reducing this form of victimization. Schools that tolerate bullying increase the level of bullying as well as the risk that an unusual act of retaliation will occur.

RESEARCH ON VIOLENCE CLOSELY RELATED TO SCHOOL RAMPAGES

In looking in detail at some of the characteristics of the school rampages, one can think about them as more or less similar to specific categories of violence. In this section, we examine research on the phenomena of mass murder, public rampages and work violence, and murders followed by suicides or crime sprees that look as though they were designed at least in part to produce a "suicide by cop."

Mass Murders

Mass murder is a very rare event. Defining a mass murder as an incident with four or more victims, one analysis of FBI Supplementary Homicide Report Data from 1976 through 1995 found 483 mass murders involving nearly 700 offenders and over 2,300 victims (Fox and Levin, 1998). This amounts to less than 1 percent of the more than 400,000 homicides committed during that period.

That analysis also compared mass murders with single-victim murders and noted a number of noteworthy differences. Mass murders are more likely to occur in small town or rural settings (43 percent) compared with single-victim murders (34 percent). Mass murders are not concentrated in the South, unlike single-victim murders. Mass murders are more likely to involve firearms (78 percent) than single-victim crimes (66 percent). And 40 percent of mass murders are committed against family members and almost as many involve other victims acquainted with the perpetrator, such as coworkers; this is more pronounced for mass murders than for single-victim crimes.

Mass murderers are usually older than single-victim murderers. While more than half of all single-victim homicides occur during an argument between the victim and the offender, it is relatively rare for heated disputes to escalate to mass murders (23 percent). Many mass murders are committed to cover up other felonies, such as armed robbery (39 percent). However, in the FBI data, an equal number of mass murders have unspecified circumstances (39 percent) because these crimes involve a wide array of motivations, including revenge.

Fox and Levin (1998) argue that a majority of mass killers have clear-cut motives—especially revenge—and their victims are chosen because of what they have done or what they represent. Most commonly, the mass killer seeks to get even with people he knows—with his estranged wife and all her children or the boss and all his employees. The more specific and focused the element of their revenge, the more likely it is that the outburst is planned and methodical. Also, the more specific the targets of revenge, the less likely it is that the killer's rage stems from extreme mental illness. These observations on the nature of mass murderers fit well with the revenge motives and premeditated attacks described by the available research on youth school rampage shooters.

Levin and Fox (1996) developed a typology of mass murderers that divides revenge into three categories: (1) individual-specific, in which the offender targets particular people, most often an estranged spouse and children; (2) category-specific, in which the murders are motivated by hatred for particular groups or categories of people; and (3) nonspecific, in which killing is precipitated by the offender's paranoid perceptions of society. In the authors' view, the final two categories involve acts that are primarily instrumental—mass murders inspired by profit, such as contract hit men, armed robbers shooting witnesses, and murders that result from acts of terrorism.

These researchers also suggest a range of factors associated with the commission of mass murder that cluster into three types, reflecting themes of the 1993 NRC report on violence.

The factors are described as: (1) predisposers—long-term and stable preconditions that become incorporated into the personality of the killer; (2) facilitators—conditions, usually situational, that increase the likelihood of a violent outburst but are not necessary to produce that response; and (3) precipitants—short-term and acute triggers or catalysts. Predisposers include frustration and externalization of blame.

A critical condition for frustration to result in violent aggression is that the individual perceives others are to blame for his personal problems. The mass killer sees himself not as the perpetrator but as the victim. It may take years for this frustration to build; hence, mass killers are usually older. Given both long-term frustration and an angry, blameful mind set, certain situations or events can precipitate or trigger violent rage.

In most instances, the killer experiences a sudden loss or the threat of a loss, which from his view is catastrophic. The loss typically involves an unwanted separation from loved ones or termination from employment (Fox and Levin, 1998). Books, manuals, and magazines may provide technical guidance in committing mass murders. Anecdotal evidence on copycat mass killings is highly suggestive. A rash of schoolyard slayings, beginning with Laurie Dann's May 1988 shooting at Winnetka, Illinois, and ending with Patrick Purdy's January 1989 attack in Stockton, California, suggests the possibility of a fad element in which mass killers inspire each other (Fox and Levin, 1998).

With respect to likelihood, mass killers are frequently isolated from sources of emotional support. Many are cut off from the very people who could have supported them when times got tough. Some live alone for extended periods of time. Others move far away from home, experiencing a sense of anomie or normlessness. Of course, it is important to recognize that most people who feel angry, hopeless, and isolated do not commit mass murder; in many cases, they simply do not have the means. The availability of firearms is important as a facilitator to mass murder (Fox and Levin, 1998).

Public Rampages and Workplace Violence

Since schools are work settings for the faculty, staff, and administrators, youth school rampage shootings could be viewed as an extreme form of workplace violence. The U.S. Occupational Safety and Health Administration uses a three-part typology to describe the range of workplace violence: Type I, violence by people unrelated to the workplace (e.g., robbery); Type II, acts committed by people who are related in some way to an employee (e.g., domestic assault); Type III, violence between employees (includes the revenge killer).

In 1998, there were 709 workplace homicides in the United States, 4 percent of the 16,910 homicides committed that year (Bureau of Labor Statistics, 1999). In 1998, homicide was the second leading cause of death at work, accounting for 12 percent of 6,026 occupational deaths (highway accidents was number one at 24 percent). Like homicides generally, workplace homicides have been declining in recent years. Since the Department of Labor began collecting data in 1992, the number held steady at about 1,050 per year (0.9 per 100,000 workers) through 1994, and it has since fallen each year since 1995, reaching a seven-year low of 709 (0.5 per 100,000 workers) in 1998. The Bureau of Labor Statistics (1997) data also suggest that 80 percent of the homicides committed at work were committed during a robbery or commission of some other crime. Disgruntled coworkers, clients, or personal acquaintances (husband, ex-husband, boyfriend, ex-boyfriend, relative) committed the remaining 20 percent.

The U.S. Postal Service (USPS) experienced a widely publicized series of homicidal rampages by disgruntled current or former employees. In 1986, letter carrier Patrick Henry Sherrill killed 14 coworkers and himself at the Edmond, Oklahoma, Post Office. This was the first instance of a worker of "going postal," and some believe that it spawned a series of copycat crimes that lasted over the next several years. For example, five years later in 1991, letter carrier Thomas McIlvane killed four coworkers and himself at the Royal Oak, Michigan, Post Office. From 1986 to 1999, 29 workplace homicide incidents involved postal employees either as victims or offenders (U.S. Postal Service Commission, 2000). There were 54 homicide victims in these incidents, including 48 postal employees. Nonemployees were responsible for 14 of the postal worker homicide incidents. The motives were varied in these incidents; six were robberies, and others involved a dispute over debt, anger over mail not delivered when expected, and intimate relationships.

The U.S. Postal Service Commission on a Safe and Secure Workplace (2000) closely examined the 15 incidents of workplace homicides committed by current or former employees since 1986. Guns were used in all 15 incidents, and all incidents had a single offender. The motives included robbery, actual and desired intimate relationships, and workplace disputes. The commission observed that 14 of 15 perpetrators had troubled histories of violence, mental health problems, substance abuse, or criminal convictions. Five exhibited behavior prior to employment that should have excluded them from being hired.

Six homicide incidents committed by postal workers involved multiple victims; five of them involved retribution for personal matters, such as a spurned intimate relationship, or work-related troubles, such as termination from job or perceived unfair treatment. Of these five killers, three committed suicide at the end of the event. One mass murder was committed to cover up a robbery by a postal employee who was a cocaine abuser; he did not kill himself. The commission reported that in some cases of homicides committed by postal workers, there were warning signs. Three perpetrators had been fired for threatening behavior. In several cases, managers, coworkers, union officials, physicians, or counselors mistakenly assessed the perpetrators as unlikely to commit violence despite warning signs; in at least two cases, managers did not even report the threats to the Postal Service's Inspectional Services.

In its broader assessment of the risks of violence in Post Office facilities, the U.S. Postal Service Commission (2000) concluded that "going postal" is a myth. Postal workers were no more likely to physically assault, sexually harass, or verbally abuse their coworkers than employees in the national workforce. Postal employees were only a third as likely as those in the national workforce to be victims of homicide at work.

The levels of violence throughout the American workplace are unacceptably high: in 1999, 1 in 20 workers was physically assaulted, 1 in 6 was sexually harassed, and 1 in 3 was verbally abused (U.S. Postal Service Commission, 2000). One researcher suggests that bullying is a large problem in all workplaces and leads to the violent victimization of the person being bullied (Barron, 2000). Postal workers are no more likely to be the victims of these crimes than other U.S. workers. However, given the publicity generated by the Post Office slayings, postal workers are more fearful than employees in the national workforce about violence in their workplace. Postal workers are six times more likely to believe that they are at greater risk to be the victim of violence from coworkers.

Homicide Followed by Suicide of the Perpetrator

Suicide following homicide is fairly rare. In the United States, city-level studies suggest that only a small fraction of homicides lead to the perpetrator's suicide. In Philadelphia, one study found that 4 percent of homicides lead to the perpetrator's suicide (Wolfgang, 1957). In Chicago, another study found that less than 2 percent of homicides lead to the perpetrator's suicide (Stack, 1997); the author speculates the perpetrators suffer from frustrated personal relationships, ambivalence, jealousy, morbid jealously, separation, helplessness, depression, and guilt. In Canada, where national data are available, homicide offenders commit suicide in about 10 percent of the cases; as the tie between the offender and victim is closer, the probability that the offender will commit suicide increases (Gillespie et al., 1998). The probability of suicide increases with offender's age and education, is higher when the offender uses a gun, and is higher when the victim is female (Gillespie et al., 1998). Homicide-suicide offenders, particularly incidents involving multiple victims, share some common elements with the youth school rampage shooters.

A review of the available research on homicide-suicides found that more than 90 percent of the perpetrators were male (Felthous and Hempel, 1995). When the victims are immediate members of the family or children, the proportion of female killers increases. The predominance of male perpetrators is not so lopsided when homicide-suicides occur in the context of intimate homicides, with some studies reporting as many as 50 percent of the killers as female in this context. The review also found that firearms are used more frequently in homicide-suicides than in spousal homicides alone. In studies in which perpetrators of homicide-suicides are compared with perpetrators of homicide alone, depression was far more common among the homicide-suicide perpetrators. The authors speculate that a core element of the motivation may be loss (Felthous and Hempel, 1995). The individual cannot perceive enduring life without a key element (such as their spouse, family, or job) and cannot bear the thought of the other person(s) carrying on without him, so he forces them to join him in death. When the loss is one of self-esteem, believed to be caused by nameless others or society, a violent, destructive, annihilistic blaze may seem the best final solution.

In the typology of homicide-suicides developed by these authors, two overlapping categories are very similar to youth school rampage shooters

(Felthous and Hempel, 1995). In the adversarial homicide-suicide, the event typically involves a disgruntled employee who feels depressed and bitter over his compromised employment or interpersonal conflicts that led to his dismissal. Although there is often a small kernel of truth to his complaints, the offender usually develops a persecutory delusion that specific individuals conspired to harm him.

In the pseudo-commando homicide-suicide, the offender selects a public place where many people can be slaughtered at once, and then he kills indiscriminately people with whom he has no relationship, formal or informal. The pseudo-commando can attack random people or may target some group of people who share a common characteristic. He brings to the scene powerful weapons, perhaps a small arsenal, and plenty of ammunition. The apparent lack of an escape plan is consistent with the presumption that the offender expects to die himself. The pseudo-commando typically forces the police to kill him. The perpetrator ensures his death by passive or active suicide, but he kills as many people as possible in a last-stand "blaze of glory." The pseudo-commando may be embittered, angry, and resentful; if mentally disturbed, excessive suspicion and paranoid thinking are features of the disturbance. Examples include Charles Whitman's mass slaying of 14 from a University of Texas tower; James Huberty's massacre of 21 people in a McDonald's restaurant in San Ysidro, California; George Hennard's slaughter of 22 people at Luby's Cafeteria in Killeen, Texas; and Marc Lepine's massacre of 14 women at the University of Montreal.

As suggested by the pseudo-commando offender, some suicidal individuals force police officers to kill them. This form of suicide is known as "suicide by cop," a term used by law enforcement officers to describe an incident in which a suicidal individual intentionally engages in life-threatening and criminal behavior with a lethal weapon or what appears to be a lethal weapon toward law enforcement officers or civilians specifically to provoke officers to shoot the suicidal individual in self-defense or to protect citizens (Hutson et al., 1998).

Research suggests that this behavior is uncommon but hardly rare. A review of files of officer involved shootings investigated by the Los Angeles County sheriff's department from 1987 to 1997 revealed that suicide by cop accounted for 11 percent of all officer-involved shootings and 13 percent of all officer-involved justifiable homicides (Hutson et al., 1998). The characteristics of the suicidal individuals included: 70 percent had a prior arrest or conviction, 65 percent abused alcohol or drugs, 39 percent were involved in domestic violence or a domestic dispute, and 63 percent had a psychiatric history (Hutson et al., 1998). Based on a review of five studies, authors of one study suggest that about 10 percent of police deadly force incidents involve suicide-by-cop situations (Homant and

Kennedy, 2000). These incidents rarely involve additional victims. Reviewing 123 cases of suicide by cop, Homant and his colleagues (2000) found that in 7 percent of the incidents, one or more people were killed by the perpetrator, and in another 7 percent of the incidents, one or more people were seriously wounded by him.

In the rare events that involve victims, the suicide-by-cop perpetrators share some characteristics with the youth school rampage shooters. According to one researcher, some suicidal individuals may prefer suicide by cop because they may see themselves as victims and set up the situation to prove it (Foote, 1995). He also points out a similarity to murder-suicide cases and warns that such individuals "may not hesitate to kill another to accomplish his or her death." Another researcher suggests that some may be seeking a final catharsis of inner rage by acting out a fantasy of dying in a shootout with police, taking as many people as possible (Gilligan, 1996). In their close review of 143 suicide-by-cop incidents, Homant and Kennedy (2000) identify 31 percent of the cases as involving direct confrontation, in which the subject plans ahead of time to attack police in order to be killed by them. One subtype of this category, the "kamikaze attack" (3.5 percent), involves the use of deadly force to attack a police station or a group of officers in spectacular fashion.

CONTAGION MECHANISMS

Criminologists have long noted that certain types of antisocial acts occur in waves. These are variously termed crime fads (Sutherland and Cressey, 1970), contagious violence (Sears et al., 1985), and deviant epidemics (Turner and Killian, 1987).

However, very little is known about the precise mechanisms that produce these clusters of violent incidents. This question has been most seriously and directly addressed in a literature that examines what are commonly called copycat crimes.

Copycat Crimes

Copycat crimes are very difficult to research. Most studies of the phenomenon rely on anecdotal evidence, which strongly suggests that the copycat phenomenon exists. For example, one study identifies a number of case studies, ranging from four youths who shot and wounded two Las Vegas police officers and who were alleged to be motivated by Ice-T's "Cop Killer" song to a 16-year-old California boy who killed his mother and admitted that he got the idea from the movie *Scream* (Surette, 2002). This study of serious and violent juvenile offenders found that one-quarter reported that they have attempted a copycat crime (Surette, 2002). The juveniles who believed that the media and close friends particularly influenced their behavior also reported copycat behaviors.

Media coverage of terrorist events is believed to motivate copycat terrorist acts (Poland, 1988). An extensive review of the relationship between terrorism and the media concludes that while other factors are probably at least equally important, media coverage is sufficient to lead to acts of imitative behavior (Schmid and DeGraaf, 1982). According to one researcher, media coverage has two effects (Surette, 1990). First, anecdotal evidence indicates that coverage encourages false threats and pseudo-copycat reactions. For example, a May 1981 bombing at New York's Kennedy Airport was followed by over 600 threats the following week. Second, real copycat events follow in significant numbers in a process called "contagion" in the terrorist literature.

Much anecdotal evidence is available that such events as hostage bank robberies, hijackings, and airline bomb plantings occur in clusters (Schmid and DeGraaf, 1982; Livingstone, 1982). Claims of contagion have also been made about larger-scale incidents of violence, including racial disturbances (Spilerman, 1970), disorders in schools (Ritterband and Silberstein, 1973), political violence (Hamblin et al., 1973), and military coups (Li and Thompson, 1975; Midlarksy, 1970). One study shows that successful hijackings in the United States generated additional hijacking attempts (Holden, 1986). There were no contagion effects of unsuccessful hijacking attempts in the United States or any effects on U.S. hijacking attempts outside the country.

In general, the effect of the media on crime seems to be more qualitative (affecting criminal behavior) than quantitative (affecting the number of criminals) (Huesmann, 1982; Comstock, 1980; Donnerstein and Linz, 1995). Offenders seldom cite the media as a motivating influence. While there are positive correlations between watching violent media and aggressive behavior, people do not become aggressive or violent solely from watching television or violent movies (Philips, 1982a, 1982b; Garofalo, 1982; National Institute of Mental Health, 1982). One study reported that only 12 percent of inmates in their study cite the media as a cause in their criminality, ranking it second to last behind all other possible factors except for "too much junk food" (Pease and Love, 1984). However, the media were endorsed by 21 percent of inmates as a source of information about crime techniques; the media ranked fourth in developing techniques, behind "myself," friends, and fellow inmates. These researchers also conclude that except for isolated cases of mentally ill individuals, copycat offenders possess a criminal intent to commit a particular crime before they copy a particular technique.

Another study suggests that the media can encourage and instruct criminal behavior through priming processes (media-portrayed behaviors activating a network of associated ideas), desensitization to violence, and attitude changes, as well as the direct imitation of behavior (Berkowitz, 1984). For individuals who rely heavily on the media for worldly information and escape and have a tenuous grasp of reality, the influence can be significant. The available evidence supports the contention that predisposed at-risk individuals who are primed by media crime characterizations are the primary agents of copycat crime (Comstock, 1980).

One conceptual model shows copycat crime as resulting from the interaction of factors in four areas: the initial crime, media coverage, social contextual factors, and copycat criminal characteristics (Surette, 1990). The model denotes a process in which particular, usually highly newsworthy and successful initial crimes and criminals (after interacting with media coverage) emerge as candidates to be copied. The pool of potential copycat criminals is affected by media coverage and other social context factors, such as norms regarding deviance and violence, the existence of social conflicts, the number of opportunities available to employ a copycat crime technique, the nature and pervasiveness of media coverage, and the size of the preexisting criminalized population. The author concludes that copycat crime appears to be a persistent social phenomenon prevalent enough to influence the total crime picture mostly by influencing crime techniques rather than criminal motivations (Surette, 1990).

Suicide Clusters

The risk of an individual's committing suicide may increase as the number of suicides in his or her peer group or community increases or as the number of suicide reports or publicity increases in the media (Gould, 1990). Both anecdotal accounts and epidemiological research indicate that significant clustering does occur, but it does not account for a large proportion of total youth suicides.

An analysis of National Center for Health Statistics (NCHS) mortality data indicates that clusters of completed suicides occur primarily among adolescents and young adults, but even in this age group such clusters account for no more than 5 percent of all suicides (Gould et al., 1990a, 1990b). Further analyses of the mortality data reveal that significant time-space clustering occurred among teenagers (ages 15–19) and that these outbreaks of suicide occurred more frequently than expected by chance alone. Moreover, the significant clustering of suicide occurred primarily among teenagers and young adults, with minimal effects beyond age 24. Clustering was two to four times more common among adolescents and young adults than among other age groups.

Public health researchers have argued that suicide clusters are caused in part by social contagion (Robbins and Conroy, 1983; Davidson and

Gould, 1989). However, the hypothesized mechanism of suicide contagion has not been well defined. Grief, especially prolonged or unresolved grief over the death of a loved one, is known to be a risk factor for suicide (Bunch et al., 1971; Hagin, 1986). There are reports that the risk of suicide may be higher among grieving relatives of suicide victims (Cain and Fast, 1972). Cluster suicides appear to be multidetermined, as are other suicides, but imitation and identification are factors hypothesized to increase the likelihood of cluster suicides. Among susceptible individuals, the route of exposure may be direct (i.e., close friendship with a suicide victim or observing a suicidal act) or indirect (i.e., watching television news coverage of a prominent person's suicide or hearing about a suicide by word of mouth).

In the context of geographically localized suicide clusters, however, there seems to be a third ingredient in contagion, in addition to grief and imitation (O'Carroll, 1990). In some suicide clusters, the tendency to glorify suicide victims and to sensationalize their deaths has frequently fostered a community-wide preoccupation, even a fascination, with suicide. The resulting highly charged emotional atmosphere is believed by many to have contributed to causing suicide. Individual susceptibility (e.g., preexisting mental health problems, family history of suicidal behavior) is probably a major influence on the individual's motivation (Gould, 1990). The occurrence of a suicide or report in the media can also increase the knowledge of how to perform the behavior.

A large body of research literature suggests that various types of suicides and murder-suicides increase following other well-publicized suicides and murder-suicides (Philips 1978, 1979, 1980; Bollen and Philips, 1982). David Philips (1986) reported that suicides increase after reported suicides of famous people and after fictional suicides have occurred in daytime television soap operas. His methodology has been criticized (Baron and Reiss, 1985) and his findings have not always been replicated (Kessler and Stipp, 1984). However, the existence of imitative suicides is supported by more informal evidence and seems to be widely accepted (Platt, 1987; Clarke and Lester, 1989).

Fictional events on television do not seem to have a large effect on imitative suicide. Little evidence suggests that fictional suicides have a grave negative impact on the general population. However, nonfictional events must be viewed differently. Research suggests that teens are more susceptible to imitative suicide when exposed to news stories about suicides; however, imitation theory only accounts for a very small percentage of total teen suicides (Philips and Carstensen, 1986). Kessler and his colleagues (1988) replicated and extended the Philips and Carstensen (1986) study and suggest that only celebrity suicide stories were associated with an increase in teen suicides. However, recent data collected from the Seattle Medical Examiner's Office and from the Seattle Crisis Center suggests that, although there was a significant increase in suicide crisis calls, there was no significant increase in completed suicides following the suicide of rock star Kurt Cobain in 1994 (Jobes et al., 1996). The lack of an apparent copycat effect in Seattle may be due to various aspects of the media coverage associated with the event that portrayed the act as tragic, selfish, and ultimately wasteful.

CONCLUSIONS

This review of research undergirds our investigation in Part I of six specific incidents of lethal school violence, including four that are considered school rampages. The research reviewed here has directed attention to the following broad classes of potential causes of the violence:

1. Macrosocial structures affecting the communities in which the rampages occurred.
2. Stable characteristics of the offenders that make them unusually susceptible to committing such acts. These characteristics are produced by some combination of social factors operating on individuals, and individual inheritances and experiences that they have.
3. Microsocial processes that create the social dynamics that make it important for the offenders to act violently, directing their violence towards more or less particular targets and enabling the action to be taken.
4. The failure of control mechanisms at the family, community, and institutional level that should have been successful in preventing and controlling the events.

Our interest tends to focus more on the last three of these classes of causal factors for several reasons. First, many of the macrosocial conditions implicated in causing high levels of violence are not present in the communities that experienced school rampages. These communities do experience poverty, discrimination, and alienation, and there are some weaknesses in the quality of the schools. But the communities as a whole seem better off in terms of these characteristics than the communities that have experienced higher levels and different kinds of lethal youth violence.

Second, both the cases of inner-city violence and the school rampage shootings seem to be caused as well as shaped by important microsocial processes that swept the offenders and their victims up in powerful circumstances that drove them to their specific acts. One can look at the events and imagine that if they had gone somewhat differently, the episode might have been prevented from occurring at all.

Third, the school rampage shootings in particular suggest a potentially important role for some kind of mental illness in the individual as a relatively important cause of the violence. Mental illness is apparent in suicide attempts and in symptoms of depression. It is also present in the fact that the grievances that the perpetrators seemed to feel, and the targets they chose to attack, seemed incomprehensible to others in the communities in which they occurred. The exaggerated sense of victimization and the arbitrary choice of victims most strongly suggest the presence of mental illness.

The more we looked at the cases of school rampages, the more they looked like other kinds of rampages rather than other kinds of youth violence or other kinds of school violence. Given the trends in this form of violence, it also seemed important to keep our eyes open to the possibility that there were contagion mechanisms operating during that period to generate the cluster of these events observed in the United States and in the world. In undertaking the case studies, the important goal was to conduct them in ways that would reveal whether these surmises have any supporting evidence.

NOTE

[1]Although one can separate out these different levels of analysis, treating the factors as more or less stable characteristics of either social groupings or individuals, the fact of the matter is that factors at one level influence factors at the next level—either up or down—and that many of these factors change over time both for individuals and for social groupings.

11

Response Strategies: Observations on Causes, Interventions, and Research

The most important issue for policy officials in responding to school rampages is finding the means to prevent them. One thing that makes such efforts particularly difficult is that no place seems to be immune from such events. Another is that they are exceedingly rare: only 35 of the nation's 116,910 elementary and secondary schools have experienced a multiple victim shooting over the last decade. This makes it difficult for communities and schools to maintain their vigilance and their preparedness over long periods of time, during which the threat seems to be both remote and receding. This suggests that in order for prevention measures to work, they have to be important for other more immediate and more persistent purposes.

A PREVENTION PARADIGM

The public health community has developed a useful way of thinking about the prevention of injuries, both intentional (crimes) and unintentional (accidents). They distinguish three different classes of preventive efforts: primary, secondary, and tertiary prevention. Not only does the public health community make analytic distinction among these different forms of prevention,[1] but they also are strong advocates for particular kinds of prevention. Generally speaking, they would prefer primary prevention over any of the others—particularly if the primary prevention instruments are both inexpensive to use and entirely effective. The preference for this sort of prevention is rooted in the extraordinary success of immunizations against disease. Immunization has many of the ideal qualities of a primary preventive instrument: it is very low in cost, can be widely distributed, and is extremely effective.

It is difficult, however, to import this model into the world of social behavior, especially delinquency. Because of the complexity of developmental and other factors that produce these behaviors, these interventions—whether primary, secondary, or tertiary—can be very costly rather than inexpensive, especially since they must be sustained over time. Research has shown that prevention efforts to date afford quite limited protection.

Cost may cause society to depart from its preference for protecting everyone and concentrate on protecting those places, people, and circumstances that seem at particularly high risk. However, because the interventions are less reliable, society may still have to cope with incidents that were not prevented. In short, the more expensive and less reliable the interventions, the more the society will be forced to employ a portfolio of preventive activities that favor tertiary over secondary and primary preventive instruments. One can even imagine situations in which the cost and unreliability of primary and secondary instruments are so great as to make tertiary prevention the best approach to dealing with a problem. In these cases, an ounce of prevention would be worth much less than a pound of cure.

This report suggests that many different factors can potentially lead to lethal violence or school rampages, including structural community variables, the ordinary processes of child development, the risks created when children become alienated from adults, cultural influences, the stable characteristics and motivations of the individuals who become offenders, and microsocial situational processes that, although surely influenced by individual characteristics and larger community forces, take on a life of their own in the give and take of interpersonal and group interaction.

Since each of these factors is a potential cause, each is a potential target of preventive efforts. And one could attack these different causes in either a general, primary prevention effort or in a more selective secondary prevention mode. For example, one could seek to modify structural conditions across all communities in the country, or only those that are judged to be at particularly high risk. One could treat every youth as a potential offender, or concentrate on those considered to be at particularly high risk. One could treat every instance of status threat and degradation in school as something to be managed, or focus only on those situations that seemed to occur in high-risk communities and schools involving high-risk individuals.

The lethal violence and school rampages are spread across different kinds of communities, different kinds of schools, different kinds of youth, and different kinds of situations. So it may be difficult to rely much on certain secondary prevention instruments, such as profiling, especially for the rampage shootings. For example, in our cases, school administrators and teachers could not easily distinguish high-risk schools, youth, or situations from low-risk ones.

However, secondary prevention in the form of uncovering and responding to plans for rampages can be highly effective and quite targeted. Tertiary preventive instruments are—for lethal violence and school rampages—insufficiently preventive. Communities need to be able to do more than simply respond after the fact. It seems clear that design of successful interventions will require a more sophisticated research base than currently exists.

With this preliminary cautionary discussion in mind, we examine particular preventive ideas that are widely discussed or have been embraced by communities that have either suffered these tragedies or been galvanized into action by the experience of those who did.

POTENTIAL TARGETS OF PREVENTIVE EFFORTS

Creating a Profile of Likely Shooters

One widely discussed preventive idea is to develop methods to identify likely offenders in instances of lethal school violence or school rampages. If they could be identified, then a secondary preventive instrument could be developed to focus on those who are at high risk of committing such offenses.

The difficulty is that looking at the relatively stable and visible characteristics of youth generally does not help to find likely offenders. The offenders are not that unusual; they look like their classmates at school. This has been an important finding of all those who have sought to investigate these shootings. Most important are the findings of the United States Secret Service, which concluded (Vossekuil et al., 2000:5):

There is no accurate or useful profile of "the school shooter."

- Attacker ages ranged from 11–21.
- They came from a variety of racial and ethnic backgrounds. In nearly one-quarter of the cases, the attackers were not white.
- They came from a range of family situations, from intact families with numerous ties to the community to foster homes with histories of neglect.
- The academic performance ranged from excellent to failing.
- They had a range of friendship patterns from socially isolated to popular.

• Their behavioral histories varied, from having no observed behavioral problems to multiple behaviors warranting reprimand and/or discipline.

• Few attackers showed any marked change in academic performance, friendship status, interest in school, or disciplinary problems prior to their attack. [Note: the cases in this report suggest something different here.]

• Few of the attackers had been diagnosed with any mental disorder prior to the incident.

While the shooters in our cases demonstrated less variability on some of these risk factors, they were, for the most part, not distinguishable from many of their peers on those factors. Trying to make these kinds of distinctions would result in many errors, of two types: mistakenly classifying many schools, individuals, or situations as high risk and wasting resources on circumstances that were not going to produce instances of lethal school violence, and mistakenly classifying many communities, schools, individuals, and situations as safe when in fact they were risky and might well produce instances of violence. This finding should prevent communities from moving quickly to a form of prevention that seeks to identify the shooters before they shoot. Not only would such efforts be ineffective, but they would also unnecessarily stigmatize a large number of adolescents as a threat to their classmates. The combination of high cost, ineffectiveness, and unfairness rules out this particular line of attack.

Alternatively, one could look for important changes in a student's status and behavior. What is notable about the instances of youth violence is that they seem to be propelled forward from the realm of potential to concrete action by small, sudden changes in circumstances: a gambling dispute became an urgent, collectible matter; failure was keenly felt in trying to achieve a certain standing in a peer group; making a date with a girl became impossibly humiliating. It is not that the youth were so unusual; it is that they found themselves in circumstances that felt to them very threatening and impossible. The challenge is to find the kids who feel this way.

This is no easy task. It requires close, continuous monitoring, and it is hard to imagine relations between adults in and out of the school that could be close enough to accomplish this. But the kids may know when something dramatic has changed in another student's status at school. In most of the shootings, other kids knew about the "beef," or the shooter gave explicit warnings or made explicit threats. The fact that rampage shootings are so rare suggests that there is some success in spotting and thwarting them, and that communities have been somewhat successful alerting adults and children to the need to take explicit or implied threats seriously.

The seriousness of this kind of threat has altered the norms that exist in schools. One hopes that students themselves are now concerned about their own safety, and they take threats of shootings and rampages seriously. Joined with efforts to draw adults and youth closer together and a commitment to keeping schools safe from lethal incidents, some kind of preventive cover can reduce even further the already low probability that a school will be victimized by a lethal shooting or a rampage.

Improving Security Arrangements at Schools

An important question is whether and how mobilization to oppose violence in the schools, of the kind just described, might be aided by the installation of specialized security arrangements, including metal detectors, fences, identity badges, and the hiring of various kinds of security specialists.

It is easy to understand the opposition to such measures. They can change the look and feel of the school from a learning community to an anxiety-ridden prison. They can draw attention to the fact that the social relations have so frayed that schools have to rely on technology and rules instead of human relationships and values to provide security. Such measures can distract students, faculty, and administrators from the important work of teaching, including teaching the idea of what it means to be a responsible school citizen. All these drawbacks make the use of such measures a last resort.

Still, in some circumstances, such measures are a welcome first step in restoring a sense of security. Case authors John Hagan, Paul Hirschfield, and Carla Shedd note that after the city-wide installation of metal detectors in Chicago, no further shootings occurred in Chicago schools. We would need evidence from experimental studies to conclude that metal detectors in schools could end either lethal shootings or school rampages, and one would want to look closely at other effects of the metal detectors on school culture and performance. But it seems reasonable for the citizens of Chicago to think that their schools were made at least a bit safer through the installation of metal detectors.

The schools in the case studies that turned to hiring school police officers or resource officers believed that to be a useful program. It is not that the officers patrolled the schools and deterred shootings. Rather, they became a symbol and a rallying point for students and faculty who were concerned about security. They also served as a communication channel for students to report information about threats to school security. Discussions about whether and how to use such security measures could have a salutary effect on the creation of an effective normative system to enhance safety in the school, even if specific measures were not adopted.

Other measures considered or used in the cases seem more suspect. Putting up fences or other measures designed to prevent outsiders from threatening the schools was not productive, because in all of the cases in which this was done, the threat came from inside the school, from members of the school community. The fact that the fences would not have been successful in preventing the incidents that occasioned their construction reveals the fundamentally irrational nature of this enterprise.

Increasing Weapons Security to Keep Guns from Youth

We cannot conclude this report without briefly discussing the issue of youth access to guns. All instances of lethal violence documented in the cases were committed by youth armed with rifles and handguns who were breaking current gun laws in addition to substantive criminal laws. Both state and federal laws prohibit children of this age from possessing or carrying guns without adult supervision, and many federal and state statutes are designed to prevent children from being able to acquire weapons. This includes bans against selling to minors. It is also true that even the most fervent champions of the rights of citizens to own guns have stood for keeping them away from unsupervised children. Nevertheless, research has found that more than half of privately owned firearms are insecurely stored (National Institute of Justice, 1997). Insecure storage of firearms was clearly a factor in four of these six cases.

Despite all of this effort to keep guns from children, the committee was somewhat astounded at the ease with which the young people in these cases acquired the weapons they used. Only in the Jonesboro case were the most powerful weapons in the home of one of the boys too well secured for them to access. But it was easy to defeat the security measures of another relative and get hold of a powerful semiautomatic rifle with a scope. In general, it was easy for these young teens to circumvent both law and informal controls designed to deny them the weapons they used in their crimes.

The committee notes that, in the Edinboro case, a gun was used by a citizen to end the rampage incident and prevent the shooter from harming additional victims or himself. This also happened in the rampage shooting in Pearl, Mississippi, in 1997.

In the committee's view, there is much useful work to be done through both law and custom to deny guns to children. Recognizing the fact that both law and community sentiment are against unsupervised children possessing and using guns, the committee believes that denying access to unsupervised youth should continue to be an important national goal.

Dealing More Effectively with Adolescent Status Concerns at Home, in Schools, and in the Community

One message that comes through loud and clear in the cases is that adolescents are intensely concerned about their social standing in their school and among their peers. For some, their concern is so great that threats to their status are treated as threats to their very lives and their status as something to be defended at all costs.

In addition to the dread and fear that loss of status of any kind inspires in some adolescents, of equal importance is how hard it is for them to form accurate estimations of their standing. They are easily confused by relatively small events, easily imagining that "everyone hates them" or that "life isn't worth living" on the basis of what may later look like relatively small setbacks.

It is important for siblings, parents, teachers, guidance counselors, youth workers, and employers to be vigilant in noticing when these threats to an adolescent's status occur and to be active in helping them deal with their status anxieties. Young people need some places where they feel valued and powerful and needed—that is part of the journey from childhood to adulthood. If they cannot find paths that make them feel this way, or they find the paths blocked by major threats, they will either retreat or, in the case of lethal shootings and rampages, strike back against those who seem not to value them, or are threatening them, or are blocking their way. Holding spaces and pathways open for them may be an important way of preventing violence.

The best way to prevent lethal school shootings and school rampages is to create communities that are committed to the safety and healthy development of their youth. The value of this work is not measured solely in its success in preventing lethal school violence and school rampages. To become and remain a nation that creates equal opportunity for all, that creates the conditions under which individuals can make the most of their talents in whatever pursuit interests them, communities must help young people get to the starting line of adult life with health, vitality, and confidence. Communities that cannot keep their children safe from lethal violence, that endure conditions in which those reaching for adult status and competence in schools cannot be safe, are failing to protect the American dream.

RESEARCH

Improved basic research is needed to support theories about the scope and nature of lethal school violence. There is already a large body of research on violence among urban youth, much of it emphasizing such risk factors as the influence of a violent environment, belonging to criminal juvenile gangs, and illegal gun carrying. But there is almost no research on violence among suburban and rural youth inside or outside schools. Longitudinal research is needed to follow up the perpetrators of school shootings, both those who remain in prison and those who are released. In addition, case studies of other kinds of rampages involving adults are needed as a point of comparison with student rampage shootings. Finally, while some research has followed children who have been victims of school rampage shootings in the past, further longitudinal research is needed to understand the impact of school shootings on both perpetrators and victims, including secondary victims.

Two Types of Violence

As the committee worked with both the cases and the data, it seemed that there might be two quite different strands of lethal violence. In one strand, the lethal violence seemed to emerge from a fairly specific conflict among individuals, in which all the parties to the conflict were aware of the problem and all the parties felt vulnerable to a physical threat or attack. It also seems clear that others outside the conflict—the audience—knew about it and understood its logic and patterns. Because this kind of violence seemed most typical of the inner-city school cases—as one would expect in communities in which violence had become highly prevalent—we began calling this the "inner-city form of violence." In a second strand, the lethal violence also emerged from a conflict in which the persons who committed the offenses felt that they were under threat and needed to protect themselves. But there are some important differences in the nature of the conflict. In this second strand of violence, the conflict was much less specific and differentiated than in the first strand. The "beef" was not with a particular person over a particular issue; it involved some abstract idea of "the school" or "the other kids who don't respect me." There can be many particular individuals who embody this abstract enemy. A second important distinction between this kind of lethal violence and the first kind is that not all those who were potential targets of the violence knew they were involved in a conflict with the person who did the shooting. Because this seemed to be the pattern that was characteristic of the suburban and rural shootings we examined closely, we came to call this the suburban-rural pattern.

It is the committee's strong view that additional research is necessary to investigate the question of whether there are two different strands of lethal school violence and, if so, whether they are correlated with urban on one hand, and suburban and rural on the other. If these two different patterns of lethal school violence exist, and if they are in fact correlated with the characteristics of the communities in which the violence occurred, then these trends should be revealed in the aggregate statistics on lethal violence in schools. However, because detailed case studies have not been developed for all the instances of lethal school violence, for most of them we remain uncertain both about the motivations of the offenders and the relationships that existed between them and their victims. All we can see is the gross pattern of victimization in the attacks: the number of individuals who were killed and injured and the most superficial account of the characteristics of victims that might give us a clue about their relationship to the offender. It is quite possible that, while rampage shootings showed up in suburban and rural schools that had not experienced much if any "inner-city style" violence, there was some portion of rampage "suburban-rural style" violence that added to the burdens of inner-city schools that were already struggling with violence.

Nonlethal Violence and Bullying

Lethal violence in any school is rare, but nonlethal violent crime in schools is much more common. The occurrence of frequent bullying is a very serious problem among students in grades 6–10. Such bullying, which can range from physical assaults to verbal harassment and threats, has been shown to have lifelong social and psychological consequences for victims. Moreover, despite attempts to control it, some of the most serious behavior takes place below the radar screen of responsible adults.

The committee recommends that research be conducted on the nature and causes of school violence, including seriousness of behavior, motivation of perpetrators, and the role of recognized gangs, crews, and cliques or informal social groups inside and outside the school in both crime and other antisocial behavior, such as serious bullying. The consequences of both lethal and nonlethal school violence for students and adults not directly victimized is also an important area of inquiry.

Gun Carrying

Virtually every case of lethal school violence that has occurred since 1992 has involved the use of a firearm. In our review of these cases and other research on lethal youth violence, the committee found that illegal gun carrying by youth crosses racial and class boundaries, and that a substantial number of boys—particularly those becoming involved with gangs—illegally carry firearms, at least sometimes, at young ages. In the cases studied in this report, the boys carried guns to school for protection or to enhance their status among their peers. Few youth carry a firearm all or most of the time over a long period, and most who carry one for protection stop when there is no longer a threat.

Because illegal gun carrying increases with age, it is important to develop research that will inform prevention or deterrence strategies. The sources of the firearms carried and used in the cases in this volume were friends, especially gang friends, and parents or friends, from whom they were stolen. It was extremely easy for these boys to get weapons, and in most of the cases, they were familiar with their use.

The committee recommends that a program of research be developed to further examine illegal gun carrying by adolescents, especially carrying a gun to school. This research should examine the circumstances and motivations related to illegal gun carrying, the sources of and ease of access to guns, socialization to illegal gun use, and the relationship, if any between legal and illegal gun use by adolescents.

Individual Risk Factors for Violent Behavior in School

The extensive and sound knowledge base on risk factors for delinquency and violent behavior, which can be used to design prevention programs, would not have helped identify the young people in most of these cases as high risk. Most of the shooters in the cases studied were not thought to be at high risk by the adults around them. While some of them had one or two risk factors, none had the multiple, high-risk factors described in longitudinal studies of delinquency that normally presage violent acts.

However, among the eight shooters in this set of cases, the specter of developing mental illness surfaced for five of the boys. In most cases, other than attention deficit disorder, the presence of serious mental illnesses can be difficult to detect in youth ages 11–15. This is particularly true of those who encounter high levels of violence in their everyday lives. In its examination of these cases, the committee found symptoms of mild and severe depression, stress disorders, personality disorders, and developing schizophrenia in these youth. Suicidal thinking was a prominent feature in all of the suburban and rural cases in this study and in many such cases studied by others. Greater research attention is needed to determine how and when mental illness begins to develop in young adolescents, how the social environment inside or outside school contributes to the development of pathology, and how it becomes manifest in the behavior of youth in this age group.

The committee concludes that empirical research is needed to measure the prevalence of developing mental illness in young adolescents. We recommend that public health surveillance research methods be applied to the identification of risk factors and signs or symptoms of developing, serious mental illness in children in grades 6–10. Such a research effort might include the development of new, culturally appropriate instruments to measure the effects of exposure to both actual violence and virtual or vicarious (media) violence on the individual development of young adolescents and preteens. Research on the nature and operation of contagion mechanisms that may lead to copycat behavior, especially focusing on suicidal behavior, is also an important avenue of scientific inquiry. Evaluation studies are needed on the effects of suicide prevention programs with this age group.

Workplace Violence

In the cases presented in this volume, three of the slain victims were teachers. Violent crimes—over 0.5 million in a five-year period in the mid-1990s—and thefts—over 1 million—committed against teachers in middle school and high school have become a serious problem. In addition to the harm caused to the victims directly, crimes against teachers, especially when committed by students, undermine adult authority and severely compromise the safety of students at school.

The committee recommends that a program of research be designed and conducted on the nature and causes of crimes committed against teachers in middle school and high school. This research should examine the seriousness of incidents; the identity, status in school, and motivations of the attackers; the effect on the learning environment and social relationships in the school, including the effect on school order and discipline; the individual consequences of these attacks on victims, other teachers, and students; and the effect of these crimes on the school system's ability to retain and recruit qualified teachers.

Community Research

The committee found important structural differences in the kinds of communities described in the cases in this study. The urban neighborhoods were characterized by social and physical conditions that created a milieu for the development of youth delinquency and violence. These include a high degree of social and economic disinvestment in the neighborhoods of the shooters, especially the withdrawal of community services; population change resulting in neighborhood hypersegregation; poor housing stock; social disorganization, characterized by the presence of criminal juvenile gangs; and high levels of violent crime. The suburban and rural cases did not evince these community structural conditions and in fact were demographically the opposite—thriving economically, having a high degree of social capital, and mostly free of crime and violence. This is true of the rampage shooting cases not only in this study, but also of rampage shootings in general.

The environmental similarity across most of the cases is the presence of rapid social change leading to possible instability, even when the changes are positive ones. While many parents successfully supervise their children in these kinds of situations, it is clear, in both the urban and rural and suburban cases, that parents had a poor understanding of their children's exposure to changing community conditions and involvement in social situations, including at school. It is important to the development of prevention efforts to have a better understanding of social and structural features of communities where shooting rampages have taken place, and of parental supervision of children's activities in all communities where school shootings occur.

The committee recommends that research be conducted on the effects of rapid change in increasingly affluent rural and suburban communities on youth development, socialization, and violence. Such research should include the effects of new industries, such as manufacturing and service industries, and accompanying jobs on the community and its long-term residents; the effects of an influx of many new residents on social class structure, organizations, and institutions, including schools; changes in zoning or differences in the quality of housing, playgrounds, or schools in neighborhoods; and the presence or absence of community conflict among different economic and social groups and whether such conflicts affect youth behavior. In addition, the committee recommends that research be conducted on parental styles of supervision for youth in grades 6–10 when parents are at work or when their children are away from home. To the extent that parents rely on the schools to supervise youth in this age group, it also is important to better understand the effectiveness of various supervisory roles and styles of teachers and other adults, such as administrators, counselors, volunteers, and security personnel.

Security and Enforcement

The shooters in the cases were arrested almost immediately after the shootings without further incident, and the police were able to respond in an appropriate and timely manner. In no instance did the police have previous warning of the shooting. This is in contrast to the circumstances of other school rampages. In the Columbine incident, which involved two older teenage offenders and much more elaborate planning, there was far more confusion about what was taking place and why, once the police arrived on the scene. And there was some evidence that reports to police about threats to a student, including threats posted on a web site, had been ignored. In addition, over the past two years there has been some experience with preventing planned attacks when there is information that permits the potential problem student or group to be fairly precisely targeted.

The committee recommends that evaluation studies be conducted on police response to rampages and on new protocols that have been developed to uncover and respond to plans for rampages in schools. It would also be useful to develop detailed case studies of failed or thwarted incidents, not focusing on police action alone, but including other details as to why the plans were made and what caused them to be disrupted. In addition, most of the schools in these cases adopted new security measures, such as deploying metal detectors, security guards, and police resource personnel and building perimeter fences. The committee recommends that these security efforts be evaluated in terms of their impact on the learning environment and on the overall safety of the school.

NOTE

[1]Some make distinctions along some time dimension that run from conditions or events that are antecedent to the injury to those that follow the injury. Others make distinctions among the different kinds of prevention on the basis of the probability that the condition or event that is the focus of preventive effort will lead to the injury. Still others associate the types of prevention with their ultimate effectiveness: primary prevention essentially eliminates the risk of injury across the general population; secondary prevention reduces but does not eliminate the risk among a subsection of the population; and tertiary prevention consists of efforts to mitigate the damage once primary and secondary prevention have failed to prevent the injury.

References

Note: This reference list covers the report of the panel. References for the case studies appear within the respective chapters.

Anderson, E.

1994 The code of the streets. *Atlantic Monthly* (May):81–94.

Anderson, M., J. Kaufman, T.R. Simon, L. Barrios, L. Paulozzi, G. Ryan, R. Hammond, W. Modzeleski, T. Feucht, and L. Potter

2001 School-associated violent deaths in the United States, 1994–1999. *Journal of the American Medical Association* 286(21):2695–2702.

Baird, A.A., S.A. Gruber, D.A. Fein, L.C. Mass, R.J. Steingard, P.F. Renshaw, B.M. Cohen, and D.A. Yurgelun-Todd

1999 Functional magnetic resonance imaging of facial affect recognition in children and adolescents. *Journal of the American Academy of Child and Adolescent Psychiatry* 38(2):195–199.

Baron, J.N., and P.C. Reiss

1985 Same time, next year: Aggregate analyses of the mass media and violent behavior. *American Sociological Review* 50:347–363.

Barron, O.

2000 Why workplace bullying and violence are different: Protecting employees from both. *Security Journal* 13:63–72.

Bennett, W.J., J.J. DiIulio, and J.P. Walters

1996 *Body Count: Moral Poverty and How to Win America's War Against Crime and Drugs.* New York: Simon and Schuster.

Berkowitz, L.

1984 Some effects of thoughts on anti- and pro-social influences of media events: A cognitive-neoassociation analysis. *Psychological Bulletin* 95:410–417.

Blumstein, A.

1995 Youth violence, guns and the illicit-drug industry. *The Journal of Criminal Law and Criminology* 86(1):10–36.

Blumstein, A., and R. Rosenfeld
1998 Explaining recent trends in U.S. homicide rates. *The Journal of Criminal Law and Criminology* 88(4):1175–1216.
Bollen, K A., and D. Phillips
1982 Imitative suicide: A national study of the effects of television news stories. *American Sociological Review* 47:802–809.
Bunch, J.G., B. Barraclough, B. Nelson, and P. Sainsbury
1971 Suicide following bereavement of parents. *Social Psychiatry* 6:193–199.
Bureau of Labor Statistics
1997 Fatal Workplace Injuries: A Collection of Data and Analysis. Washington, DC: U.S. Department of Labor.
1999 Census of Fatal Occupational Injuries, Summary. Press release. Washington, DC: U.S. Department of Labor.
Cain, A.C., and I. Fast
1972 The legacy of suicide: Observations on the pathogenic impact of suicide on marital partners. In *Survivors of Suicide*, A.C. Cain, ed. Springfield, IL: Charles C. Thomas.
Clarke, R.V., and D. Lester
1989 *Suicide: Closing the Exits.* New York: Springer-Verlag.
Comstock, G.
1980 New emphases in research on the effects of television and film violence. In *Children and the Faces of Television: Teaching, Violence, Selling*, E.L. Palmer and A. Dorr, eds. New York: Academic Press.
Cook, P.J,. and J.H. Laub
1998 The unprecedented epidemic in youth violence. Pp. 27–64 in *Youth Violence*, Crime and Justice, vol. 24, M. Tonry and M.H. Moore, eds. Chicago: University of Chicago Press.
Davidson, L., and M.S. Gould
1989 Contagion as a risk factor for youth suicide. In *Report of the Secretary's Task Force on Youth Suicide: Risk Factors for Youth Suicide*, U.S. Department of Health and Human Services, ed. Washington, DC: U.S. Department of Health and Human Services.
DiIulio, J.J., Jr.
1995 Why violent crime rates have dropped. *The Wall Street Journal* (September 6):A19.
Dodge, K.A., G. Pettit, C. McClaskey, and M. Brown
1986 Social competence. *Monograph of the Society for Research in Child Development* 58:213–251.
Donnerstein, E., and D. Linz
1995 The media. In *Crime*, J.Q. Wilson and J. Petersilia, eds. San Francisco: ICS Press.
Elliott, D.S., B.A. Hamburg, and K.R. Williams
1998 Violence in American schools: An overview. Pp. 3–28 in *Violence in American Schools: A New Perspective*, D.S. Elliott, B.A. Hamburg, and K.R. Williams, eds. Cambridge: Cambridge University Press.
Fagan, J. and D.L. Wilkinson
1998 Guns, youth violence, and social identity in inner cities. Pp. 105–188 in *Crime and Justice*, vol. 24. Chicago: University of Chicago Press.
Farrington, D.
1993 Understanding and preventing bullying. In *Crime and Justice: A Review of Research*, vol. 17, M. Tonry, ed. Chicago: University of Chicago Press.

Farrington, D.P., and R. Loeber
1999 Transatlantic replicability of risk factors in the development of delinquency. Pp. 299-329 in *Historical and Geographical Influences on Psychopathology*, P. Cohen, C. Slomkowski, and L.N. Robins, eds. Mahwah, NJ: Lawrence Erlbaum Associates.

Federal Bureau of Investigation
2000 *The School Shooter: A Threat Assessment Perspective*. Quantico, VA: U.S. Department of Justice.

Felthous, A., and A. Hempel
1995 Combined homicide-suicides: A review. *Journal of Forensic Sciences* 40:846–857.

Foote, W.
1995 Victim-precipitated homicide. In *Lethal Violence 2000: A Sourcebook on Domestic, Acquaintance, and Stranger Aggression*, H.V. Hall, ed. Kamuela, HI: Pacific Institution of Conflict.

Fox, J. A.
2001 *Uniform Crime Reports [United States]: Supplementary Homicide Reports, 1976–1999* [Computer file]. ICPSR version. Boston, MA: Northeastern University, College of Criminal Justice [producer], 2001. Ann Arbor: Inter-university Consortium for Political and Social Research [distributor], 2001.
1996 *Trends in Juvenile Violence: A Report to the United States Attorney General on Current and Future Rates of Juvenile Offending*, March. Washington, DC: Bureau of Justice Statistics.

Fox, J.A., and J. Levin
1998 Multiple homicide: Patterns of serial and mass murder. In *Crime and Justice: A Review of Research*, vol. 23, M. Tonry, ed. Chicago: University of Chicago Press.

Garofalo, J.
1982 Crime and the mass media: A selected review of research. *Journal of Research in Crime and Delinquency*, 18:319–350.

Gilligan, J.
1996 *Violence: Reflections of a National Epidemic*. New York: Vintage Books.

Gillespie, M., V. Hearn, and R. Silverman
1998 Suicide following homicide in Canada. *Homicide Studies* 2:46–63.

Gottfredson, D.C.
2001 *Schools and Delinquency*. New York: Cambridge University Press.

Gould, M.S.
1990 Suicide clusters and media exposure. In *Suicide Over the Life Cycle: Risk Factors, Assessment, and Treatment of Suicidal Patients*, S. Blumenthal and D. Kupfer, eds. Washington, DC: American Psychiatric Press.

Gould, M.S., S. Wallenstein, and M. Kleinman
1990a Time-space clustering of teenage suicide. *American Journal of Epidemiology* 131:71–78.

Gould, M.S., S. Wallenstein, M. Kleinman, P.O'Carroll, and J. Mercy
1990b Suicide clusters: An examination of age-specific effects. *American Journal of Public Health* 80:211–212.

Graham, H.D., and T.R. Gurr
1969 *Violence in America: Historical and Comparative Perspectives*. Washington, DC: U.S. Government Printing Office.

Gurr, T.
1989 Historical trends in violent crime: Europe and the United States. In *Violence in America: Volume I*, T. Gurr, ed. Newbury Park, CA: Sage Publications.

Hagin, A.D.
1986 Understanding suicide. In *Suicide Prevention and Caregiving*, S. Daniels and D. Pass, eds. Milwaukee, WI: National Funeral Directors Association.
Hamblin, R.J., R.B. Jacobsen, and J.L. Miller
1973 *A Mathematical Theory of Social Change*. New York: Wiley-Interscience.
Holden, R.
1986 The contagiousness of aircraft hijacking. *American Journal of Sociology* 91:874–904.
Homant, R., and D. Kennedy
2000 Suicide by police: A proposed typology of law enforcement officer-assisted suicide. *Policing: An International Journal of Police Strategies and Management* 23:339–355.
Homant, R., D. Kennedy, and R.T. Hupp
2000 Real and perceived danger in police officer assisted suicide. *Journal of Criminal Justice* 28:43–52.
Huesmann, L.R.
1982 Television violence and aggressive behavior. In *Television and Behavior*, Vol. 2. U.S. Department of Health and Human Services. Washington, DC: U.S. Government Printing Office.
Hutson, H.R., D. Anglin, J. Yarbrough, K. Hardaway, M. Russell, J. Strote, M. Canter, and B. Blum.
1998 Suicide by cop. *Annals of Emergency Medicine* 32(6):665–669.
Jobes, D., A. Berman, P. O'Carroll, S. Eastergard, and S. Knickmeyer
1996 The Kurt Cobain suicide crisis: Perspectives from research, public health, and the news media. *Suicide and Life-Threatening Behavior* 26:260–264.
Kaufman, P., X. Chen, S.P. Choy, S.A. Ruddy, A.K., Miller, K.A. Chandla, C.D. Chapman, M. R. Rand, and P. Klaus
1999 Indicators of School Crime and Safety. U.S. Departments of Education and Justice. NCES 1999 - 057/NCJ – 178906. Washington, DC.
Kessler, R.C., and H. Stipp
1984 The impact of fictional television suicide stories on U.S. fatalities: A replication. *American Journal of Sociology* 90:151–167.
Kessler, R.C., L. Downey, J.R. Milavsky, and H. Stipp
1988 Clustering of teenage suicides after television news stories about suicides. *Journal of Psychiatry* 145:1379–1383.
Klein, M.
2002 Street gangs: A cross national perspective. In *Gangs in America*, 3rd edition, C. R. Huff, ed. Thousand Oaks, CA: Sage Publications.
Levin, J., and J.A. Fox
1996 A psycho-social analysis of mass murder. In *Serial and Mass Murder: Theory, Research, and Policy*, T. O'Reilly-Fleming, ed. Toronto: Canadian Scholars' Press.
Li, R.P., and W.R. Thompson
1975 The 'Coup Contagion' hypothesis. *Journal of Conflict Resolution*, 19:63–88.
Livingstone, N.
1982 *The War Against Terrorism*. Washington, DC: Heath.
Lizotte, A., and D. Sheppard
2001 Gun use by male juveniles: Research and prevention. *Juvenile Justice Bulletin*, Office of Juvenile Justice and Delinquency Prevention.
Loeber, R., and D.P. Farrington, eds.
1998 *Serious and Violent Juvenile Offenders: Risk Factors and Successful Interventions*. Thousand Oaks, CA: Sage.

McFall, R.
1982 A review and reformulation of the concept of social skills. *Behavior Assessment* 4:1–33.
McGee, J.P., and C.R. DeBernardo
1999 The classroom avenger: A behavioral profile of school based shootings. *The Forensic Examiner* 8(5)16–18.
Mellor, A.
1990 *Bullying in Scottish Secondary Schools.* Edinburgh: Scottish Council for Research in Education.
Meloy, J.R., A.G. Hempel, K. Mohandie, A.A. Shiva, and B.T. Gray
2001 Offender and offense characteristics of a nonrandom sample of adolescent mass murderers. *Journal of the American Academy of Child Adolescent Psychiatry* 40(6):719–728.
Menninger, K.
1938 *Man against Himself.* New York: Harcourt, Brace & World.
Midlarsky, M.
1970 Mathematical models of instability and a theory of diffusion. *International Studies Quarterly* 14:60–84.
Moore, M. and M. Tonry, eds.
1998 *Youth Violence.* Chicago: University of Chicago Press.
Nansel, T., M. Overpeck, R. Pilla, W.J. Ruan, S. Simons-Morton, and S. Scheidt
2001 Bullying behaviors among US youth: Prevalence and association with psychosocial adjustment. *Journal of the American Medical Association* 285:2094–2131.
National Institute of Mental Health
1982 *Television and Behavior: Ten Years of Scientific Progress and Implications for the Eighties.* Vol. 1, Summary Report. Rockville, MD: National Institute of Mental Health.
National Research Council
1993 *Understanding and Preventing Violence.* Panel on the Understanding and Control of Violent Behavior. A.J. Reiss, Jr. and J.A. Roth, eds. Commission on Behavioral and Social Sciences and Education. Washington, DC: National Academy Press.
National School Safety Center
2001 *School Associated Violent Deaths.* Westlake Village, CA: National School Safety Center.
O'Carroll, P.
1990 Suicide prevention: Clusters and contagion. In *Suicide Prevention: Case Consultations,* A. Berman, ed. New York: Springer.
Olweus, D.
1978 *Aggression in the Schools: Bullies and Whipping Boys.* Washington, DC: Hemisphere Publishing Corporation.
1992 Bullying among schoolchildren: Intervention and prevention. In *Aggression and Violence Throughout the Life Span,* R. Peters, R. McMahon, and V. Quinsey, eds. London: Sage Publications.
Pease, S., and C. Love
1984 The copy cat crime phenomenon. In *Justice and the Media,* R. Surette, ed. Springfield, IL: Charles C. Thomas.
Perry, D., L. Perry, and P. Rasmussen.
1986 Cognitive social learning mediators of aggression. *Child Development* 57:700–711.
Philips, D.
1978 Airplane accident fatalities increase just after newspaper stories about murder and suicide. *Science* 201:748–750.
1980 Airplane accidents, murder, and the mass media: Evidence towards a theory of imitation and suggestion. *Social Forces* 58:1001–1024.

1982a The behavioral impact of violence in the mass media. *Sociology and Social Research* 66:387–398.

1982b The impact of fictional television stories on U.S. adult fatalities: New evidence on the effect of mass media on violence. *American Journal of Sociology* 87:1340–1359.

1984 Natural experiments on the effects of mass media violence on fatal aggression. In *Advances in Experimental Social Psychology*, L. Berkowitz, ed. New York: Academic Press.

Philips, D.P.

1979 Suicide, motor vehicle fatalities, and the mass media: Evidence toward a theory of suggestion. *American Journal of Sociology* 84(5):1150–1174.

Philips, D., and L. Carstensen

1986 Clustering of teenage suicides after television news stories about suicide. *New England Journal of Medicine* 315:685–689.

Platt, S.

1987 The aftermath of Angie's overdose: Is soap (opera) damaging to your health? *British Medical Journal* 294:954–957.

Poland, J.M.

1988 *Understanding Terrorism: Groups, Strategies, and Responses*. Englewood Cliffs, NJ: Regents/Prentice Hall.

Reddy, M., R. Borum, J. Berglund, B. Vossekuil, R. Fein, and W. Modzeleski

2001 Evaluating risk for targeted violence in schools: Comparing risk assessment, threat assessment, and other approaches. *Psychology in the Schools* 38:157–172.

Ritterband, P., and R. Silberstein

1973 Group disorders in public schools. *American Sociological Review* 37:461–467.

Robbins, D., and R.D. Conroy

1983 A cluster of adolescent suicide attempts: Is suicide contagious? *Journal of Adolescent Health Care* 3:253–255.

Sampson, R.J.

1997 Collective regulation of adolescent misbehavior: Validation results from eighty Chicago neighborhoods. *Journal of Adolescent Research* 12(2):227–244.

Schmid, A., and J. DeGraaf

1982 *Violence as Communication*. Newbury Park, CA: Sage Publications.

Sears, D.O., J.L. Freedman, and L.A. Peplau

1985 *Social Psychology*. Fifth edition. Englewood Cliffs, NJ: Prentice–Hall.

Spilerman, S.

1970 The causes of racial disturbance: A comparison of alternative explanations. *American Sociological Review* 35:627–629.

Stack, S.

1997 Homicide followed by suicide: An analysis of Chicago data. *Criminology* 35:435–454.

Surette, R.

1990 Estimating the magnitude and mechanisms of copycat crime. In *The Media and Criminal Justice Policy: Recent Research and Social Effects*, R. Surette, ed. Springfield, IL: Charles C. Thomas.

2002 Self-reported copycat crime among a population of serious and violent juvenile offenders. *Crime and Delinquency* 48:46–69.

Sutherland, E.H., and D.R. Cressey

1970 *Criminology*. Eighth edition. Philadelphia: Lippincott Company.

Thompson, P.M., J.N. Gledd, R.P. Woods, D. MacDonald, A.C. Evans, and A.W. Toga

2000 Growth patterns in the developing brain detected by using continuum mechanical tensor maps. *Nature* 404(6774):190–193.

Turner, R.H., and L. M.Killian
1987 *Collective Behavior*. Third edition. Englewood Cliffs, NJ: Prentice-Hall.

U.S. Department of Labor
1997 *BLS NEWS: National Census of Fatal Occupational Injuries, 1996*. Washington, DC: U.S. Department of Labor.

U.S. Federal Bureau of Investigation
2001 *FBI's Supplementary Homicide Reports: 1980-1999*. Available: http://ojjdp.ncjrs.org/ojstatbb/ezashr/ [Accessed August 17, 2002].

U.S. Postal Service Commission
2000 *United States Postal Service Commission on a Safe and Secure Workplace*. Report by The National Center on Addiction and Substance Abuse, Columbia University. Washington, DC: U.S. Postal Service.

U.S. Secret Service
2000 "Deadly Lessons: School Shooters Tell Why," *Chicago Sun-Times* October 16.

Vossekuil, B., M. Reddy, R. Fein, R. Borum, and W. Modzeleski
2000 *Safe School Initiative: An Interim Report on the Prevention of Targeted Violence in Schools*, U.S. Secret Service National Threat Assessment Center. Washington, DC: U.S. Department of the Treasury.

Whitney, I., and P.K. Smith
1991 A survey of the nature and extent of bullying in junior/middle and secondary schools. Final report to the Gulbenkian Foundation. Sheffield: University of Sheffield, Department of Psychology.

Wilson, J.Q.
1995 Crime and public policy. In *Crime*, J.Q. Wilson and J. Petersilia, eds. San Francisco, CA: Institute for Contemporary Studies Press.

Wolfgang, M.
1957 An analysis of homicide-suicide. *Journal of Clinical and Experimental Psychopathology* 18–19:208–217.

Ziegler, S., and M. Rosenstein-Manner
1990 *Bullying in School*. Toronto: Board of Education.

Zimring, F.
1998 *American Youth Violence*. New York: Oxford University Press.

Appendix A

Case Study Methodology and the Study of Rare Events of Extreme Youth Violence: A Multilevel Framework for Discovery

Mercer L. Sullivan and Mindy Thompson Fullilove

The deliberations and decisions by the committee about which cases to select for intensive study and how to structure the collection and analysis of the data were undertaken in a pragmatic manner by a working group of researchers confronting an unusual and important task under a tight deadline. The workings and outcomes of that process are described in the report. This appendix connects that pragmatic process to formal methodological writings dealing with case studies and with the study of human behavior, including deviant and psychopathological behavior, in ecological context.

The nature of the phenomena to be studied necessitated a comparative case study approach by virtue of both the extreme rarity of the phenomena, in terms of numbers of occurrences, as well as the extreme severity of the behavior involved, whether that behavior is seen on any of a number of possible continua, including general violence, youth violence, or school violence. The small number of cases precluded gathering a large sample, while the severity of the violence seemed to demand in-depth scrutiny of such extraordinary events.

While single case studies and comparative studies of small numbers of cases are not suited to the same ends as more conventional statistical analyses of large samples, this approach is well suited to situations like the present one. The underlying assumptions and appropriate ends of case study approaches have been laid out with increasing sophistication in recent years, particularly in such fields as historical sociology and community studies, in which dealing with small numbers of cases is the usual state of affairs. This appendix reviews briefly some of those assumptions and that methodological literature as they relate to the current investigation of extreme violence in American schools.

Following that examination, the appendix then connects the aims and strengths of case study methodology to the study of extreme, and therefore puzzling, human behavior. The same extremity of deviance that makes such cases rare demands a multilevel conceptualization of the problem to be investigated. The emphasis on discovery of relevant factors through naturalistic methods of inquiry, which is the hallmark of case studies, leads directly to the conceptualization of human behavior as taking place in a set of hierarchically nested systems that underlies several well-established schools of thought and research methodology in psychology and psychiatry. It is precisely when things occur that seem extreme, unprecedented, and difficult to explain that one needs to cast the widest possible net to identify potentially important factors and processes. Viewing human behavior as occurring in a set of hierarchically nested systems is of great heuristic value in this situation, certainly as a way of casting a wide net and potentially as a way of actually catching something in an uncharted sea.

APPROPRIATE USES OF CASE STUDY METHODS IN SCIENTIFIC RESEARCH

Case studies have a variety of practical uses, including widespread application as teaching texts in such fields as law and business. In recent years, however, the particular and appropriate uses of cases studies as tools for scientific research have been delineated. Yin (1989) has noted that case studies, far from being suitable merely as illustrative material for quantitative analyses using larger samples, can be employed for three distinct and legitimate types of scientific ends: descriptive, exploratory, and explanatory. As an example of how even a single case study can serve explanatory ends, he cites Allison's study of the Cuban missile crisis, in which the author tests three rival hypotheses against the data (Allison, 1971).

In introducing Yin's exposition of case methods, Donald Campbell, among the most distinguished of all social science methodologists, contrasts this kind of operation with the highly touted ability of randomized experiments to exclude all rival hypotheses, without, he notes with a large dose of irony, "specifying what any of them are" (Campbell, 1989:8). Campbell invokes the effort to exclude rival hypotheses as the "core of scientific method" (p. 7). In the case of extreme school violence, for example, there may be popular theories, such as "the perpetrators are all the victims of bullying" that might be excluded on the basis of even a single case and that look even more untenable if they are inapplicable to multiple cases from a small universe.

Yin (1989:23) offers a working formal definition of a case study as: "an empirical inquiry that:

- investigates an empirical phenomenon within its real-life context; when
- the boundaries between phenomenon and context are not clearly evident; and in which
- multiple sources of evidence are used."

The second of these criteria is of particular interest in the present circumstance. In cases of extreme school violence, the question of the relevant context is both urgent and extremely problematic. Is it the psychopathology of the individual who offends? The borders of the school (and are those merely physical or to be thought of as the social field of school-related activities)? The community in which it occurs (and is that a local neighborhood or an incorporated municipality or county)? The national culture, drenched in violent media and grounded in a history of violence unique among wealthy, "civilized" nations (or subnational cultural traditions, defined by race, ethnicity or region, that tend more to legitimize violence)? With so few cases and the frighteningly suggestive but statistically unverifiable appearance of a recent upward trend, we cannot rule out any of these possibilities at the beginning of an inquiry.

The inquiry at hand, then, is a textbook example of a research situation in which case studies are not merely a default position but the proper tool for the job at hand. The boundaries between these phenomena and their contexts are about as murky as they get. The exploratory functions of the case study method are at the forefront here, and a central task is to probe for the relevant contexts and their interrelationships in these rare and extreme occurrences.

Many well-trained social scientists may be tempted at this point to raise two objections. First, how can even a meticulously researched case study prove the relevance of a factor or process to the occurrence of an outcome? Second, even if some such relationships can be established in a given case, how does one know that they have anything relevant to any other case? Both points, appropriate as they are in dealing with many common research situations, are of limited relevance here.

A case study may not prove beyond all doubt that a certain relationship exists in that particular case, but it is very likely the most trustworthy way to go about it, absent the ability to isolate the relationship in a laboratory, outside and perhaps irrelevant to the real world in which something occurs. The use of multiple sources of evidence, as opposed to strictly predefined data elements, helps to ensure construct and context validity and provides checks and balances that protect the inquiry from contamination by methodological artifacts related to reliance on single sources of evidence.

One of the most common failures in attempting to traverse the perilous gulf between correlation and causation is that of ignoring a previous causal factor that explains the association of interest. Because case studies treat the boundary between phenomenon and context as inherently problematic and consult multiple sources of evidence in order to discover as many potentially important factors as possible, the danger of ignoring important factors in this way is reduced.

Inferring relationships in a single case from aggregate patterns in a large number of cases sharing some but not all the features of the case in point is not inherently superior. Inference downward from the aggregate to the single case is not illegitimate either, but it runs the well-known risk of ecological fallacy (Robinson, 1950). The danger of generalizing inappropriately from the aggregate to the individual is no less than that of moving in the other direction (Sullivan, 1998). And yet, even a single well-documented case can refute absolutely a potentially overly broad theory such as "they are all victims of bullying."

The second objection, that a single case cannot be generalized to others without additional scientific operations, is not incorrect. Rather, it misses the point in a situation in which there are as yet few viable hypotheses to generalize. Here we return to Campbell's assertion of the centrality to the scientific method of examining rival hypotheses. Case studies are excellent tools for pruning extraneous hypotheses and generating potentially viable ones. Objecting to case studies in the name of generalization and science ignores the variety of important functions they play in scientific inquiry along with the complexity of the process of scientific generalization.

CASE STUDY, NATURALISTIC INQUIRY, AND SYSTEMATIC DISCOVERY

Case studies are inherently naturalistic, conducted by means of gathering data from within the naturally existing social fields in which the phenomena of interest are located (Lincoln and Gruba, 1985). This naturalistic perspective is well suited to the systematic discovery of both important factors, "variables" in other terminology, as well as the processes that connect these factors, or varying conditions, in the unfolding of social action (Abbott, 1992). Considering the potentially infinite number of factors that might be related to any given social phenomenon, a systematic approach to discovery of a more limited set of potentially important factors is essential (Glaser and Strauss, 1967).

Naturalistic field methods allow researchers a practical method of casting nets wide enough to minimize the chance of missing important contributing factors and yet focused enough to make it possible to do useful and timely work. This process unfolds in practice through an iterative process of conceptualization and data-gathering in which research questions and operations are continually refined as initial data analyses suggest subsequent data-gathering, ideally until the data themselves appear to reach a point of saturation, at which point subsequent data add very little to the overall picture (Strauss and Corbin, 1998). The naturalistic approach allows researchers to test any theoretical notions they have brought to the field against the perceptions of social actors in that field as well as to invite the social actors to nominate other potentially important factors. This systematic approach to discovery, while it does not eliminate the possibility that researchers may ignore important factors or impose biases as a result of preexisting theoretical predispositions, helps to minimize those risks.

While the working group of researchers on this project engaged in a minimum of discussion of these fundamental aspects of case study research, it is probably fair to say they shared many of these methodological assumptions as a result of past research experience. Prior to entering the field to collect data, they did engage in a concerted collective effort to identify a common set of theoretical categories for which data would be sought. For the most part, this effort was successful. (See the introduction to Part I for a list of the categories used. The work of Katherine Newman and her research team in generating the initial set of categories was particularly helpful for the entire working group.)

A notable aspect of the working group's ongoing effort to identify and gather data for a common set of categories across the cases is the hierarchical, nested structure of the categories, including community-level, situational, and individual-level factors (see Figure A-1). The emergence of this hierarchical, nested structure was the outcome of a pragmatic group process of defining the task at hand. It was not the result of a preexisting set of theoretical orientations shared among a group of researchers from diverse disciplinary backgrounds, including sociology, anthropology, education, and psychiatry. Yet there is ample precedent for this kind of theoretical structure in a number of fields, including the study of violence. This appendix next discusses some of the intellectual roots of this approach, not because the particular scholarly traditions to be discussed were explicitly invoked during the process, but because the theories to be discussed help make explicit key issues that this enterprise

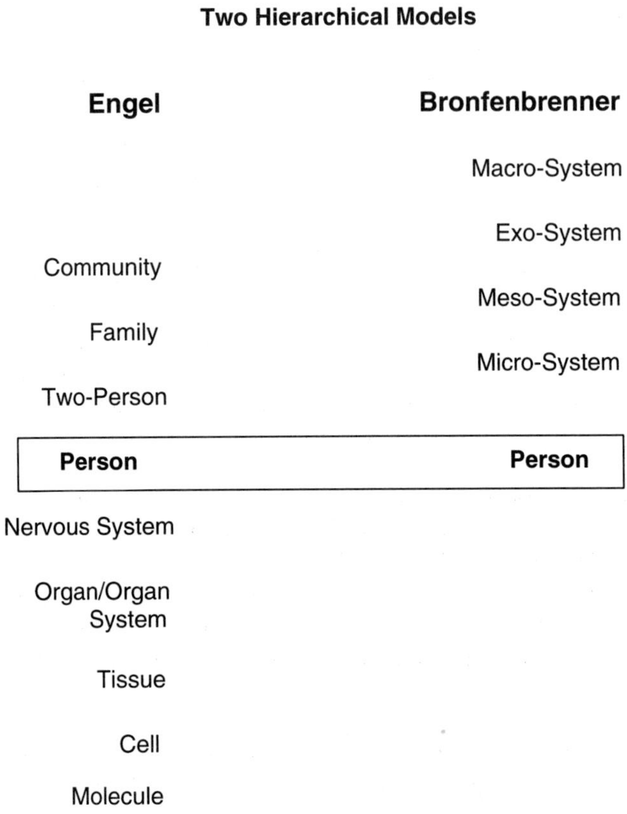

Two Hierarchical Models

Engel	Bronfenbrenner
	Macro-System
	Exo-System
Community	Meso-System
Family	Micro-System
Two-Person	
Person	**Person**
Nervous System	
Organ/Organ System	
Tissue	
Cell	
Molecule	

FIGURE A-1 Two hierarchical models.

had to confront in trying to gather and analyze data about these rare and extreme cases of violence.

HIERARCHICALLY NESTED SYSTEMS AND THE STUDY OF VIOLENCE

Previous syntheses of the research literature on violence have consistently suggested a hierarchically nested explanatory framework. For example, a comprehensive review of the literature commissioned by the National Research Council a few years ago was organized around the nested levels of community, situation, and individual (National Research Council, 1993). Violence clearly varies in frequency and type in ways correlated with each of these levels.

A great deal of recent research in the human sciences has also employed this kind of framework, guided by explicit theoretical orientations that focus attention on processes of person-environment interaction as underlying human behavior and development (Sampson and Lauritsen, 1990). The brief discussion of these traditions presented here delineates some of the key assumptions and discusses their heuristic value in guiding the present inquiry.

The conceptualization of hierarchically nested systems has undergirded important scientific breakthroughs over the past 50 years in a variety of fields, including engineering, computer science, administration, as well as developmental psychology and psychiatry (March and Simon, 1993; Simon, 1996). Basic principles of systems theory have become invaluable for understanding and designing a wide range of systems, both mechanical and social. Among these principles are such ideas as the following:

- Each level of a hierarchically nested system operates according to principles unique to that level.
- Lower levels are constrained by higher levels.
- Levels otherwise operate independently, unless occurrences at one level upset operations at another level.

In Herbert Simon's (1996) well-known illustration of these principles, for example, it does not matter for the navigation of a ship whether the compass used is mechanical or electronic, so long as the compass supplies sufficiently accurate results. The internal workings of the two types of compass, in contrast, are completely dissimilar. At a certain point, however, a demand for greater accuracy at the level of the ship, as a result of more precise steering systems, may demand greater accuracy than the mechanical compass can provide. At this point, the internal workings of the compass may need to be changed completely, along with the articulation of operations at the level of the compass with those at the level of the ship (Simon, 1996).

Two of the case study authors, the authors also of this appendix, independently found it useful to make use of hierarchical theories of this kind in order to analyze the data gathered in their case studies. One of these theories is drawn from clinical psychiatry (Engel, 1980), the other from developmental psychology (Bronfenbrenner, 1979, 1992).

The two theories share some common assumptions about how human behavior unfolds within a set of hierarchically nested ecological levels. Neither is a prescriptive theory; both were designed as guides for practice, clinical treatment in the case of Engel's (1980) work and psychological research in that of Bronfenbrenner (1979, 1992). The main overlapping points of these theories are discussed here, not because they were explicitly employed across the case studies—which they were not—but because they highlight some of the intellectual assumptions reflected in the template of factors arrived at by the working group as a guide for collecting and analyzing data across the case studies.

Figure A-1 is an interpretive comparative representation of the theories of Engel and Bronfenbrenner. The left-hand column lists the hierarchical levels delineated by Engel, from molecules at the bottom to community at the top. The right-hand column lists Bronfenbrenner's levels, from person at the bottom to what he calls "macro-system" at the top. In the middle is the pivot of interpretive comparison, showing "person" as the common element of the two theories. As the figure illustrates, Bronfenbrenner's theory is purely social and psychological and does not deal with biological levels of organization underneath the level of the person. Engel's theory is, as he names it, biopsychosocial theory.

Engel's theory was developed for clinical application and is widely used in the training of psychiatrists and other medical practitioners. The theory emphasizes the embeddedness of biological functioning in social context and the continuous interaction of psychological and organic functions at the level of the person. Engel notably emphasizes both that one knows enough to know that this embeddedness exists, while, at the same time, one knows that one does not know very much about how it works. His theory is therefore fully grounded in empirical science and in no way a misty invocation of alternative medical systems based on anything other than empirical scientific premises. He therefore stresses the need to consider all potential factors at various levels of hierarchical embeddedness (within the practical limits of time and available information) when dealing with a case of medical need. As is also true of the case studies undertaken here, the great stimulus for casting as wide a net as is practical under the given circumstances is the perplexing acknowledgment that one does not know what it is that one needs to know.

Bronfenbrenner's work has been dedicated to expanding knowledge of the interaction of person and environment in the process of human development. Building on the foundational work in this area by Vygotsky (1978) and Lewin (1935), especially Lewin's fundamental theorem that development is a function of the interaction of person and environment, Bronfenbrenner proposes a hierarchical conceptualization of the social environment, divided into the formally defined levels listed in the right-hand column of Figure A-1. This appendix is not the place to rehearse Bronfenbrenner's schema in detail, but, briefly, his conception of micro-system is of the immediate environment of the developing child. The meso-system is the set of relationships of the different social systems directly experienced by the child; the exo-system is the larger community embracing meso-system components; and the macro-system is the most embracing system of all, that of the national culture in which communities are nested.

Figure A-1 presents an interpretive comparison of Engel's and Bronfenbrenner's models. Bronfenbrenner's micro-system is staggered in between Engel's two-person and family levels, to indicate that Bronfenbrenner's category of micro-system probably embraces the more fine-grained distinction between dyadic and familial relationships made by Engel. Similarly, Engel's community level seems to include both the meso- and exo-system levels of Bronfenbrenner. Finally, Engel stops at the level of community, while Bronfenbrenner goes up one additional level, to that of national culture.

Discussing what of these systems is more correct would be futile, both for present purposes and almost certainly from the perspectives of their originators. They were always intended for heuristic rather than prescriptive purposes, and they are remarkably similar in theoretical assumptions, structure, and intent. They share two fundamental assumptions: the embeddedness of behavior in the social environment and the hierarchically nested structure of that social environment. They are frameworks for solving problems by gathering appropriate data in an efficient manner, acknowledging the challenge of the unknown, refusing to accept inappropriate limits of possibly important information, and setting problems in their real-world contexts.

For all these reasons, these frameworks inform this set of case studies in an appropriate and useful manner. It is hardly surprising that the theoretical premises of this work are highly congruent with the research template that emerged from the work of the committee and the case study researchers to coordinate collection and analysis of data. The practical needs of the research group were shaped by the extraordinary nature of the phenomena to be studied, the lack of an extensive base of previous research on these phenomena, and the problematic nature of the boundaries between phenomena and their contexts. The research design that emerged was one that approached the problem as one involving multiple levels of context requiring field inquiry that could investigate different levels of context and their interrelationships in the emergence of these events of extraordinary violence.

The fact that these are case studies on highly unusual phenomena makes this kind of multilevel, comparative case study approach particularly appropriate. Both case study methodology and multilevel ecological frameworks are potentially useful for a wide range of research problems. Bronfenbrenner's theory, for example, was designed for the study of normal child development, across societies and developmental stages, while case study methods have long been used to study both ordinary and extraordinary phenomena (Hamel, 1993; Rabin and Becker, 1992; Yin, 1989). The lack of knowledge about these kinds of rare events, however, makes the multilevel, case study approach particularly useful.

Systematic, multilevel, comparative case studies, in this instance, are far more than simply a fallback methodological option, undertaken only because of the frustrating lack (from a scientific point of view) of a sufficiently large number of cases for statistical analysis. The challenge here is more daunting than that. Not only do we lack a large-*N* database; we also do not know what kinds of questions we should be asking when confronted with such rare, puzzling, and ostensibly irrational behavior. Given this need to define questions, even if, horrifically, we suddenly had large numbers of cases, it would be prudent to undertake these kinds of case studies at this stage of inquiry.

WHAT KINDS OF CONCLUSIONS ARE APPROPRIATE?

Given the methodological issues discussed to this point, what kinds of conclusions would it be reasonable to expect might be drawn from a group of case studies such as those found here? No attempt is made here to frame those conclusions. Rather the question here is what kinds of conclusions might be appropriate, whether from this volume itself, on the part of readers of the case studies, or in terms of directions for future work in this area.

First, at least in the opinion of the authors of this appendix, expectations should not be confined solely to the framing of hypotheses. As discussed earlier, case studies as instruments of scientific research can be descriptive, exploratory, or explanatory. The descriptive and exploratory functions should certainly come to pass, but the explanatory is not out of the question. The most likely form of that would probably be negative case refutation of overly broad or simplistic generalizations, such as "they were all victims of bullying." Only one case is needed for that.

Beyond that, the small universe of these extremely serious incidents presents the possibility that finding one or more negative cases might cast serious doubt on a number of potential generalizations of interest. Negative explanatory findings can be highly useful in times of sudden and emotionally charged attention to a problem, during which many explanations are likely to be and should be tried out. Unfortunately, some theories can become highly popular in the absence of systematic testing of any kind. The range of data assembled for these case studies is unique at the present time, making these studies a distinctive and important resource. For these reasons, the potential explanatory use of these studies should not be ruled out.

Negative case analysis also plays a central role in exploratory, hypothesis-generating activity of case study analysis and of qualitative research generally (Lindesmith, 1968; Becker, 1998). The process involves the iterative trying out of hypotheses on data, followed by successive refinements of the hypotheses in an effort to make them fit more and more cases. These cases should provide ample grist for that kind of effort, both here and in the future.

The multilevel structure of the data gathered across these cases presents unique opportunities for generating hypotheses, for it invites thinking and perhaps future research about the interactions across levels. Are some ecological levels more salient than others, for example? Do conditions at one level seem to constrain processes at a lower level, as systems theory would suggest? If so, how does that seem to happen?

These kinds of questions may appear overly ambitious, and perhaps they are. If the kind of theory-building engaged in is the kind that aims to establish necessary and causal conditions for the occurrence of a certain kind of phenomenon, then asking these questions of these data is indeed inappropriately grandiose.

Recent writings on case study methodology, however, have argued that the search for causality is not the only possible or useful social scientific enterprise. If one presses hard on the conditions needed to establish causality, the number of truly successful causal demonstrations in social science diminishes very rapidly. Becker has contrasted the search for understanding "the real complexity of historical cases" with the effort to find "relationships between variables in a universe of hypothetical cases." He calls the relationships established in this kind of enterprise "conjunctures" rather than "causes" and advocates the search for such conjunctures as an eminently worthwhile scientific goal (Becker, 1992:208).

In a similar vein, Ragin has noted that finding a relationship in a particular case is interesting and important, even if other possible combinations of antecedents or independent variables might have produced the same outcome (Ragin and Becker, 1992). Such findings, which take on a triangular form in bivariate scatterplots, have often been dismissed in conventional social science thinking as uninterpretable, because they do not fit the expectations of linear causal modeling. But finding one kind of relationship, albeit not the only possible one, surely represents an advance over not having found any kind of relationship, or over having found every possible kind of relationship, which amounts to the same thing.

Extraordinary acts of violence in schools cannot be ignored just because they are too few in number to be subjected to rigorous statistical analysis. Theories about how they happen have arisen and will continue to do so. Establishing causal processes leading to these rare and heinous outcomes is not the only scientific approach possible in the service of the search for prevention and control. Case studies like those presented here are essential, appropriate, and scientific tools for use in this search. Only by carefully analyzing those patterns that do exist in the unfolding of these occurrences, using the full range of data available, can we make headway.

REFERENCES

Abbott, A.
1992 "What do cases do? Some notes on activity in sociological analyses." Pp. 53–82 in *What Is a Case? Exploring the Foundations of Social Inquiry*, C.C. Ragin and H.S. Becker, eds. Cambridge: Cambridge University Press.

Allison, G.T.
1971 *Essence of Decision-Making: Explaining the Cuban Missile Crisis*. Boston: Little, Brown.

Becker, H.S.
1992 "Causes, conjunctures, stories, and imagery" in *What Is a Case? Exploring the Foundations of Social Inquiry*, C.C. Ragin and H. S. Becker, eds. Cambridge: Cambridge University Press.

Becker, H.S.
1998 *Tricks of the Trade: How to Think About Your Research While You Are Doing It*. Chicago: University of Chicago Press.

Bronfenbrenner, U.
1979 *The Ecology of Human Development*. Cambridge, MA: Harvard University Press.

Bronfenbrenner, U.
1992 "Ecological Systems Theory." in *Six Theories of Child Development*, R. Vasta, ed. Philadelphia: Jessica Kingsley Publishers.

Campbell, D.T.
1989 "Foreword." in *Case Study Research: Design and Methods*, Robert K. Yin, ed., Newbury Park, CA: Sage Publications.

Engel, G.L.
1980 "The Clinical Application of the Biopsychosocial Model." *American Journal of Psychiatry* 137:535–544.

Glaser, B.G., and A. Strauss
1967 *The Discovery of Grounded Theory*. Chicago: Aldine.

Hamel, J.
1993 *Case Study Methods*. Newbury Park, CA: Sage Publications.

Lewin, K.
1935 *A Dynamic Theory of Personality*. New York: McGraw-Hill.

Lincoln, Y.S. and E. G. Guba
1985 *Naturalistic Inquiry*. Newbury Park: Sage Publications.

Lindesmith, A.R.
1968 (1947) Method and Problem. in *Addiction and Opiates*, A. R. Lindesmith, ed. Chicago: Aldine.

March, J.G., and H.A. Simon
1993 *Organizations*. Cambridge, MA: Blackwell.

Ragin, C.C., and H.S. Becker (Eds.)
1992 *What Is a Case? Exploring the Foundations of Social Inquiry*. Cambridge: Cambridge University Press.

Robinson, W.S.
1950 Ecological Correlation and Behavior of Individuals. *American Sociological Review* 15:351–357.

Sampson, R.J., and J.L. Lauritsen
1990 Violent Victimization and Offending: Individual-, Situational-, and Community-Level Risk Factors." Pp. 1–115 in *Understanding and Preventing Violence: Social Influences*, A.J. Reiss and J.A. Roth, eds. Washington, DC.: National Academy Press.

Simon, H.A.
1996 *The Sciences of the Artificial*. Cambridge, MA: MIT Press.

Strauss, A., and J. Corbin
1998 *Basics of Qualitative Research*. Thousand Oaks, CA: Sage Publications.

Sullivan, M.L.
1998 Integrating Qualitative and Quantitative Methods in the Study of Developmental Psychopathology in Context. *Development and Psychopathology* 10:377–393.

Vygotsky, L.S.
1978 *Mind in Society: The Development of Higher Psychological Processes*. Cambridge, MA: Harvard University Press.

Yin, R.K.
1989 *Case Study Research: Design and Methods*. Newbury Park: Sage Publications.

Appendix B

Biographical Sketches

COMMITTEE MEMBERS

MARK H. MOORE *(Chair)* is the Daniel and Florence Guggenheim professor of criminal justice policy and management at the Kennedy School of Government and Director of the Hauser Center for Nonprofit Institutions at Harvard University and founding chair of the Kennedy School's Committee on Executive Programs. He is the faculty chair of the school's Program in Criminal Justice Policy and Management. Professor Moore also was chair of the National Research Council's Panel on Alcohol Policy, and has served on several other committees of the National Research Council. His research interests include public management and leadership, criminal justice policy and management, and the intersection of the two. He is the author of numerous books, including most recently: *Creating Public Value: Strategic Management in Government*; *Buy and Bust: The Effective Regulation of an Illicit Market in Heroin*; *Dangerous Offenders: The Elusive Targets of Justice*; and (with others) *From Children to Citizens: The Mandate for Juvenile Justice*; and *Beyond 911: A New Era for Policing*.

PHILIP J. COOK is the ITT/Sanford professor of public policy at Duke University. An economist, criminologist, and public health researcher, Professor Cook has been a member of the Duke University faculty since 1973. He has twice served as director of the Sanford Institute of Public Policy, from 1985–89, and again from 1997–99. Professor Cook is a fellow of the American Society of Criminology, and has been a research associate with the National Bureau of Economic Research since 1996. He has served on a total of five different expert panels and committees at the National Academies. He is author of several books, including: *Gun Violence: The Real Costs* with Jens Ludwig; *The Winner-Take-All Society* with Robert H. Frank; and *Selling Hope: State Lotteries in America* with Charles T. Clotfelter.

THOMAS A. DISHION is a research scientist and associate professor in the Department of Counseling Psychology at the University of Oregon. His research focus falls within the broad area of developmental psychopathology. He is interested in understanding how children's social interactions with parents, peers and siblings influence stability and change in developmental trajectories. More recently he has studied the role of discourse topic in organizing the affective exchanges among children and their peers with respect to the development and escalation of antisocial behavior and drug use. He is also interested in applying knowledge of developmental processes to the design of preventive and clinical interventions that reduce conflict and distress in families. He is currently principal investigator of a prevention trial targeting multiethnic families in an urban setting, with the goal of promoting successful adaptation during the adolescent transition as well as reducing problem behavior such as violence and drug abuse. He co-authored the volume *Antisocial Boys*, and his most recent book for parents is *Preventive Parenting with Love, Encouragement and Limits*, co-authored with Scott Patterson.

DENISE C. GOTTFREDSON is a professor of criminal justice at the University of Maryland. She received a Ph.D. in social relations from the Johns Hopkins University, where she specialized in sociology of education. Dr. Gottfredson's research interests include delinquency and delinquency prevention, and particularly the effects of school environments on youth behavior. She has contributed to the literature of school-based crime prevention by testing specific strategies and more recently by summarizing the literature. Her earlier evaluations include Project PATHE, an environmental approach to delinquency prevention; a three-year organization development intervention to reduce violence and related problem behaviors in two troubled Baltimore City junior high schools; and a three-year effort in eight Charleston, South Carolina, middle schools aimed at altering school and classroom environments to reduce student misbehavior. Her recent and current work includes a report to the U.S. Congress on what works, what doesn't work, and what is promising in school-based crime prevention; a recent National Study of Delinquency Prevention in Schools; and a recent book on school-based delinquency prevention.

PHILIP B. HEYMANN is the James Barr Ames professor of law at Harvard Law School, and director of the Harvard Law School Center for Criminal Justice. He studied philosophy at the Sorbonne under a Fulbright scholarship, and served as law clerk to Justice Harlan, United States Supreme Court, October Term, 1960. He has held government policy-level positions within the Department of State, as deputy assistant secretary for the Bureau of International Organizations, and executive assistant to the Under-secretary of State; and within the Department of Justice, where his most recent appointment was as deputy attorney general. His research interests include international law, especially prosecution and court procedures, investigations, violence, and terrorism. He currently serves on the National Research Council's Committee on Law and Justice. Professor Heymann received his law degree in 1960 from Harvard University.

JAMES F. SHORT, JR., is professor emeritus, Washington State University. He was director of research (with Marvin Wolfgang) of the National Commission on the Causes and Prevention of Violence (1968–69), and a member of the National Research Council Committee on Law and Justice, and that committee's Panel on the Understanding and Control of Violent Behavior. He served as editor of the American Sociological Review and as president of the American Sociological Association, the Pacific Sociological Society, and the American Society of Criminology. His most recent book is *Poverty, Ethnicity, and Violent Crime*. Currently he is a member of the U.S. Academic Advisory Council of the National Campaign Against Youth Violence, and advisory committees for the National Consortium on Violence Research, and the National Youth Gang Center. His honors include the Edwin H. Sutherland Award of the American Society of Criminology, the Bruce Smith Award from the Academy of Criminal Justice Sciences, the Paul W. Tappan Award from the Western Society of Criminology, and the Guardsmark Wolfgang Award for Distinguished Achievement in Criminology. He is the namesake for the James F. Short, Jr., Best Article Award, created by the American Sociological Association Section on Crime, Law, and Deviance.

STEPHEN A. SMALL is a professor of human development and family studies at the University of Wisconsin-Madison and Human Development and family relations specialist for the University of Wisconsin-Extension. He received a Ph.D. in developmental psychology from the Department of Human Development and Family Studies at Cornell University in 1985. Professor Small's work is primarily focused on adolescent development, parenting, program development and evaluation, and action-oriented research methods. His research in Wisconsin has addressed a range of issues, including adolescent risk-taking, youth violence, positive youth development, mental health, sexuality, drug use, parent-child relations, and building organizational and community capacity. He is currently completing a book entitled *Bridging the Gap Between Research and Action*, which is aimed at helping social science scholars make their work more relevant to policy and practice. Professor Small has served as a member of the Wisconsin State Legislature's Special Committee on Teen Pregnancy Prevention and the National Research Council's Forum on Adolescence. He has testified on federal drug policy for the U.S. Senate and on youth issues for the Wisconsin legislature.

LEWIS H. SPENCE served as deputy chancellor for operations for the New York City Board of Education, the nation's largest school system. He oversaw the school system's budget and financial operations, information systems, human resources, school facilities, student safety, and food and transportation. He also served as special advisor to the president of the College Board, developing middle and high school integrated instructional programs. Mr. Spence currently holds the office of commissioner, Department of Social Services, Boston, Massachusetts.

LINDA A. TEPLIN is professor of psychiatry and behavioral sciences at Northwestern University Medical School, with joint appointments in the Institute for Policy Research, Sociology Department, and School of Education and Public Policy. She is currently conducting the first large-scale longitudinal study of mental health needs and outcomes of juvenile detainees, funded by a consortium of federal agencies and private foundations. Professor Teplin serves as a member of the National Policy Committee, American Society of Criminology, and the Children and the Law Committee, American Judicature Society. She has been commissioner on the American Bar Association Commission on Mental and Physical Disability Law and their Commission on Lawyer Assistance Programs. She also serves on the Children's Mental Health and Juvenile Justice Initiative of the National Mental Health Association. She has received numerous awards, most recently the 2001 Bernard P. Harrison Award of Merit from the National Commission on Correctional Health Care.

CASE AUTHORS, CONSULTANTS, AND STAFF

GINA ARIAS is a Columbia University graduate with a dual master's degree in international affairs and public health. She has worked in Africa in the field of public health and is currently working with Alianza Dominicana, a community-based organization serving the predominantly low-income Latino population in Washington Heights, New York City. In her spare time she is active in various progressive social justice organization.

ANTHONY A. BRAGA is a senior research associate in the Program in Criminal Justice Policy and Management of the Malcolm Wiener Center for Social Policy at Harvard University's John F. Kennedy School of Government, and a visiting fellow at the National Institute of Justice, U.S. Department of Justice. His research focuses on developing problem-oriented policing strategies to control violent crime hot spots, disrupt drug markets, and reduce firearms violence. He has served as a consultant on these issues to the Rand Corporation; the National Academies; U.S. Department of Justice; U.S. Department of the Treasury; Bureau of Alcohol, Tobacco, and Firearms; Boston Police Department; and the New York Police Department. He holds a master's degree in public administration from Harvard University and a Ph.D. in criminal justice from Rutgers University.

WILLIAM H. DEJONG is a professor of social and behavioral sciences at the Boston University School of Public Health. Professor DeJong also serves as director of the U.S. Department of Education's Higher Education Center for Alcohol and Other Drug Prevention at Education Development Center, Inc., in Newton, Massachusetts. Trained as a social psychologist with a Ph.D. from Stanford University, Dr. DeJong is the author of nearly 300 professional publications in the diverse fields of mass communications and health promotion, criminal justice, and social psychology. His major area of interest is the development and testing of health communications in the areas of alcohol and tobacco control, drunk driving prevention, and STD/HIV prevention.

JOEL C. EPSTEIN is a senior associate at Education Development Center, Inc. in Newton, Massachusetts. He is the author of *A Parent's Guide to Sex, Drugs, and Flunking Out: Answers to the Questions Your College Student Doesn't Want You to Ask*, as well as numerous monographs and articles on crime control. Mr. Epstein is a consultant to colleges, schools, and attorneys on campus safety and liability.

CYBELLE FOX is a Ph.D. candidate in sociology and social policy at Harvard University and a doctoral fellow in the Harvard Multi-Disciplinary Program on Inequality and Social Policy. Previously a policy analyst for the Urban Justice Center and a research associate for the Institute for Children and Poverty, her other research interests center on race, immigration and the politics of redistribution. Her graduate work is supported by a National Science Foundation Graduate Research Fellowship.

MINDY THOMPSON FULLILOVE is professor of clinical psychiatry and public health at Columbia University, a research psychiatrist with New York State Psychiatric Institute, and co-director of the Community Research Group. She holds a master's degree in nutrition and an M.D. degree from Columbia University. Fullilove's research focuses on the intersection of disease and setting, with a particular emphasis on the psychology of place. She has completed a longitudinal study of housing resettlement in Central Harlem, involving the experiences of 10 families, and an interview study of over 100 men and women to understand their experiences of the violence epidemic in their neighborhood. She serves on the National Task Force on Community Preventive Services, a nonfederal task force preparing evidence-based guides to public health practice. She has received grants from the United Hospital Fund, the Centers for Disease Control, and the Robert Wood Johnson Foundation. She is a current member of the National Research Council's Board on Children, Youth, and Families.

ROBERT E. FULLILOVE III is the associate dean for community and minority affairs and associate professor of clinical public health in sociomedical sciences at the Joseph L. Mailman School of Public Health of Columbia University. He currently co-directs the Community Research Group at the New York State Psychiatric Institute and Columbia University. His research is focused on the impact of drug treatment programs on the lives of men and women addicted to crack cocaine and other drugs. He has authored numerous articles on HIV/AIDS, minority health, substance abuse, and mathematics and science education. Dr. Fullilove has served on the Board on Health Promotion and Disease Prevention at the Institute of Medicine since 1995 and in 1998 was appointed to the Advisory Committee on HIV and STD Prevention at the Centers for Disease Control; in the summer of 2000 he became the chair of this committee. He also serves on the editorial board of the journal *Sexually Transmitted Disease* as well on the editorial board of the *Journal of Public Health Policy.*

ROB T. GUERETTE is a doctoral student at Rutgers University-Newark in the School of Criminal Justice and is a fellow at the Eagleton Institute of Politics, Rutgers University-New Brunswick. He has worked on projects in collaboration with the National Research Council, U.S. Department of Justice Office of Community Oriented Policing Services (COPS), British Home Office Research Directorate, and the New Jersey Department of

Probation and Parole. His research interests include crime and public policy, juvenile violence, problem oriented policing, situational crime analysis and prevention, and crime offending patterns.

JOHN HAGAN is John D. MacArthur professor of sociology and law at Northwestern and senior research fellow at the American Bar Foundation. He is co-author with Holly Foster of "Youth Violence and the End of Adolescence." His most recent book is *Northern Passage: American Vietnam War Resisters in Canada.* He is the co-editor with Karen Cook of the *Annual Review of Sociology* and the criminology editor of the *Journal of Criminal Law and Criminology.* Professor Hagan's principal research and teaching interests encompass crime, law and the life course. His book, *Mean Streets: Youth Crime and Homelessness* with Bill McCarthy received the 1998 C. Wright Mills Award from the Society for the Study of Social Problems and in 1997 he received the Ed-win H. Sutherland Award from the American Society of Criminology.

DAVID J. HARDING is a Ph.D. candidate in sociology and social policy at Harvard University, a doctoral fellow in the Harvard Multi-Disciplinary Program on Inequality and Social Policy, and a recipient of a National Science Foundation graduate fellowship. His research interests include urban sociology, urban policy, stratification, and qualitative and quantitative methodology.

THOMAS E. HART is a master patrolman with the Portsmouth, New Hampshire Police Department. He has 20 years of law enforcement experience, including work as a police academy instructor and department training officer. In addition, Mr. Hart is the owner and principal consultant for Emerald Griffin, LLC, a small security company that specializes in emergency response planning for schools, business security, and women's self-defense.

PAUL HIRSCHFIELD is a graduate student in Northwestern University's sociology department. He has participated in evaluations of the Comer School Development Program, and residential mobility programs, and in studies of racial disparities in drug enforcement and the impact of felony disenfranchisement laws. He has also worked and consulted for several research and policy organizations, including the National Network for Youth, the Sentencing Project, the Urban Institute, the Children and Family Justice Center, and Human Rights Watch. His dissertation research focuses on how involvement in the juvenile justice system impacts school performance.

PETER L. McFARLANE is currently the principal of the Hugo Newman College Preparatory School located at 370 West 120th Street in Harlem, New York. He received his advanced degree from Teachers College, Columbia University, with a focus on school restructuring and its impact on urban schools. During his tenure at the Hugo Newman College Preparatory School he has successfully led this school's removal as a School Under Registration Review (SURR) and facilitated an increase in academic achievement for four consecutive years. He is currently a member of Kappa Delta Pi International Honor Society, the Council of Supervisor and Administration, the National Association of Elementary School Principals, the National Association of Supervision and Curriculum Development, the Dramatist Guild and The Authors League of America Inc. Dr. McFarlane was recently honored by the New York City Board of Education as an outstanding educator representing his school district as Principal of the Year. Dr. McFarlane continues his scholarly work on school reform models and the complexities of the principalship.

MICHELLE AUCOIN MCGUIRE is on the staff with the Division of Behavior and Social Sciences and Education in the National Research Council/National Academy of Sciences. Her research contributions include publications in Creole and theoretical linguistics. She has a B.A. in English literature and an M.A. in applied linguistics both from the University of South Florida. She received her Ph.D. in linguistics from the University of Chicago in 2002.

BRENDA McLAUGHLIN is a research assistant with the Committee on Law and Justice at the National Research Council. She has previously worked on projects on juvenile crime, policing, and improving data and research for drug policy. Ms. McLaughlin received a B.A. in sociology and Spanish from Juniata College in 1997, and an M.A. in sociology from American University in 1999.

JAL MEHTA is a Ph.D. candidate in sociology and social policy at Harvard University, and a doctoral fellow in the Harvard Multi-Disciplinary Program on Inequality and Social Policy. A National Science Foundation graduate fellow, Mehta's previous work focused on the role of social-psychological mechanisms in perpetuating social stratification and sponsoring social mobility, and on racial differences in achievement. His research interests include combining normative political theory on justice with empirical research on poverty and inequality.

KATHERINE S. NEWMAN is the Malcolm Wiener professor of urban studies at Harvard University's John F. Kennedy School of Government and the dean of social science at the Radcliffe Institute for Advanced Study. She is the author of several books on middle-class economic insecurity, including *Falling From Grace* in 1988 and *Declining Fortunes* in 1993. Her 1999 book, *No Shame in My Game: The Working Poor in the Inner-City,* focused on the job search strategies, work experiences, and family lives of African American and Latino youth and adults in Harlem. The book won the Sidney Hillman Book Prize and the Robert F. Kennedy Book Award for the year 2000. Newman's next book, *A Different Shade of Gray: Mid-Life and Beyond in the Inner City,* will be published in January of 2003.

MOISES NUNEZ is a research associate of the Community Research Group at the New York State Psychiatric Institute and Columbia University, and is a 2001 graduate of Hampshire College.

CAROL PETRIE, study director, serves as staff director of the Committee on Law and Justice, National Research Council, a position she has held since 1997. Prior to her work there, she was the director of planning and management at the National Institute of Justice, responsible for policy development and administration. In 1994, she served as the acting director of the National Institute of Justice during the transition between the Bush and Clinton administrations. Throughout a 30-year career she has worked in the area of criminal justice research, statistics, and public policy, serving as a project officer and in administration at the National Institute of Justice and at the Bureau of Justice Statistics. She has conducted research on violence, and managed numerous research projects on the development of criminal behavior, policy on illegal drugs, domestic violence, child abuse and neglect, transnational crime, and improving the operations of the criminal justice system.

ERICKA PHILLIPS is a research associate of the Community Research Group and is currently enrolled in the School of General Studies at Columbia University, where she is planning to major in sociology.

LECIA QUARLES was a project assistant with the Committee on Law and Justice from December 2000 to March 2002. Prior to that, she served with the Office of Administration, National Research Council. She has attended Bethune-Cookman College, Daytona Beach, Florida, and Park College-Barstow, California. She is pursuing a bachelor's degree in social psychology.

WENDY D. ROTH received her master's degree in sociology from Oxford University and was formerly a researcher at the National Centre for Social Research in London, where her work involved youth offending and welfare-to-work policy evaluations. She is currently a Ph.D. candidate in sociology and social policy at Harvard University and a fellow of the Multidisciplinary Program in Inequality and Social Policy. Her research includes studies on the racial and ethnic identities of Latino immigrants and the socialization of multiracial children.

CARLA SHEDD is a doctoral student in the Sociology Department at Northwestern University. She earned her bachelor's degree from Smith College in economics and African-American studies in May 2000. Carla is primarily interested in urban sociology, with specific research interests in crime, community, race, and poverty. She is currently a fellow for the Joint Center for Poverty Research. She will also serve as a summer fellow for the National Consortium on Violence Research and the Law and Social Science Program with Northwestern Law School and the American Bar Foundation in 2002–2003. She is currently working on research projects with John Hagan, Mary Pattillo, and Dan Lewis.

MERCER L. SULLIVAN is an urban anthropologist and associate professor in the School of Criminal Justice at Rutgers University. He has written extensively about delinquency and youth crime, the male role in teenage pregnancy and parenting, and community development efforts in inner-city neighborhoods. Professor Sullivan has conducted many studies with inner-city adolescents, using comparative ethnographic data analysis to explore the role of neighborhood and other social context features in adolescent development. He has also worked with other social scientists in interdisciplinary projects integrating qualitative and quantitative approaches to the study of these problems. Professor Sullivan was a member of the 1998 National Research Council Panel on Juvenile Crime Prevention, Treatment, and Control. He is a member of the National Consortium on Violence Research and the Selection Committee for Faculty Scholars Program of the William T. Grant Foundation. He is the editor of the *Journal of Research in Crime and Delinquency*, and the author of the book, *Getting Paid: Youth Crime and Work in the Inner City.* He holds a Ph.D. in anthropology from Columbia University.

RODRICK WALLACE is a research scientist in the Department of Mental Health Epidemiology at the New York State Psychiatric Institute. He received his PhD in physics from Columbia University in 1977. His research has focused on the ecological effects of civic neglect of cities. He is co-author, with Deborah Wallace, of *A Plague on Your Houses: How New York Was Burned Down and National Public Health Crumbled*, published in 1999. He was the recipient of a Robert Wood Johnson Health Policy Investigator Award.

Index

A

Abstract grievances, 4, 337
Academic performance, 5, 38, 45, 75, 88, 103, 104, 107, 136, 148, 177, 202, 204, 257
Access to weapons, 5, 254, 256, 307–308
  bomb-making instructions, 47, 50
  Heath High School shooter, 140, 145–146
  Heritage High School shooter, 34–35, 42, 48, 52–53, 68
  Parker Middle School shooter, 81–82, 84, 93
  prevention of school violence and, 7, 93, 335
  research needs, 337–338
  Tilden High School shooter, 179
  Westside Middle School shooters, 106, 109, 117, 130
  See also Guns
Adolescent development, 249, 256–257
  status anxieties, 336
Adolescent mass murderers, 303–304
Alexander, Eric, 212
Alford plea, 153
Ambrose, Leonard G., 97
Amer, Christina, 108
Anderson, Robert, 198, 203
Atlas.ti software, 201
Attention deficit disorder, 42–43, 258

B

Barnes, Crystal, 108
Beck, Carol, 201, 203, 211, 212, 227
Benedek, Elissa, 150
Bentley, Jason, 198, 204, 214, 231–233. *See also* Thomas Jefferson High School (East New York, New York) shootings
Bentley, Jermaine, 202
Berger, Robert H., 216
Betts, Ashley, 108
Biopsychosocial research, 199
Blundell, Mae Dean, 41
Bond, Bill, 132, 142
Boraten, Edrye May, 73, 81
Breen, Michael, 157
Brooks, Jenna, 108
Brooks, Natalie, 108
Buckley, Brendan P., 204
Bullying and taunting, 94, 317
  definition and characteristics, 316, 317
  harmful effects, 317
  life-span development and, 317
  prevalence, 316–317
  preventive interventions, 67, 124, 317–318
  research needs, 338
  by shooter, 104, 115, 116
  shooter as target of, 6, 49, 67, 115, 116, 146–147, 308
Byrne, Thomas, 187, 188–189, 191

C

Cameron, Shawn, 204
Carneal, Ann, 135, 143, 144
Carneal, John, 135–136, 143, 144
Carneal, Kelly, 133, 135, 137, 143, 144, 148
Carneal, Micheal, 135–138. *See also* Heath High School (Paducah, Kentucky) shooting
Carroll, Patrick, 206, 221, 222
Case study findings
  contagion mechanisms, 259–260
  cultural factors, 253–255
  responses to incidents, 260–264
  school characteristics, 255–256
  shooter characteristics, 256–259
  significance of, 15, 311–312, 351, 353–354, 360–362
  similarities and differences among, 249–252, 266–283
  socioeconomic factors, 252
  See also specific case
Case study methodology, 14–15, 17–18, 23, 199
  appropriate use, 8, 24, 351, 352–354, 360–362
  case selection, 2, 3, 13, 18–20, 351
  cross-case analysis, 247–248
  data sources, 22–23, 26–27, 71–72, 101, 132–133, 199–202
  defining characteristics, 353
  generalizability, 353–354
  goals, 10–11, 17
  naturalistic approach, 354–355
  negative case analysis, 360–361
  systems approach to behavioral research, 352, 355–360
  template design, 20–22
Castillo, Sheila, 185
Causes of school violence, 4, 309, 310, 328–329
  adult–child relationships, 152–153, 235–236, 253–254
  bullying and taunting, 67, 116, 146–147
  challenges in identifying, 285
  classification, 328
  community characteristics and, 19, 252, 328
  copycat incidents, 39, 41, 97, 259–260, 325, 326
  cross-case analysis, 248
  cultural factors, 253–255
  dynamic/situational factors, 311–312, 314
  exposure to media violence, 117–118, 130, 151–152, 253–255, 315–316, 325
  family problems, 114–116, 143–145
  gun culture, 116–117, 145–146, 254
  Heath High School shooting, 143–153
  Heritage High School shooting, 41, 67–68
  interpersonal conflict, 202, 205, 225
  mental health of shooter, 39–41, 114, 150–151
  multifactorial, 247–248, 331
  research needs, 8, 338
  school characteristics, 255–256
  shooter characteristics, 256–259, 309
  socioeconomic factors, 252
  structural, 311, 312
  student beliefs, 65
  Thomas Jefferson High School shootings, 202, 205, 237
  Tilden High School shooting, 163
  Westside Middle School shooting, 113–120, 129–130
  See also Goals of shooter
Chicago, Illinois
  history of school violence, 164
  See also Tilden High School (Chicago, Illinois) shooting
Christian Institute of Human Relations, 96
Civil law suits, 63, 97, 128, 155–157, 262–263
  against gun manufacturers, 117, 128
Clark, Charles, 150
Coleman, James S., 160, 238
Columbine High School (Littleton, Colorado) shooting, 1, 25, 234, 260, 341
  influence on Heritage High School shooter, 39, 41, 49–52, 56–57, 67
Committee to Study Youth Violence in Schools, 1–2, 10–11
Community characteristics, 2–3, 5–6, 18, 21
  aftermath of Heath High School shooting, 154–160
  aftermath of Heritage High School shooting, 64–67
  aftermath of Tilden High School shooting, 184
  aftermath of Westside Middle School shooting, 127–129
  attitudes toward civil suits filed by victims' families, 156–157

attitudes toward shooter after incident, 64–65, 154, 155
attitudes toward shooter's family after incident, 155–156
attributions of causality for shootings, 261–262
causes of school violence related to, 19, 252, 328
criminal and violent behavior, 31–33, 92
criminal justice system, 33–34
cross-case comparison, 268–271
demographics, 28, 29, 89–90, 133, 134, 169–171, 172
entertainment and recreation, 29, 32–33, 37, 90–92
forms of school violence related to, 18–19
gun culture, 34–35, 66, 93, 116–117, 145–146, 209, 221, 226, 230
Heath High School, 133–134, 152–153
Heritage High School, 25, 27–34
interventions to prevent violence, 7, 92–93
involvement with school after incident, 191, 203, 212–214
motivations of shooters and, 249–252
obstacles to violence prevention, 227–228
Parker Middle School, 88, 89–92
race/ethnicity, 169–170, 171, 172
rapid social change, 6, 27–28, 29, 30–31, 66, 169–170, 207–210, 230–231, 252, 341
religiosity, 30, 90, 102, 135
research needs, 340–341
resident self-image, 29–30
responses to shooting incidents, 261
shooter characteristics and, 256
socioeconomic, 28, 29, 31, 90, 133–134, 163, 169–172, 175, 224
street violence, 203
Thomas Jefferson High School, 202, 204–205, 207–210, 212–214, 217–222, 229, 230–233, 237
Tilden High School, 169–172
urban confrontation scenarios, 222–225
urban ecological transition, 217–221
Westside Middle School, 101–102
youth–adult relationships, 6, 7–8, 119–120, 152–153, 160, 227–228, 231–233, 235–236, 237–239, 253–254, 256, 263–264
youth exposure to violence, 228, 229
See also Suburban and rural schools; Urban schools
Consequences of violence, 3–4, 22
community actions and beliefs, 64–67, 127–129, 154–155
definition of school violence, 13
Heath High School shooting, 132, 138–139, 154–155, 159–160
Heritage High School shooting, 25, 55–56, 62–67
long-term, 159–160
mortality/morbidity, 1, 55–56, 62, 70, 73, 101, 202–203
Parker Middle School shooting, 97
resident self-image, 29–30
suicidal behavior of shooter, 25, 27, 61, 62
for teachers, 126, 127, 160, 340
threats against school, 123
victim mental health, 62–63, 125–127, 159–160
Westside Middle School community, 122–123
See also Responses to violent incidents
Conti, Joseph P., 74
Conyers, Georgia. *See* Heritage High School (Rockdale County, Georgia) shooting
Copycat incidents, 5
conceptual model, 326
contagiousness of violence, 3
cross-case analysis, 259–260
Heritage High School shooting as, 39, 41, 49–52, 56–57, 67
influence of Columbine High School shooting, 1, 25, 39, 41, 49–52, 56–57, 67, 234, 260, 341
influence of urban youth violence epidemic, 316
mass murders, 319–320
media role in, 325–326
Parker Middle School shooting as, 97
prevalence, 324–325
prevention, 309
proliferation of, 68
research needs, 340
statistical evidence, 296–297
suicide clusters, 326–328
Westside Middle School, 118
workplace violence, 320–321

Corley, Sharron, 226
Cornell, Dewey, 150
Corporal punishment, 43–44, 48, 107
Cosby, Bill, 212–213
Course of events, 20–21, 305
cross-case comparison, 250–251, 266–267
Heath High School shooting, 140–143
Heritage High School shooting, 25, 52–57
initial press reports in Tilden High School shooting, 164, 165–167, 168
Parker Middle School shooting, 70, 72–74, 76–77, 80–81
shooter behavior after shooting, 45–47, 77, 110–111
Thomas Jefferson High School shootings, 200, 202–203, 205–206, 216
Tilden High School shooting, 179–180
Westside Middle School shooting, 108–113
Craft, Jeffrey, 96
Criminal behavior
characteristics of violent juvenile offenders, 313
Heritage High School community, 31–33
national homicide patterns, 297–301
Parker Middle School community, 92
shooter history, 39, 48–49, 105, 136, 177, 182–183, 205, 257
suspected co-conspirators of shooter, 74, 142–143
Tilden High School community, 175, 176
Criminal justice process
competency of juveniles in, 121
confidentiality issues, 165–166
cost of defense, 84
cross-case analysis, 262–263
eligibility for parole and probation, 57, 61
Heath High School shooting, 153–154
Heritage High School shooting, 26–27, 34, 57–62
juveniles tried as adults in, 34, 57–58, 59–60, 74, 121, 153, 165
media coverage and, 166–167, 168, 181
mental health care in prison, 99
mental health of defendant, 39–40, 57, 58–61, 62, 68–69, 74–75, 79, 121, 150, 153, 215–217, 258
Parker Middle School shooting, 70, 74, 79
prosecution and sentencing of shooter(s), 6–7, 25, 34, 57–62, 120–122, 153–154, 166–167, 168, 178–184, 262
public perception of leniency in, 7, 121–122, 127–128, 153–154
response to violent incidents, 6–7
Rockland County, Georgia, 33–34
Thomas Jefferson High School shootings, 214–217
Tilden High School shooting, 163, 164, 166–167, 168, 178–184
victim participation in prosecution and sentencing, 63
Westside Middle School shooting, 120–122
Crist, Patricia M., 73, 74, 96

D

Daley, Richard, 163, 164, 185, 186, 187, 193, 194
Definition of school violence, 11–13, 287–288
Delusional thinking, 78–79
Depression
classification of mental illness in legal proceedings, 58
in shooter, 39, 47, 305
Dinkins, David, 199, 206, 211, 213, 236
Donaldson, Greg, 201, 204, 223, 224–225
Drug and alcohol use, 256
crack cocaine, 314
in Heritage High School, 32–33, 37
in Parker Middle School community, 92
by shooter, 48, 75, 84
in Thomas Jefferson High School community, 209, 221–222
youth violence and, 314, 315

E

East New York, New York. *See* Thomas Jefferson High School (East New York, New York) shootings
Ecological mapping, 201
Ecological psychology assessment, 41
Edinboro, Pennsylvania. *See* Parker Middle School (Edinboro, Pennsylvania) shooting
Egitto, Francis X., 217

Emotional functioning
family of shooter, 44, 45–47
shooter characteristics, 50–51, 78, 85, 105–106, 115, 204–205
of shooting victims, 62–63
trauma-related disorders, 234
Engel, George, 199

F

Family structure and functioning
community attitudes toward perpetrator's family, 155–156
cross-case analysis, 248
divorce, 41–42, 103
emotional functioning, 44, 45–47
Heath High School shooter's, 135–136, 143–145, 155
Heritage High School shooter's, 25, 41–44
interactions after shooting, 45–47
marital relations, 84
Parker Middle School shooter's, 76, 78, 83–85
protective factors, 92–93
recognition of mental problems, 97–98
residential moves, 42–43, 103
shooter characteristics, 5, 25, 41–44, 76, 78, 83–85, 103, 104–107, 114–116, 135–136, 171–172, 228, 258–259, 305, 307
sibling relations, 43, 76, 84, 144, 148
supervision and discipline, 43–44, 103, 105, 106–107, 114, 144–145
Thomas Jefferson High shooters', 204
Tilden High School shooter's, 171–172
victims' families, 62, 154
in violent urban environment, 228
Westside Middle School shooter's, 103, 104–107, 114–116
Federal Bureau of Investigation (FBI), 308–309
Fernandez, Joseph, 199, 204, 211, 212
Fletcher, Justin, 73, 76, 80–81, 86–87
Forms of school violence
conceptual classification, 288–291
definition of school violence, 11–13, 287–288
research needs, 337–338
trends, 2, 4–5
in urban *vs.* rural/suburban communities, 5, 7, 18–19, 249–252, 289
Fresh, Doug E., 204
Frustration, 319

G

Gang activity/affiliation, 33, 118, 226, 253, 313
alliances between gangs, 173
racial conflict and, 173, 174–175, 178
Tilden High School shooting and, 75, 164, 165, 172–174, 176, 177, 178, 183, 184
in youth homicide epidemic of 1990s, 313, 315
Gillette, John J., 70, 71, 76, 80, 83, 97
Glover, Dewaun, 178, 179
Goals of shooter, 307
to achieve group acceptance, 148–149
cross-case comparison, 268–269
to get attention, 59, 118
in Heath High School shooting, 146–150
media speculation about, 56
in Parker Middle School shooting, 80
revenge against despised groups, 49, 147–148
revenge for teasing and bullying, 67, 116, 146–147
shooters' statements, 56, 76–77, 143
suicide by cop, 59, 323–324
in Tilden High School shooting, 180
in urban *vs.* rural/suburban communities, 249–252
in Westside Middle School shooting, 112, 113
Golden, Andrew, 101, 106–108, 115–116, 118
See also Westside Middle School (Jonesboro, Arkansas) shooting
Golden, Doug, 106
Greene, Bob, 188
Guns
civil suits against manufacturers of, 117, 128
community attitudes, 34–35, 66, 93, 116–117, 145–146, 209, 221, 226, 230
control efforts, 222
reasons for possessing, 35
research needs, 337–338
school security policies, 66–67
shooter's enthusiasm for, 44–45, 49, 117
student attitudes and experiences, 210, 226

trigger locks, 93, 117
use in peer conflicts, 37–38
youth homicide epidemic of 1990s and, 313, 314, 315
See also Access to weapons
Gutmann, Ann, 216, 217

H

Habib, Joseph, 165, 166, 168, 179, 181, 182
Hadley, Nicole, 137–138, 139, 148, 149
Hamburg, Margaret, 210
Hard, Kelly, 139
Hawkins, Yusef, 205
Hayes, Lonnie, 223, 225
Heath High School (Paducah, Kentucky) shooting, 20
causes, 143–153
civil suits arising from, 155–157
community characteristics, 133–134, 152–153
consequences for community, 154–155
course of events, 140–143
data collection, 132–133
mortality/morbidity, 132, 138–139
prosecution and sentencing, 153–154
school characteristics, 134–135
school security response, 157–159
shooter characteristics, 135–138
suspected co-conspirators, 142–143
Henderson, Jermaine, 203
Heritage High School (Rockdale County, Georgia) shooting, 20
causes, 41, 67–68
community characteristics, 25, 27–31, 64–67
consequences, 30, 62–67
course of events, 25, 52–57
data collection, 26–27
gun ownership and, 34–35, 68
influence of Columbine shooting, 49–52, 56–57, 67
overall crime rate of region, 31–33
school characteristics, 38–39
security measures before, 38–39
shooter characteristics, 25, 39–49, 67–68
student characteristics, 35–38
trial and sentencing, 34, 57–62, 68–69
Herring, Paige, 108
Hill, John, 215
Holm, Hollan, 139, 160

I

Impellizzeri, Irene, 213
Infoshare, 201
Internet, 47, 50, 136–137, 144
Irushalmi, Bruce, 201
Irving, Whitney, 108

J

Jackson, Jesse, 184–185, 194
Jacobs, Jane, 222–223
Jacobs, Jennifer, 108, 112
James, Jessica, 138
Jenkins, Melissa, 139, 160
Johnson, Mitchell, 101, 102–106, 108, 114–115, 117–118. *See also* Westside Middle School (Jonesboro, Arkansas) shooting
Johnson, Scott, 103
Johnson, Stephanie, 108, 110
Jonesboro, Arkansas. *See* Westside Middle School (Jonesboro, Arkansas)
Juvenile Justice Reform Act, 34
Juviler, Michael, 214

K

Kaltenbach, Tim, 153
Keene, Craig, 139
Kimbaugh, Ted, 188, 194
Knox, George, 174
Kuby, Ron, 215, 216
Kunstler, William, 215, 216, 217

L

Lambie, Brittany, 108
Lawson, Delondyn, 163, 164, 165, 168, 179
Lawson, Linda, 164, 168, 181
Lee, Spike, 213
Literature review, 13–14, 302
contagion mechanisms, 324–328
mass murders, 303–304, 318–320
school rampage shootings, 303–309
suicide of homicide offenders, 322–324
violence in general, 309–312
workplace violence, 320–322
youth violence, 313–316
Loners, 36, 44, 45, 48, 303, 305
Lowery, Richard, 210

M

Mass murderers, 303–304, 318–320
Masters, David A., 73, 89
McCall, Carl, 204
McGowan, Tristan, 108
McIlvane, Thomas, 320–321
Media coverage, 29–30, 63
  community resentment of, 64, 121–122, 155
  copycat crimes and, 325–326
  criminal justice process and, 166–167, 168, 181
  cross-case analysis, 260
  exposure to media violence as cause of violence, 117–118, 130, 151–152, 253–255, 315–316, 325
  historical consistency, 295
  imitative suicide and, 327–328
  incident response, 56, 122
  initial press reports in Thomas Jefferson High School shooting, 214–215
  initial press reports in Tilden High School shooting, 164, 165–167, 168
Medley, Lena, 201
Mental health
  access to care, 66, 99
  assessment and diagnosis, 40–41, 47, 93–94
  causes of rampage shootings, 329
  competency of juveniles, 121
  detection of potential problems, 93–95, 97–98, 125, 159
  development of psychosis in shooter after trial, 151
  ecological psychology approach to assessment, 41
  Heath High School shooter, 138, 150–151
  Heritage High School shooter, 26–27, 39–41, 49–50, 57, 58–61, 62, 67
  legal issues, 57, 58–61, 68–69, 121, 153, 215–217, 258
  Parker Middle School shooter, 74–75, 76, 77–80, 97–98
  provision of counseling services after incident, 65–66, 125–126, 203
  psychological problems of victims, 62–63, 125–127, 159–160
  research needs, 339–340
  shooter characteristics, 5, 42, 257–258, 305, 306
  Westside Middle School shooters, 114
Merski, Robert, 95
Moore, Ian, 198, 205, 206–207
Moral panic, 193–195
Music, 47–48, 51–52, 75, 117–118, 315

N

Nation, Sidney, 57, 60, 61
National Center for the Analysis of Violent Crime, 308–309
National Organization of Victim Assistance, 125
National School Safety Center, 291
National Threat Assessment Center, 306–308

O

O'Brien, John S., 79
O'Connor, Kathleen, 151
Ojiste, Lance, 216

P

Paducah, Kentucky. *See* Heath High School (Paducah, Kentucky) shooting
Palmisano, Michael M., 70, 97
Parker Middle School (Edinboro, Pennsylvania) shooting, 20
  causes, 97–98
  community characteristics, 89–92
  course of events, 70, 72–74, 76–77, 80–81
  data collection, 71–72
  responses to, 92–96
  school characteristics, 70–71, 88–89
  shooter characteristics, 70, 74–80, 83–87
  suspicion of conspiracy, 74
  trial and sentencing, 70, 97
  victims and witnesses, 70, 76–77, 80–83
Peer relations
  adolescent information processing deficits, 316
  association with delinquent peers, 48–49
  attitudes toward shooter after incident, 64–65
  between shooters, 108, 109, 113
  causes of school violence, 4, 202, 205
  conflict between groups, 37
  conflict between individuals, 37–38
  gaining and protecting reputation, 226

Heath High School shooter, 136, 137, 147–149
Heritage High School, 35–38
individuals aware of shooter's plans, 74, 81–83, 111, 140, 307, 308
interventions to improve, 213
loners, 36, 44, 45, 48, 303
Parker Middle School shooter, 74, 85–87
sexual behavior, 29–30
shooter characteristics, 5, 6, 39, 43, 44, 45, 48–49, 74, 85–87, 104, 107, 116, 150–151, 259
student groups and cliques, 6, 35–37, 49
suspected co-conspirators of shooter, 74, 142–143
threat assessment, 308–309
Westside Middle School shooters, 104, 107
youth–adult relationships and, 235–236
Pillars of Good Character, 94
Police, 66, 67
data sources, 26–27
gang interaction, 173
incident response, 54–55, 110–111, 167–168, 341–342
in schools, 89, 95–96, 124–125, 158, 186–187, 211, 263, 334
suicide by cop, 59, 323–324
youth perceptions of, 32–33
Pope, Elijah, 203
Porter, Candice, 108, 112
Prevention of violent incidents, 7–8
access to guns and, 7, 93, 335
challenges in designing interventions for, 285, 330
character education programs, 94, 98, 192
community-level interventions, 7
community standards as obstacle to, 227–228
consideration of dynamic/situational factors, 312
costs, 331
cross-case analysis of strategies for, 263–264
fostering student relations for, 213
Heath High School interventions, 89
identification of likely offenders, 332–334
improving youth–adult relationships for, 92–93, 160, 235–236, 237–239, 256, 263–264
intervention with bullying behavior, 67, 124, 317–318
interventions targeting adolescent development, 336
mental health interventions, 40–41, 68, 93–94
monitoring for problems in students, 93–95, 98, 125, 333
Parker Middle School interventions, 89
primary interventions, 330, 331
public health model, 330–331
research needs, 8
school shooter threat assessment, 308–309
secondary/tertiary interventions, 330, 331–332
in smaller schools, 212
Thomas Jefferson High School interventions, 211–214, 236
thwarted events since Columbine High School shooting, 296
United for Safety Project, 213–214
weekend retreats, 213, 236
Westside Middle School interventions, 123–125
See also Security systems and practices

R

Race/ethnicity
in Chicago gang conflict, 173, 174–175, 178
Heath High School demographics, 134
Heritage High School community, 29, 35, 36
Tilden High School community, 169–170, 171, 172
Rampage violence, 4, 7, 288
existing research, 303–309
international incidents, 295–296
trends, 4–5
Religion/spirituality
to foster child–adult relationships, 119
Heath High School community, 135
Heritage High School community, 30
Parker Middle School shooter, 75
violence and, 30
Westside Middle School community, 102
Westside Middle School shooters, 103–104
Research needs, 8
contagion mechanisms of school violence, 340

developmental effects of exposure to violence, 339–340
effects of community change, 341
effects of school shootings for perpetrators, 337
effects of school shootings for victims, 337
failed attempts of school violence, 342
gun access and use, 338–339
individual risk factors for school violence, 339–340
nonlethal school violence, 337–338
parenting styles, 341
police response to school shootings, 342
prevalence of adolescent mental illness, 339
school security systems, 342
suicidal behavior/ideation among adolescents, 340
types of violence in urban schools *vs.* suburban/rural schools, 337–338
violence among suburban and rural youth, 336–337
workplace violence against teachers, 340
Responses to incidents, 22
attitudes toward shooter, 64–65, 154
cross-case analysis, 260–264, 278–283
moral panic analysis, 193–195
Parker Middle School community, 92–96, 99
policy formulation, 4, 22, 263–264
preparedness for, 102
principal's immediate actions, 190–192, 203
provision of counseling services, 65–66, 125–126, 154–155, 191, 203
school security reforms, 7, 55, 66–67, 95–96, 123–125, 157–159, 184–196, 211–212, 236, 262–263
Thomas Jefferson High School community, 203, 212–214
Tilden High School shooting, 184–190, 192–196
victim memorial, 71, 128, 154, 159
victim's funeral, 203–204, 206–207
Revenge, 49, 67, 116, 147–148, 306, 307
classroom avenger model, 304–306
Rights of Passage Program, 192
Risk factors, 310
dynamic/situational, 311–312
research needs, 339–340
school shooter threat assessment, 308–309
shooter characteristics, 5, 256–257, 332–334
structural, 311, 312
Ritalin, 42–43, 52
Rockdale County, Georgia. *See* Heritage High School (Rockdale County, Georgia) shooting
Romantic relationships of shooters, 45, 75, 87, 112, 115, 137–138, 139, 259
Rural schools. *See* Suburban and rural schools
Russell, John, 215, 216

S

Sadoff, Robert L., 74, 75, 76, 77, 78, 79, 82, 84, 87
Scaletta, Rick, 96
Schaberg, Shelley, 139, 160
Schetky, Diane, 150
Schneider, William, 58, 59–60
School characteristics, 22
as cause of school violence, 255–256
corporal punishment, 107
cross-case comparison, 272–273
culture of bullying, 317–318
gang activity, 178
Heath High School, 134–135
Heritage High School, 38–39, 66–67
immediate actions in response to shootings, 190–192, 203
interventions to prevent violence, 93–95, 98, 157–159
parental involvement, 135, 238–239
Parker Middle School, 70–71, 88–89
post-incident trauma counseling, 65–66, 125–127, 154–155, 203
reforms following Tilden High School shooting, 184–190, 192–196
rule setting and enforcement, 123, 125–126, 135, 163, 189–190
teacher–student relationships, 119–120
Westside Middle School, 102, 116
See also Security systems and practices
School resource officers, 89, 95–96, 124–125, 158, 334
School violence, generally
in aggregate patterns of violence, 285–286, 289–291, 297, 299–301

classification, 288–291
existing research, 303–309
international incidents, 295–296
operational definition, 11–13, 287–288
research needs, 8, 336–342
trends, 1, 9–10, 11, 291–297, 300–301
Secret Service, 306–308
Security systems and practices
effectiveness of, in preventing school violence, 334–335
exclusion of students, 192, 196, 212
Heath High School, 157–159
Heritage High School, 38–39
incident response, 55, 190–191
metal detectors, 66, 163, 164, 167, 184–185, 186–189, 190, 191, 193, 194, 195, 210–211, 236, 263–264, 334
Parker Middle School, 89, 95–96
political context, 193–194
reforms of, after violent incidents, 7, 55, 66–67, 95–96, 123–125, 157–159, 184–196, 211–212, 263
research needs, 341–342
school resource officers, 124–125, 158
student identification tags, 158
Thomas Jefferson High School, 211–212, 236
Tilden High School, 163, 184–190, 192–196
Westside Middle School, 123–125
See also Prevention of violent incidents
Self-image, 305, 336
Sex differences, 5
Sexuality/sexual behavior
Heritage High School, 29–30, 37
taunting of shooter, 146
Sharpe, Daryl, 198, 202–204
Sherrill, Patrick Henry, 320
Shooter characteristics, 5, 21
academic performance, 5, 38, 45, 75, 88, 103, 104, 107, 136, 148, 177, 202, 204, 257
age, 256–257
as cause of school violence, 256–259, 309, 328, 329
childhood abuse experiences, 103, 115
classroom avengers, 304–306
cross-case analysis, 249, 256–259, 272–276
diversity, 307, 332–333
drug and alcohol use, 48, 75, 84
emotional functioning, 50–51, 85, 116
enthusiasm for guns, 44–45, 49, 106, 305
family structure and functioning, 5, 25, 41–44, 76, 78, 83–85, 103, 104–105, 106–107, 114–116, 135–136, 171–172, 228, 258–259, 305, 307
Heath High School shooting, 135–138
Heritage High School shooting, 25–26, 39–49, 67–68
history of antisocial behavior, 38, 48–49, 104, 105, 107, 136, 138, 177, 256, 257
mental health, 5, 26–27, 39–41, 68, 77–80, 97–98, 114, 257–258, 305, 306
Parker Middle School shooting, 70, 74–80
peer relations, 5, 6, 39, 43, 44, 45, 48–49, 74, 85–87, 104, 107, 136, 137, 259
preparatory actions, 276–279, 306, 307
recreational and entertainment activities, 47–48, 75, 103–104, 117–118, 136–137, 305
relationships with teachers, 119–120
religious beliefs/practices, 75, 103–104
risk factors, 5, 256
romantic experiences, 45, 75, 87, 112, 115, 137–138, 139, 259
social status, 252, 259, 336
as target for preventive intervention, 332–334
Thomas Jefferson High shootings, 204–205
threat assessment, 308–309
Tilden High School shooting, 169–170
Westside Middle School shooting, 102–108, 114–116
Sinkler, Tyrone, 198, 205
Situation analysis, 199
Small schools, 212
Smith, Marlon, 206
Societal violence, 2–3, 9, 249
avoidance strategies, 225–226
contagion effects, 297
Heritage High School community, 30
literature review, 309–312
in marginalized urban communities, 231
media violence, 117–118, 130, 137, 151–152, 253–255, 315–316, 325
national trends, 175, 297–300
school violence in context of, 289–291, 297, 299–301
Thomas Jefferson High School community, 203, 204–205, 209, 221–222, 229, 230–231, 233–234

Tilden High School community, 174–175, 176
urban confrontation scenarios, 222–226
violent video games, 118, 130, 136–137, 151, 315
in workplace, 320–322, 340
youth exposure, 228, 229
Solomon, Anthony, 41
Solomon, Anthony B., Jr., 25, 39–49. *See also* Heritage High School (Rockdale County, Georgia) shooting
Statements and writings of perpetrators
after shootings, 45–47, 56, 79, 112, 120–121, 140, 146
before shootings, 47, 50–52, 79, 81–83, 85, 111, 143, 305–306, 307, 308
description of incident, 53–54, 76–77
during incident, 73, 142
regarding co-conspirators, 142–143
suicide note, 84
Steger, Kayce, 139
Steward, Hazel, 178, 188, 189, 190–191, 196
Strand, James A., 73, 77, 93
Stuck-Lewis, Denise, 74
Student characteristics
group identification, 35–36, 137, 147–148
Heritage High School, 35–38
loners, 36
recreational activities, 32–33, 37, 91–92, 96
relationships with adults, 6, 7–8, 119–120, 160, 227–228, 253–254, 256, 263–264
Suburban and rural schools, 2–3
attributions of causality for shootings, 261–262
causes of violence in, 4
community characteristics, 5–6, 340
consequences of violence in, 3–4
forms of violence in, 7, 18–19, 289
influence of urban youth violence epidemic, 314–315, 316
motivations of shooters, 250–252
peer relations, 6
research needs, 337–338
responses to shooting incidents, 261, 263
shooter characteristics, 256, 257
Suicidal behavior/ideation, 25, 27, 36, 46, 51–52, 54, 61, 62, 82–83
contagion mechanisms, 326–328
early manifestations in shooter, 42, 48, 76, 147
as evidence of mental illness, 58
following commission of homicide, 322–324
Heath High School shooter, 140, 142, 147, 151
legal considerations, 58–59
research needs, 340
shooter characteristics, 258
suicide by cop, 59, 323–324
Summers, Warren, 45–46
Sumpter, Khalil, 198, 204, 214–217, 234, 235–236. *See also* Thomas Jefferson High School (East New York, New York) shootings
Suttles, Gerald, 172

T

Teich, Stephen, 216–217
Thetford, Lynette, 108
Thomas Jefferson High School (East New York, New York) shootings, 20, 198–199
anti-violence programs after, 212–214, 236
community characteristics, 202, 204–205, 207–210, 217–222, 229, 230–233, 237
course of events, 200, 202–203, 205–206, 216
data collection, 199–202
precipitating events, 202, 205
prosecution and sentencing, 214–217
security reforms after, 211–212
shooter characteristics, 204–205
student perceptions of, 234–235
victims, 202–203
youth–adult relationships, 227–228, 235–236, 237–239
See also Bentley, Jason; Sumpter, Khalil
Thomas theorem, 199
Thompson, Jesse, 202
Thrasher, Frederick, 174
Tilden High School (Chicago, Illinois) shooting, 20
arrest of perpetrator, 167–168
causes, 163
community characteristics, 169–178
course of events, 179–180
gang activity and, 164, 165, 172–174, 175, 176, 177, 178–179, 183
initial media coverage, 164, 165–167, 168
mortality/morbidity, 163, 165

prosecution and sentencing, 164, 166–167, 168, 178–184
school reforms following, 184–190, 192–196
school violence preceding, 193–194, 195
Tury, Jacob, 73, 81
Tyson, Cicely, 213

U

United for Safety Project, 213–214
Urban schools, 2–3
attitudes toward weapons, 210
attributions of causality for shootings, 261
case study selection, 19–20
causes of violence in, 4
community characteristics, 5–6, 340
consequences of violence in, 4
forms of violence in, 5, 7, 18–19, 289
motivations of shooters, 249–250
peer relations, 6
responses to shooting incidents, 261, 263
shooter characteristics, 256, 257

V

Vallas, Paul, 189, 190, 194
Varner, Britthney, 108
Victims of school violence, 10
attitudes toward shooter after incident, 64–65, 97
community trauma, 125–127
Heath High School shooting, 138–139
Heritage High School shooting, 62–63
mortality/morbidity, 1, 55–56, 62, 70, 73, 108, 110, 132, 138–139, 163, 165, 202–203, 249
Parker Middle School shooting, 70, 76–77, 80–81, 83, 97
psychological problems, 62–63, 125–127, 159–160
selection of, by shooter, 76–77, 80, 81, 112–113, 142
teachers as, 340
Thomas Jefferson High School shootings, 202–203
Tilden High School shooting, 163, 164, 165
Westside Middle School shooting, 108, 110, 112–113
Video games, 118, 130, 136–137, 151, 315

W

War themes, 303–304, 323
Washok, James J., 73
Weitzel, William, 150
Weprin, Saul, 211
Westside Middle School (Jonesboro, Arkansas) shooting, 20, 82, 97
adjudication, 120–122
causes, 113–120, 129–130
community characteristics, 101–102
course of events, 108–113
data collection, 101
effects on community, 122–129
mortality/morbidity, 101, 108
shooter characteristics, 102–108, 114–116, 119–120
victim characteristics, 108, 112–113
White, Joseph, 163, 164, 165, 166, 167–168, 169–170, 171, 174, 177, 178–184, 196
White, Karen, 167–168, 171–172, 174, 181, 182
Wilding behavior, 226–227
WiseSkills, 94
Woodard, Gretchen, 101, 103
Woodard, Terry, 103
Wooten, Priscilla, 213
Workplace violence, 320–322, 340
Wozniak, Eric, 73, 77, 80
Wright, Kenneth, 183
Wright, Shannon, 104, 108
Wurst, Andrew Jerome, 70, 74–80. *See also* Parker Middle School (Edinboro, Pennsylvania) shooting
Wurst, Catherine, 74, 83
Wurst, Jerome J., 83

Y

Youngblood, Johnny Ray, 207, 209, 233
Youth violence, generally
characteristics of offenders, 313
homicide epidemic of 1990s, 313–315
literature review, 313–316
school violence in context of, 289–291, 299–300
social causes, 313–314
trends, 9, 298–299, 313–314

Z

Zemcik, Robert, 73, 81